Finite Elements in BIOMECHANICS

WILEY SERIES IN
NUMERICAL METHODS IN ENGINEERING

Consulting Editors
R. H. Gallagher, *College of Engineering,*
University of Arizona

and

O. C. Zienkiewicz, *Department of Civil Engineering,*
University College Swansea

Rock Mechanics in Engineering Practice
Edited by K. G. Stagg and O. C. Zienkiewicz

Optimum Structural Design: Theory and Applications
Edited by R. H. Gallagher and O. C. Zienkiewicz

Finite Elements in Fluids
Vol. 1 Viscous Flow and Hydrodynamics
Vol. 2 Mathematical Foundations, Aerodynamics and Lubrication
Edited by R. H. Gallagher, J. T. Oden, C. Taylor and O. C. Zienkiewicz

Finite Elements for Thin Shells and Curved Members
Edited by D. G. Ashwell and R. H. Gallagher

Finite Elements in Geomechanics
Edited by G. Gudehus

Numerical Methods in Offshore Engineering
Edited by O. C. Zienkiewicz, R. W. Lewis, and K. G. Stagg

Finite Elements in Fluids
Vol. 3
Edited by R. H. Gallagher, O. C. Zienkiewicz, J. T. Oden, M. Morandi Cecchi and C. Taylor

Energy Methods in Finite Element Analysis
Edited by R. Glowinski, E. Rodin and O. C. Zienkiewicz

Finite Elements in Electrical and Magnetic Field Problems
Edited by M. V. K. Chari and P. Silvester

Numerical Methods in Heat Transfer
Edited by R. W. Lewis, K. Morgan and O. C. Zienkiewicz

Finite Elements in Biomechanics
Edited by R. H. Gallagher, B. R. Simon, P. C. Johnson and J. F. Gross

Finite Elements in BIOMECHANICS

Edited by
R. H. Gallagher
B. R. Simon
P. C. Johnson
J. F. Gross

University of Arizona, Tucson, Arizona

A Wiley–Interscience Publication

175 YEARS OF PUBLISHING
1807 1982

JOHN WILEY & SONS
Chichester · New York · Brisbane · Toronto · Singapore

Library of Congress Cataloging in Publication Data:
Main entry under title:

Finite elements in biomechanics.

(Wiley series in numerical methods in engineering)
'A Wiley–Interscience publication.'
Includes index.
1. Human mechanics—Mathematics—Congresses.
2. Finite element method—Congresses. I. Gallagher, Richard H. II. Series.

QP303. F47 612′.76 81-13084

ISBN 0 471 09996 1 AACR2

British Library Cataloguing in Publication Data:

Finite elements in biomechanics.—(Wiley series in numerical methods in engineering)
1. Human mechanics
1. Gallagher, R. H.
612′.76 QP303

ISBN 0 471 09996 1

Typeset by Macmillan India Ltd, Bangalore and printed by Vail-Ballou Press, Inc., Binghamton, U.S.A.

Contributing Authors

K. -N. AN — *Orthopaedic Biomechanics Laboratory, Mayo Clinic, Rochester, MN 55907, U.S.A.*

G. BERGMANN — *Orthopaedische Klinik und Poliklinik der Freien Universitaet Berlin, D-1000 Berlin, 33, Germany*

M. P. BIENIEK — *Department of Civil Engineering and Engineering Mechanics, Columbia University, New York, NY 10027, U.S.A.*

T. D. BROWN — *Department of Orthopaedic Surgery, University of Pittsburgh, PA 15261, U.S.A.*

E. Y. S. CHAO — *Orthopaedic Biomechanics Laboratory, Mayo Clinic, Rochester, MN 55901, U.S.A.*

G. W. CHRISTIE — *Theoretical and Applied Mechanics Department, The University of Auckland, New Zealand*

I. C. CLARKE — *Department of Orthopaedics, Orthopaedic Hospital – University of Southern California, Los Angeles, CA 90007, U.S.A.*

A. B. FERGUSON — *Department of Orthopaedic Surgery, University of Pittsburgh, PA 15261, U.S.A.*

F. H. FU — *Department of Orthopaedic Surgery, University of Pittsburgh, PA 15261, U.S.A.*

T. A. GRUEN — *Department of Orthopaedics, Orthopaedic Hospital – University of Southern California, Los Angeles, CA 90007, U.S.A.*

W. C. HAYES — *Orthopaedic Biomechanics Laboratory, Beth Israel Hospital and Harvard Medical School, Boston, MA 02215, U.S.A.*

R. R. HOSEY — *Center for Materials Research, College of Engineering, University of Iowa, Iowa City, IA 52242, U.S.A.*

R. HUISKES — *Laboratory for Experimental Orthopaedics, University of Nijmegen, The Netherlands*

J. D. JANSSEN — *Division of Applied Mechanics, Department of Mechanical Engineering, Eindhoven University of Technology, The Netherlands*

A. D. KARAKAPLAN — *Department of Civil Engineering and Engineering Mechanics, Columbia University, New York, NY 10027, U.S.A.*

D. W. KELLY — *Civil Engineering Department, University of Wales, Swansea, SA2 8PP, U.K.*

R. KOELBEL — *Orthopaedische Klinik und Poliklinik der Freien Universitaet Berlin, D-1000 Berlin, 33, Germany*

G. C. LEE — *Faculty of Engineering and Applied Science, State University of New York at Buffalo, Amherst, NY 14260, U.S.A.*

B. M. LEVINE — *Department of Orthopaedic Surgery, University of Pennsylvania, Philadelphia, PA 19104, U.S.A.*

Y. K. LIU — *Center for Materials Research, College of Engineering, University of Iowa, Iowa City, IA 52242, U.S.A.*

I. C. MEDLAND — *Theoretical and Applied Mechanics Department, The University of Auckland, New Zealand*

Y. C. PAO — *Department of Engineering Mechanics, University of Nebraska-Lincoln, NE 68588, U.S.A.*

G. T. RAB — *Department of Orthopaedics, University of California – Davis School of Medicine, Sacramento, CA 95817, U.S.A.*

S. RAMASWAMY — *Orthopaedic Biomechanics Laboratory, Beth Israel Hospital and Harvard Medical School, Boston, MA 02215, U.S.A.*

A. ROHLMANN — *Orthopaedische Klinik und Poliklinik der Freien Universitaet Berlin, D-1000 Berlin, 33, Germany*

E. RYBICKI — *Department of Mechanical Engineering, University of Tulsa, Tulsa, OK 74104, U.S.A.*

A. SARMIENTO — *Department of Orthopaedics, Orthopaedic Hospital – University of Southern California, Los Angeles, CA 90007, U.S.A.*

W. H. SHIELDS — *Oceanic Division, Westinghouse Corporation, Baltimore, MD 21404, U.S.A.*

R. SKALAK — *Department of Civil Engineering and Engineering Mechanics, Columbia University, New York, NY 10027, U.S.A.*

T. J. SLOOFF — *Laboratory for Experimental Orthopaedics, University of Nijmegen, The Netherlands*

B. SNYDER — *Department of Orthopaedic Surgery, University of Pennsylvania, Philadelphia, PA 19104, U.S.A.*

R. L. SPILKER — *Department of Materials Engineering, University of Illinois at Chicago Circle, Chicago, IL 60680, U.S.A.*

R. R. TARR — *Department of Orthopaedics, Orthopaedic Hospital – University of Southern California, Los Angeles, CA 90007, U.S.A.*

H. TÖZEREN — *Department of Engineering Sciences, Middle East Technical University, Ankara, Turkey*

N. T. TSENG — *Faculty of Engineering and Applied Sciences, State University of New York at Buffalo, Amherst, NY 14260, U.S.A.*

D. L. VAWTER — *School of Engineering Science and Mechanics, Georgia Institute of Technology, Atlanta, GA 30332, U.S.A.*

M. E. WAY — *Biomedical Engineering Program, Carnegie-Mellon University, Pittsburgh, PA 15213, U.S.A.*

S. O. WILLE — *Institutt for Informatikk, University of Oslo, Blindern, Oslo, 3, Norway*

O. C. ZIENKIEWICZ — *Civil Engineering Department, University of Wales, Swansea, SA2 8PP, U.K.*

Contents

Preface

On 18–20 February 1980, the first International Conference on Finite Elements in Biomechanics was held in Tucson, Arizona. Sponsored, in part, by the U.S. National Science Foundation and the College of Engineering of the University of Arizona, the conference featured nearly sixty papers, including eight invited feature lectures. The Proceedings of the conference, which were published under the auspices of the office of Special Professional Education of the College of Engineering, contained some forty-five contributed papers, in abbreviated form, and brief summaries of the talks of the invited lecturers. Following the conference the organizers (the editors of this book) asked certain of the authors of the contributed papers to contribute longer versions of their papers for inclusion in this book. The choices were made in the interests of having a coherant treatment of the topics to be covered in the book. The latter were defined principally by the subject matter examined by the invited lecturers, whose contributions are given in full for the first time in this book.

The book contains eighteen chapters in all, covering five major areas. Two papers are devoted to rather general discourses on biomechanics and the role that the finite element method plays in this developing field. Two chapters are concerned with bio-fluid mechanics, five deal with soft tissue mechanics, eight of the chapters examine orthopaedics problems, and crash injury analysis of the head is covered in one chapter. The remainder of this Preface is devoted to a thumbnail sketch of each of these chapters.

One of the practical impediments to progress in biomechanics is the relative scarcity of researchers and practitioners who are educated in both mechanics and medicine. One of the few individuals who has such a background is Dr. George Rab, the author of Chapter 1. He takes the opportunity, in this chapter, to look at the difficulties of collaboration of engineers and physicians, and of the intrinsic difficulties of biomechanical analysis. He looks at both the failures and successes that have been achieved in orthopaedic analysis by finite elements, and presents some thoughts for the future.

While Dr. Rab has a vantage point that is primarily one of the physician, Zienkiewicz and Kelly (Chapter 2) take the view of the finite element specialist. They place the finite element method in the context of the broadest class of numerical analysis procedures and describe its advantages. They explain the key features of the finite element method and the most significant developments of its basic theory.

Chapter 3 is likewise a review article, but its focus is concentrated upon biological flow analysis, which is also the topic of the subsequent chapter. In Chapter 3, Skalak and Tözeren look at past and present accomplishments, with particular attention being given to microcirculation. Wille, in Chapter 4, constructs a finite element formulation of the axisymmetric Navier–Stokes equations. These are employed in analyses of two models of arterial abnormalities, from which conclusions are drawn about various characteristics of the blood flow.

Lee and Tseng, in Chapter 5, set the stage for a series of four chapters on topics in finite element analyses of soft tissue mechanics. After reviewing the significance of the finite element method as a procedure for analysis of a variety of soft tissue problems, they recount their experiences in applying the method in lung tissue studies. Vawter and Shields (Chapter 6) and Karakaplan, Bieniek, and Skalak (Chapter 7) likewise study the lung. Chapter 6 describes a finite element model for the determination of the macroscopic behaviour of the lung parenchyma. It includes the effects of non-linear elastic behaviour, finite deformation, and interfacial tension. The specific problem studied is the deformation of the lung due to its own weight.

Another finite element analysis of mechanical processes in lung parenchyma is presented by Karakaplan, Bieniek, and Skalak in Chapter 7. The lung structure is modelled as a three-dimensional system of membranes in which the basic unit consists of a cluster of alveoli surrounding an air duct. The computational procedure developed in this work consists of the finite element modelling of the system of alveolar walls and the liquid surfactant effect, formulation of the system of equations, and a method of their solution for various loading conditions. The results of this analysis include data on the stress and strain conditions in the solid tissue, changes in the alveolar geometry under loading, definition of factors which determine the mechanical behaviour of the lung, and the theory of mechanical testing of specimens of parenchyma.

Pao (Chapter 8) deals with finite element analyses of the heart. The author recapitulates much of the prior finite element work on this body organ, discusses modern techniques for the measurement of the geometry and material properties of the heart, and presents the results of recent finite element analyses of the left ventricle of the heart. Finite element analyses of bioprosthetic heart valves are described by Christie and Medland in Chapter 9. Their finite element representation characterizes the heart valve leaflets as hyperelastic, incompressible membranes which are capable of undergoing finite strains.

Rybicki (Chapter 10), in a review article, describes the background of analysis of orthopaedics problems, including the early approaches based on classical methods of structural analysis. He discusses the components of any analysis of an orthopaedic problem, such as the goals of the analysis, the approximations that are made, and the needed material property measurements. This chapter is intended as a philosophical basis for the subsequent seven chapters on specific aspects of orthopaedics problems. It includes, however, a description of other work in the field that is not covered in the subsequent chapters.

The analysis, by the finite element method, of external fixation devices for the treatment of open bone fractures, is presented by Chao and An in Chapter 11. Analyses with both the use of beam elements and by means of finite element codes are discussed. Among other things, the chapter demonstrates that effective biomechanical analyses can be performed with widely distributed finite element programs.

Hayes, Snyder, Levine, and Ramaswamy, Chapter 12, study the behaviour of trabecular bone of the patella by use of finite element analytical models. These models are developed on the basis of microstructural analyses of patellar geometry and literature values for loading and material properties. They demonstrate that the predicted stresses in trabecular bone are highly sensitive to the specification of trabecular anisotropy; other key features of structural behaviour are also elucidated.

A two-dimensional finite element model of the femoral head is formulated by Brown, Way, Fu, and Ferguson in Chapter 13. Their objective is the study of femoral head necrosis and of factors in the normal growth patterns which might be implicated in the onset of Legg–Perthes disease. Spilker (Chapter 14), in a departure from the finite element approach used by all others in the orthopaedic studies of this book, uses the hybrid stress approach to study the invertebral disc. This approach has a special appeal for problems involving a solid enclosing an incompressible viscous fluid, which is the case here.

Huiskes, Janssen, and Slooff (Chapter 15), introduce not only two- and three-dimensional finite element models for the stress analysis of artificial joint fixation problems, but also a classical formulation. The latter provides rough design guidelines that are simple to use. The finite element models then enable first-order evaluations of specific stem designs. Tarr, Clarke, Gruen, and Sarmiento (Chapter 16) take up the difficult question of failure of total hip replacements on account of failure of the bond between the bone and the implant. They also examine two- and three-dimensional finite element representations and discuss, in some detail, the requirements for improved analytical model of the femur and prosthesis.

A combined finite element experimental study of the femur is described in Chapter 17 by Rohlmann, Bergmann, and Koelbel. The finite element representation of the bone and the endoprosthetic fixation consists of solid, brick-type elements. Because of the three-dimensionality of the representation almost 2000 distinct elements were required. The experimental study involved a strain-gauge-instrumented femur. Interesting theory–test correlations are given in this chapter.

The single paper on head injury analysis, by Hosey and Liu (Chapter 18), brings together a considerable amount of past work by these and other investigators, in a relatively large-scale and sophisticated finite element representation of the human head and neck. The model describes accurately the geometry of the skull, brain, spinal cord, and cervical column, and other components, into one single model. Special finite elements are employed for each class of component. Computations performed with use of the model are intended to delineate the complex phenomena associated with direct-contact closed head injury.

Finite Elements in Biomechanics
Edited by R. H. Gallagher, B. R. Simon, P. C. Johnson, and J. F. Gross

Chapter 1

An Orthopaedist Looks at Finite Element Analysis

George T. Rab

1.1 INTRODUCTION

Orthopaedic surgery has always been a pragmatic and clinically oriented medical speciality. However, with the development of total joint replacement over the last fifteen years, there has been an explosion of technological advance and excitement centred about this speciality that has changed its entire complexion. More clinicians than ever are entering orthopaedics with a specific basic science commitment and at least a rudimentary knowledge of mechanics. Engineers involved in materials testing, structural analysis, kinematics, and design are involved in both research and patient-related activities. The emphasis placed on the understanding of stress and strain in bone and on the behaviour of prosthetic implants has encouraged the development and application of sophisticated structural analytic techniques to orthopaedic biological systems. In many ways, finite element analysis is the most exciting and promising of these; but serious difficulties may conceivably limit its ultimate usefulness unless they are addressed and corrected. This chapter concerns the promises and problems of finite element analysis (FEA) as they are perceived by an orthopaedic surgeon.

Orthopaedists are clinicians, having a basic interest in patient care and usually an empiric interest in mechanics. The average orthopaedic surgeon will treat 20 to 40 patients per day, or perhaps perform two or three surgical procedures, which tends to use a lot of time that could otherwise be spent in study. Even the academic orthopaedist, whose job it is to provide a more basic scientific emphasis in the field, ends up with a heavy clinical load. For this reason, when they go to the literature for information, most orthopaedic surgeons read medical journals and avoid engineering literature. Very few orthopaedists have a mechancial engineering background *per se*; medicine by its nature to be empiric, and logic and technique of a procedure often mean more to the average surgeon than the basic mechanical reasons for performing it. Because most physicians have a empiric understanding of the way the body reacts, they are frequently right in their assumptions; unfortunately, they are also frequently wrong, and some disasters that have occurred in our field should have been predictable by a more academic approach. In general, the orthopaedic community needs to be shown specific areas where finite element analysis can be helpful and where it can not, so that appropriate amounts of hard science and empiric knowledge can be blended.

1.2 PROBLEMS OF COLLABORATION

As finite element analysis is increasingly employed for orthopaedic problems, orthopaedists and engineers find themselves frequent collaborators in research projects. The physician's role is usually that of advisor, providing a clinically important context for the project; the engineer's responsibility is usually that of actual modelling and software production and its interpretation. Engineers and physicians, although both scientists, may have a very different philosophy of investigation. Engineering has a directed, single-minded approach to problems, whereas physicians frequently feel that they can't wait for science to give them all the answers and they must move ahead on a empiric basis. They may feel that rigorous engineering studies take too much time, and make too many assumptions (often rendering the project of less clinical value). For this reason, communication between the two specialities has often been difficult. It is rare to get engineering and medical talents together where they can both nourish ideas, teach each other, and be a team rather than two separate people. The most important aspect of cooperation between the orthopaedist and the engineer is a true desire of each speciality to understand the problems of the other. Without this, the magic that is mandatory for useful propagation of ideas is lost.

Problems of collaboration are not small. Engineers must seek out physician collaborators who have a definite interest in learning engineering. A sensitive physician will realize he needs to learn new ways of approaching ideas, new nomenclature, and some mathematics. On the other hand, each engineer who is striving to do quality work in orthopaedics has to be willing to immerse himself into clinical areas where he will begin to get the empiric feel of biological tissue that most physicians have refined since the time that they started medical school. Without this, projects will continue to be intellectually stimulating but devoid of deep medical value except in rare instances. One source of collaboration for the engineer in an academic department is a member of the resident staff or a young faculty member who has not yet made his mark clinically. Such a person is a bit closer to some of the basic sciences in his speciality and is not yet so immersed into clinical medicine that he can not extract himself to get free for a period of time and learn something new. The engineer also needs to attend all the clinical conferences he can; frequently, he will be called to comment on points which he may not feel are answerable, but by listening and participating he can be cognizant of ideas that have great clinical relevance.

1.3 BIOLOGICAL TISSUES

Biological tissues are frustrating for engineers. They do not behave like normal materials that were studied in classical structural analysis. All the biological systems require significant assumptions and optimizations, yet there is no good way of verifying the optimizations. The systems are three-dimensional, adding a degree of complexity that puts finite element work in a very difficult position. The basic knowledge that even astute clinicians possess is primarily based on two-dimensional thinking, and yet this may have little relevance to a three-dimensional human body and the way it behaves. Nevertheless, a large number of current finite element projects are two-dimensional and almost all of our knowledge of joint loading is two-dimensional.

It may be argued that we really have little knowledge of the loading conditions for any joint. The hip is an interesting example of this. The hip joint has been fascinating to orthopaedic surgeons as well as to engineers because of the unique anatomy, blood supply, and the unique diseases that occur in it. It forms the nidus of a large number of common clinical problems, and it was the site of the first major successful implant surgery in orthopaedics. Yet even in two dimensions, we do not understand the exact geography of forced transmission across the joint. We may be able to define an approximation of the three-dimensional forces in a stance phase of gait, but when we examine normal gait cycles or analyse any type of activity, we can get completely lost. Before accurate finite element work can be developed to a high degree, we must return to the science of joint mechanics and learn much more. Most of the exquisite mechanisms of the body do not easily lend themselves to assumptions and opitimizations.

1.4 SKELETON AS A LABORATORY

Fortunately for those engineers dealing in orthopaedics, the skeleton is an excellent physiological laboratory. X-rays, despite their problems, give a window through which to look at the skeleton. It has been known for many years that form in the skeleton follows function. The patient who has polio grows up with a limb that is smaller than the normal, because muscle force is less; the patient who has a fracture that heals with angulation will eventually fill in the concave side of the fracture and resorb the convex side, because of transmission of stress; the patient who is heavy will have larger bones; the patient who is aged may have the same quantity of bone but the bone itself will expand. These are all manifestations of Wolff's law: bone will remodel along the lines of stress; therefore, we have excellent approximation of stress within the skeleton simply by examination of the skeleton itself. Finite element work that demonstrates close correlation between lines of stress and trabecular pattern lends more credence to the analysis itself.

One area that is an excellent physiological laboratory is the growth plate in children. The growth plate is made of cartilage, and is the site at which growth occurs at the end of the bone. Cartilage is relatively weak in shear. The structure and orientation of the cartilage growth plate is an excellent indication of the relative amounts of shear and compressive force, averaged over time, being applied to the skeleton in three dimensions at that site. As an example, the growth plate of the proximal femur is conical and convoluted so that it can resist shear forces of one to two times body weight in several different planes, whereas the growth plate of the proximal tibia is flat and smooth, indicating the lack of shear on the tibial plateau. This correlates well with the findings that the tibial component of total knee replacement arthroplasty need not be grossly convoluted in order to remain fixed, at least on a short-term basis. The physiological behaviour of bone can frequently be assessed by examining the structure of the bone itself.

In the last few years, the development of computerized tomography has provided us with information on three-dimensional structure and on localized material densities within bone. These may be correlated with material properties so that our anatomic information can be refined. New revolutionary radiographic equipment is currently being developed which will further improve our understanding of three-dimensional bone and soft tissue anatomy, and radiologic imaging of skeletal systems to obtain material data and three-dimensional anatomic data will be supportive of future efforts in finite element method applications to orthopaedics.

1.5 FAILURES AND DISAPPOINTMENTS

The ultimate determinant of success in any type of medical research, be it clinical or basic: is 'does it help the patient?' Many studies have been done which, despite meticulous attention to detail, have simply not been relevant. This does not mean that a irrelevant project is irrevelant to the person doing it, but it does mean that in a biological field, purely theoretical studies remain theoretical. They take second place to studies which have importance to clinical science. Frequently we do not know what the ultimate result of the studies will be, and I don't advocate eliminating pure theory. However, the persistence of highly technical, highly theoretical studies without clinical goals in mind can be detrimental. Many of the finite element studies of the last five years have melted into obsurity because they were either poorly designed, or not received as pertinent to any recognized problem, or there were too many assumptions made as to loading and boundary conditions. If loading conditions are grossly simplified, especially if the analysis is done in two dimensions, there can be very little translation from basic science to the clinical reality of practice.

The computer (which is the engineer's closest ally in the finite element work and produces so much information) can become the reason for doing the research rather than the method by which the research is accomplished. The computer is also expensive, and there seems to be general feeling that an expensive project is better than an inexpensive one. We gear ourselves to this type of an attitude, even as scientists. In a quest for grant money and in a quest for new ideas, we sometimes forget what it is that we are trying to achieve. Computer expenses will continue to grow as more and more analyses are done in three dimensions, and one must be cautioned not to look at the computer as the answer. The answer is going to come in characterizations of biological systems.

1.6 SUCCESSES OF FINITE ELEMENT ANALYSIS IN ORTHOPAEDICS

I do not mean to imply that the finite element method has been a failure in all of its applications to orthopaedics. There have been several areas where FEA has had extensive clinical impact; a large number of the authors of these papers have contributed to this volume. Early two-dimensional characterization of stress in the proximal femur has been particularly relevant to the subtrochanteric area, and helped explain on a theoretical basis why we in orthopaedics see so many failures of fixation devices in this portion of the skeleton. There have been excellent descriptions of stem breakage in total hip arthroplasty, and stem breakage proved to be a major clinical problem which was unanticipated in the early history of total hip surgery. Finite element work has provided us with a much sounder theoretical basis for cementing techniques in total joint prosthesis, for positioning of prosthesis, for methods of reaming bone, and for some of the designs of our prosthetic stems. Yet, although much work in stem design has been assisted by FEA, the same deductions could have easily been made by empiric evaluation of the data and did not really require finite element analysis.

Parametric studies using FEA have been extremely important in our understanding of bone, cement and prosthetic interfaces, and in decisions on how thick cement layers should be. This is particularly true in understanding the behaviour of the proximal tibial component of the total knee replacement. Whereas excellent work using the finite element method has been done, the fact remains that a poor prosthesis put in by a skilled technical surgeon will almost always work better than a well-designed prosthesis put in by a poor surgeon. This has been the very factor that has added to some of the scepticism that practising physicians feel about sophisticated engineering evaluation of biological structures. Thus, despite the encouraging advances of finite element work that have had clinical impact, the very behaviour of bone as a living substance – becoming denser or resorbing in relation to stress – can confuse even the most meticulous scientist and quickly render elegant analytic models useless. We think that our design is excellent, we think that our procedures are excellent, we think our surgical technique is excellent; and yet in a short time a joint replacement is lost because the supporting bone has resorbed since it was not subjected to the proper amount of stress needed to maintain strength.

1.7 THOUGHTS FOR THE FUTURE

As pure research, finite element work is justifiable on many levels FEA and its relationship to the biological sciences is an excellent new avenue for scientists of different backgrounds to get together; thus a young engineering Ph.D. candidate, working on a project or presenting a paper, can get insight into an area that may launch an entire career. A young physician involved in an engineering project may learn why engineers attack projects analytically instead empirically, and thus bridge an important communications gap between medical and engineering scientists. It is impossible to put a 'dollar-and-cents' figure on the value of this particular interaction.

Many of us are involved in teaching or research that depends on grant money for support. Recently, finite element analysis has been generously backed by both private and institutional funds, and I think that it is both realistic and justifiable to consider finite element studies as one method for obtaining such grant money.

Finite element analysis, more than almost any other technique in orthopaedic basic science, provides a pictorial quality that tends to attract the orthopaedist into engineering. That particular talent of finite element work should be nourished and kept alive to serve as a crossroads and training ground for engineers and physicians. FEA will also continue to attempt to provide theoretical analysis of recognized problems (such as total hip stem breakage) but engineers working in the field must recognize the limitations of trying to use the finite element method with biological materials. The true behaviour of biological materials (as opposed to the engineering materials) has not yet been described. Over the next decade, orthopaedist and engineers will need to have very close contact with bone physiologists and engineers studying bone.

In the coming years, several things will be mandatory. One is that studies will need to become three-dimensional. Two-dimensional analyses, except for specific problems, do not suffice. Secondly, we need to become more cognizant of the dynamic behaviour of joint loading as opposed to static behaviour. It is possible that because of the interest generated by finite element work we have jumped beyond our capabilities. We really have to go back a step and understand joint mechanics. We need to understand the distribution characteristics of forces across a joint surface, particularly when dealing with structural analysis of that portion of the joint close to the articular surface. We have to decide whether to model bone as an anisotropic or isotropic material. We need to find out isolated material properties of small areas of bone, determine their geographic location in the skeleton, and model them. We need to understand the behaviour of cartilage better and how it can be modelled with finite element analysis.

We need to redefine the relationship between stress and bone remodelling and between stress and trabecular patterns – this is an area where the physiologist and the clinician can be of great help to the engineering scientist. We desperately need verification studies, including animal models, instrumented implants, and mechanical stress testing of bone and joint systems. We need to correlate our long-term clinical studies with theoretical FEA studies, to check our hypotheses and confirm or deny model assumptions.

These are difficult decisions for the scientists of today working in orthopaedic applications of FEA. Shall we continue blindly with finite element work of ever-increasing complexity and elegance, only to discover it may be completely devoid of meaning in the ultimate clinical sense, or should we step back, take a hard scientific look at loading and boundary conditions and mechanical characterization of biological tissue, and verify where we are before journeying onward? The answers to these questions may determine whether we write only a brief chapter in biomechanics or whether we lay a new foundation in the scientific study of an exciting clinical field.

Finite Elements in Biomechanics
Edited by R. H. Gallagher, B. R. Simon, P. C. Johnson, and J. F. Gross

Chapter 2

Finite Elements – A Unified Problem-solving and Information Transfer Method

O. C. Zienkiewicz and D. W. Kelly

2.1 INTRODUCTION

The words 'finite element method' made their first appearance in literature of mechanics some twenty years ago. Originally introduced as a process of solving structural mechanics problems, the method was soon recognized as a general procedure of numerical approximation to all physical problems capable of being modelled by a differential equation description. At this stage it was recognized also that its roots lie in the approximation processes derived at the turn of the century and that, in its most general definition, it unifies all alternative processes of approximation, for instance, the finite difference and boundary solution (Trefftz) methods.

Clearly therefore, the range of application of the finite element methodology becomes ever wider and conferences dealing with specific fields of activity are natural. Over the years we have witnessed many such meetings ranging from structural mechanics, geomechanics, fluid mechanics, to electromagnetics. In parallel with this, numbers of meetings in which the methodology rather than the field of application predominates, have been increasing.

Both phenomena are of considerable benefit as they encourage a rapid transfer of information and development from one field to another. This occurs through commonality of the mathematics basis, software development and, more directly, physical analogy. By encouraging the same practitioner to work simultaneously (or consecutively) in different fields of application, the measure of dilletantism is counterbalanced by transfer of valuable experience, and progress is made in this manner.

Biomechanics and indeed the wider subject of bio-engineering, is no exception in this information chain. Here direct application of methodologies developed for other problem areas is often possible as the starting point, and conversely special problems encountered lead to new solutions whose usefulness elsewhere will in turn be found.

As an illustration of the first, stress analysis problems of bone or prosthesis are most obvious. Methodology and indeed, complete software existing for analysis of mechanical components or civil engineering structures can be used directly with a simple change of geometric definition. Figure 2.1 illustrates an early model (1971) of a femur which defines an analysis of just such an extension.

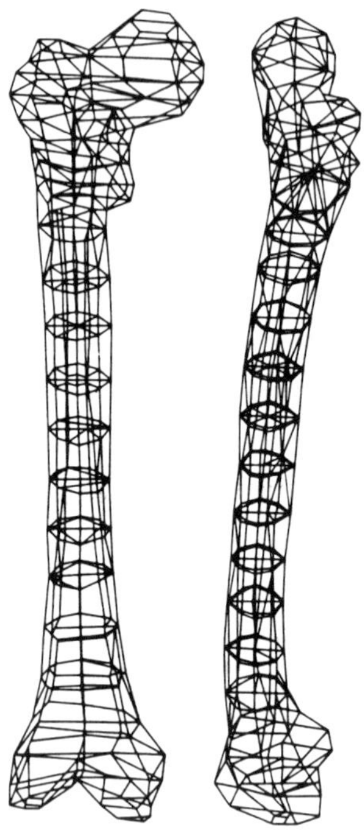

Figure 2.1 Problem of biomechanics. Plot of linear elements form only; curvature of elements omitted

Immediately, however, problems arise. The essentially two-phase nature of the bone material consisting of a solid, but porous, matrix and an interstitial fluid poses a special problem of definition. Under rapid loading the material as a whole acts in almost an incompressible manner while in slower loading phenomena, transfer of the interstitial fluid occurs giving an entirely different total response. It appears immediately that a new and specialized problem applicable only to the biomechanics area has been encountered and must be solved in that group of activity. This doubtless is true in other disciplines, e.g. in geomechanics, where an analogous situation exists and has been frequently addressed. Thus, via the medium of finite elements, a transfer of modelling knowledge is achieved and both fields benefit.

The example quoted involves indeed another connection. In both situations (i.e. the biomechanical and the geotechnical), under certain circumstances incompressible or almost incompressible behaviour occurs and standard approaches are no longer applicable. Here developments which have taken place in the mathematics of finite elements (motivated by fluid mechanics) are immediately applicable.

2.2 THE FINITE ELEMENT METHOD AND ITS RELATION TO OTHER APPROXIMATION PROCESSES

In its present form the definition of the finite element method may be made as follows:

The (generalized) finite element method is a process of approximation to the solution of a continuous function (or functions) ϕ in which:

(a) the unknown function is approximated by a series of trial functions N_i (alternatively known as basis or shape functions) and a set of unknown parameters a_i, i.e.

$$\phi \simeq \hat{\phi} = \sum_{i=1}^{n} N_i a_i = Na \tag{2.1}$$

(b) the algebraic equations from which the unknown parameters are determined are of the form

$$\int_{\Omega} G_j(\hat{\phi})\, d\Omega + \int_{\Gamma} g(\hat{\phi})\, d\Gamma = 0 \qquad j = 1, \ldots, n \tag{2.2}$$

where Ω is the domain of the problem and Γ its boundary.

Immediately the reader will observe that it is possible to subdivide the problem domain (and its boundary) into any number of subdomains (or elements) such that

$$\sum_{e=1}^{m} \Omega^e = \Omega \qquad \sum_{e=1}^{l} \Gamma^e = \Gamma \tag{2.3}$$

It follows therefore that the *assembly* of the set of equations is carried out on the lines familiar to all structural engineers who obtain their overall equilibrium equations by simply summing the contribution made by individual members of a structure. Indeed, this system of assembly is common to many other disciplines – such as those involving electric networks or assemblages of hydraulic components. It is not surprising, therefore, to observe the popularity of finite element methods due to their analogous behaviour to known systems – and the ease with which they can be adapted to existing computer software.

In the general definition given above, the trial functions can be of a form extending over the whole domain (in a manner of well-known trigonometric Fourier series), or, as is more usual, it can be defined in a piecewise way over the elements with their 'base' being narrow and associated with parameters a_i defined as nodal values belonging to only few elements. Such local definitions are particularly advantageous in the solution of the equation systems resulting from Equation (2.2), which in linear problems will take the form of a simple algebraic set

$$Ka = f \tag{2.4}$$

Now the matrix K is sparse and, if suitably arranged can take up a banded form for economical computer storage.

So far we have not mentioned the manner in which the approximation equations (2.2) are derived from the continuum statement of the problem. Most commonly this is given as a set of differential equations defined in the domain Ω and written here in a simple operator form as

$$A(\phi) = 0 \quad \text{in } \Omega \tag{2.5}$$

These together with the boundary conditions imposed on the boundary Γ as

$$B(\phi) = 0 \quad \text{on } \Gamma \tag{2.6}$$

define the problem.

The reader can visualize a typical example. For instance, the solution of the Poisson equation

$$A(\phi) = \nabla^2\phi - q = 0 \tag{2.5a}$$

in Ω with

$$B(\phi) = \phi - \overline{\phi} = 0 \tag{2.6a}$$

on Γ (where $\overline{\phi}$ are prescribed values).

To obtain the approximating equation in form of Equation (2.2) we proceed as follows. Taking a set of prescribed test (or weighting) functions W_j and $\bar{W}_j$, we write

$$\int_\Omega W_j^{\mathrm{T}} A(\hat{\phi})\, \mathrm{d}\Omega + \int_\Gamma \bar{W}_j^{\mathrm{T}} B(\hat{\phi})\, \mathrm{d}\Gamma = 0 \qquad j = 1, \ldots, n \tag{2.7}$$

i.e. we 'orthogonalize' the errors $A(\hat{\phi})$ and $B(\hat{\phi})$ with these functions.

Clearly as $\hat{\phi} \to \phi$ by Equations (2.5) and (2.6), the exact solution will be obtained for *any* test functions chosen. Equation (2.7) simply approximates this limiting case. Providing a set of trial functions of a form capable of reproducing (in the limit) the exact answer is chosen, the solution will be 'convergent' for an increasing number of parameters $a_i (n \to \infty)$.

The procedure outlined above had been descriptively named the 'method of weighted residuals' (Crandall, 1956; Finlayson, 1972) long before the advent of the finite element nomenclature, and in essence fathers most of the well-known approximation procedures–thus defining it as the 'generalized finite element' process allows us to include all the alternatives in a single framework. The variants merely depend on the choice of *the trial and test functions.*

Thus, in most conventional finite element forms, piecewise-defined trial functions are used and the weighting functions are made identical to the trial function in the problem domain, i.e.

$$N_i = W_i \tag{2.8}$$

is taken, and the method is a particular case of the Galerkin–Bubnov process (Galerkin, 1915). A typical one-dimensional trial function is shown in Figure 2.2(a) with its corresponding weighting equivalent. Many other more or less ingenious elements in one, two, and three independent variables are now available and are well described in the literature.

However, even in 'standard' finite element applications the use of the Galerkin–Bubnov process is not necessarily optimal. In recent years much attention has been focused on problems in which convective terms appear such as typified by the heat transfer equation below

$$\nabla k \nabla \phi - c\mathbf{u}^{\mathrm{T}} \nabla \phi = 0 \tag{2.9}$$

in which k is the conductivity, c the specific heat and $\mathbf{u}$ the velocity field. Here the use of non-symmetric test functions is necessary to achieve best accuracy. In Figure 2.2(b) a typical form of such test functions is indicated (Christie *et al.*, 1976; Zienkiewicz *et al.*, 1977).

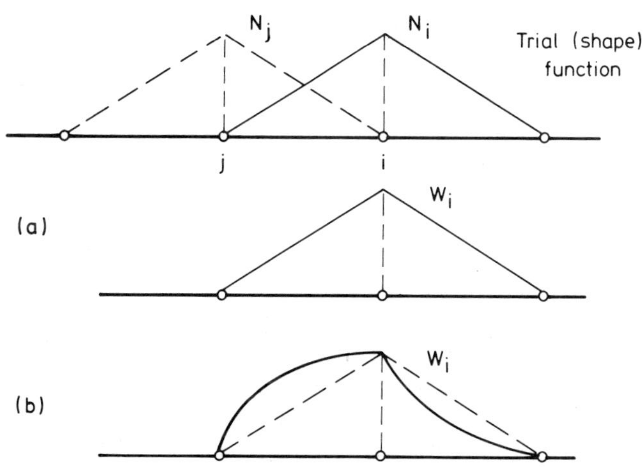

Figure 2.2 Simple linear element trial and test functions
(a) Weight functions for typical Galerkin-Bubnov process
(b) Form used for convective-diffusion problems

If functions are not locally based we revert to classical forms of approximation (e.g. Fourier series well known before the process became computerized). Today we realize that such functions are just a particular case of the general method and

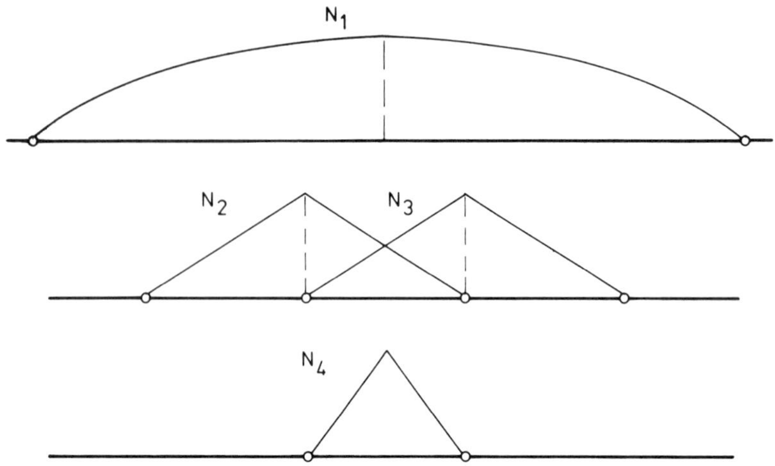

Figure 2.3 Global and hierarchical functions

often combine these with local approximations. Thus as shown in Figure 2.3, it is quite simple to define a set of 'hierarchical' functions, the superposition of which can achieve merits of both local and global approximation. Here 'elements' are no longer easily distinguished but clearly the procedures remain identical. Mote (1971) was the first to realize such possibilities, but the concept of 'hierarchical' functions was introduced earlier (Zienkiewicz *et al.*, 1970), leading to many very general adaptive refinement processes (Peano *et al.*, 1977; Babuska *et al.*, 1979). Today we are even witnessing a complete reversal of the subdivision process in which only global functions are used in a region suitably mapped into convenient rectangular or cubic domains in two- and three-dimensional problems (Delves and Hall, 1979).

While most finite element applications utilize the Galerkin–Bubnov process which in self-adjoint problems leads to symmetry of the matrix K of Equation (2.4) (and gives results identical to those obtained by the use of variational methods) it is of course possible to deliberately choose a different form of trial and test function. One typical process is that of *collocation* in which Dirac δ functions are used

$$W_j = \delta(x_j) \tag{2.10}$$

In many cases excellent approximation can be obtained by such methods (Finlayson, 1972) with globally defined trial functions. We now simply ensure that the approximation $A(\hat{\phi}) = 0$ (or $B(\hat{\phi}) = 0$) is satisfied at a selected number of domain or boundary points.

If trial functions are defined on a local basis and the collocation process is used, the finite element method becomes identical to that of 'finite difference approximation'. As an example of such an approximation the reader may consider the trivial approximation to

$$\frac{d^2\phi}{dx^2} = 0 \tag{2.11}$$

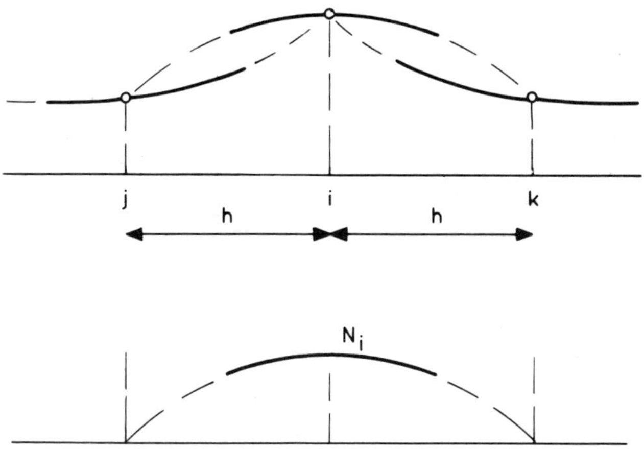

Figure 2.4 A set of discontinuous local trial functions leading to a finite difference approximation

In Figure 2.4 we show a series of (discontinuous) test functions of parabolic shape which, if weighted in the collocation manner will lead to the well-known finite difference approximation to Equation (2.11),

$$(\phi_j - 2\phi_i + \phi_k)/h^2 = 0 \tag{2.12}$$

The question of continuity of trial functions and indeed of their form necessary for convergence presents problems which are discussed elsewhere. It is, however, easy to show that the above approximation is consistent and presents a special (but perhaps cumbersome) process of deriving a well-known and useful method. This generally yields little advantage *vis-à-vis* the more general Galerkin–Bubnov methods.

Of more interest is perhaps the exploration of yet another form of trial function leading to the 'rediscovery' of the well-known 'boundary methods' (Trefftz, 1926; Banerjee and Butterfield, 1979). If the operator A of Equation (2.5) is linear and we choose *a priori* trial functions of such a form that

$$A(N_i) \equiv 0 \tag{2.13}$$

then the approximation of Equation (2.7) reduces simply to a 'boundary integral'

$$\int_{\Gamma} \bar{W}_j B(\hat{\phi}) \mathrm{d}\Gamma = 0 \tag{2.14}$$

and the dimensionality of the problem is immediately changed, necessitating fewer variables to obtain solutions of a given accuracy (Trefftz, 1926; Banerjee and Butterfield, 1979). Such boundary methods are of considerable use in some classes of problems. Realizing that they are a particular subclass of the finite element process, their merits can be appropriately exploited. Thus today it is possible to link 'standard' finite elements and 'boundary elements' in the same computer code and deal with such problems as exterior infinite boundaries or singularities (Zienkiewick, Kelly, and Bettess, 1977a, b).

The brief remarks made here are intended to illustrate the essential unity of the finite element process with its predecessors and the broad common foundation of most general approximation processes. Such a unity, however, was not immediately realized, as will be shown in the next section.

2.3 THE HISTORICAL BACKGROUND

In the mid-1950s the computer became for the first time a powerful tool for engineering design calculations, even though it had been invented more than a decade earlier. Structural computations for complex aircraft systems became commonplace and appropriate matrix methods made such calculations systematic. In this context, naturally, interest turned to the possibility of solving not only problems in which true discrete members were present, but others in which aircraft 'skin' elements were incorporated. Argyris (1960), in his early and important work on a systematic matrix presentation indicated the manner by which behaviour of panels could be approximated by the use of equivalent stiffness components. At the same time Turner *et al.* (1956) suggested how by an assemblage of such stiffnesses a true continuum could be approximated. One member of this team, Clough (1960), (realizing that a systematic method of approximation was at hand), extended such applications and for the first time suggested the name of 'finite element method' for this process. At the same time a Polish engineer, Szmelter (1959) showed how such an approximation could be derived, with locally based trial functions, and the use of an energy minimization procedure.

This early work combined with computer availability laid the foundation for the dramatic and rapid expansion of the concepts. Gallagher *et al.*, (1962) suggested extension to three-dimensional solids and non-linear behaviour, Melosh (1963) and Zienkiewicz and Cheung (1964a, b) dealt with problems of plates and shells. However, the general mathematical roots of the process were only slowly realized and the generalization to any mathematical problems for which a variational principle exists was first presented in the literature in 1965 (Zienkiewicz and Cheung, 1965).

Ergatoudis, J. G., Irons, B. M., and Zienkiewicz, O. C. (1968) Curved, isoparametric, 'quadrilateral' elements for finite element analysis, *Int. J. Solids & Struct.*, **4,** 31–42.

Finlayson, B. A. (1972) *The Method of Weighted Residuals and Variational Principles*, Academic Press.

Galerkin, B. G. (1915) Series solution of some problems of elastic equilibrium of rods and plates (Russian), *Vestn. Inzh. Tech.*, **19,** 897–908.

Gallagher, R. H., Padlog, J., and Bulaard, P. P. (1962) Stress analysis of heated complex shapes, *A.R.S. Journal*, 700–7.

Godbole, P. N., Nakazawa, S., and Zienkiewicz, O. C. (1980) Blood flow analysis by finite element method, *Proceedings of International Conference on Finite Element Methods in Biomechanics*, B. Simon (ed.), Tucson, Arizona.

Gordon, W. J. (1971) Blending-function methods of bivariate and multivariate interpolation and approximation, *SIAM J. Num. Anal.*, **8,** 158–77.

Hinton, E., Abdel Rahman, H. H., Huq, M. M. (1980) Computational models for the nonlinear analysis of reinforced concrete plates, *Proceedings of Symposium on Research in Nonlinear Structural and Solid Mechanics*, NASA-2147.

Irons, B. M. (1966) Engineering application of numerical integration in stiffness method, *J.A.I.A.A.*, **14,** 2935–7.

Irons, B. M. (1969) A conforming quartic triangular element for plate bending, *Int. J. Num. Meth. Engng.*, **1,** 29–46.

Melosh, R. J. (1963) Basis of derivation of matrices for the direct stiffness method, *J.A.I.A.A.*, **1,** 1631–7.

Mote, C. D., Jr. (1971) Global-local finite element, *Int. J. Num. Meth. Engng.*, **3,** 565–74.

Oden, J. T., and Reddy, J. N. (1976) *An Introduction to the Mathematical Theory of Finite Elements*, John Wiley & Sons Limited.

Onate, E., and Zienkiewicz, O. C. (1978) Plastic flow of axisymmetric thin shells as a non-Newtonian flow problem and its applications to stretch forming and deep drawing problems, 10th Biennial Congress of the International Deep Drawing Research Group, England.

Peano, A. G., Szabo, B. A., and Mehta, A. K. (1977) Self-adaptive finite elements in fracture mechanics, *Proc. Innovative Num. Anal. in Appl. Engng. Science*, Paris.

Strang, G., and Fix, G. J. (1973) *An Analysis of the Finite Element Method*, Prentice-Hall, p. 106.

Szmelter, J. (1959) The energy method of networks of arbitrary shape in problems of the theory of elasticity, *Proc. I.U.T.A.M. Symposium on non-homogeneity in Elasticity and Plasticity* (ed. W. Olszak), Pergamon Press.

Taig, I. C. (1961) Structural analysis by the matrix displacement method, *Engl. Electric Aviation Report No. S017.*

Taylor, C., Hinton, E., and Owen, D. R. J. (eds.) (1980) *Numerical Methods for Nonlinear Problems*, Vol. 1. Pineridge Press. (*Proceedings of International Conference on Numerical Methods for Nonlinear Problems*, Swansea, 1980).

Trefftz, E. (1926) Gegenstück zom Ritz'schen Verfahren, *Proc. Sec. Int. Congress Applied Mechanics*, Zurich.

Turner, M. J., Clough, R. W., Martin, H. C., and Topp, L. J. (1956) Stiffness and deflection analysis of complex structures, *J. Aero Sci.* **23,** 805–23.

Visser, W. (1968) The finite element method in deformation and heat conduction problems, Ph.D. Dissertation, Delft.

Zienkiewicz, O. C. (1970) From intuition to generality, *App. Mechs. Review*, March, 249–256.

Zienkiewicz, O. C., Chang, C. T., and Bettess, P. (1980) Drained, undrained, consolidating and dynamic behaviour assumptions in soils. Limits of validity, *Geotechnique* (to be published).

Zienkiewicz, O. C., and Cheung, Y. K. (1964a) The finite element method for analysis of elastic isotropic and orthotropic slabs, *Proc. Inst. Civ. Eng.*, **28,** 471–88.

Zienkiewicz, O. C., and Cheung, Y. K. (1964b) Finite element method of analysis for arch dam shells and comparison with finite difference procedures, *Proc. Symp. Theory of Arch Dams*, Southampton Univ. (Pergamon Press, 1965).

Zienkiewicz, O. C., and Cheung, Y. K. (1965) Finite elements in the solution of field problems, *The Engineer*, Sept., 597–10.

Zienkiewicz, O. C., Heinrich, J. C., Huyakorn, P. S., and Mitchell, A. R. (1977) An upwind finite element scheme for two dimensional convective transport equations, *Int. J. Num. Meth. Engng.*, **11,** 131–44.

Zienkiewicz, O. C., Irons, B. M., Campbell, J., and Scott, F. C. (1970) Three dimensional stress analysis, *Int. Un. Th. Appl. Mech. Symposium on High Speed Computing in Elasticity*, Liege.

Zienkiewicz, O. C., Jain, P. C., and Onatc, E. (1978) Flow of solids during forming and extrusion: some aspects of numerical solutions, *Int. J. Solids Structures*, **14,** 15–38.

Zienkiewicz, O. C., Kelly, D. W., and Bettess, P. (1977a) The coupling of the finite element method and boundary solution procedures, *Int. J. Num. Meth. Engng.*, **11,** 355–76.

Zienkiewicz, O. C., Kelly, D. W., and Bettess, P. (1977b) Marriage a la mode, Finite elements and boundary integrals, *Proc. Conf. Innovative Numerical Analysis in Engineering Science*, CETIM, Paris (to be published).

Zienkiewicz, O. C., Leung, K. H., Hinton, E., and Chang, C. T. (1980) Earth dam analysis for earthquakes: Numerical solution and constitutive relations for nonlinear (damage) analysis, *Design of Dams to Resist Earthquake*, Inst. Civil Eng., London.

Zienkiewicz, O. C., Onate, E., and Heinrich, J. C. Plastic flow in metal forming. I Coupled Thermal, II Thin Sheet Forming, *Applications of Numerical Methods to Forming Processes*, AMD-Vol. 28.

Finite Elements in Biomechanics
Edited by R. H. Gallagher, B. R. Simon, P. C. Johnson, and J. F. Gross

Chapter 3

Finite Elements in Bio-fluid Mechanics

Richard Skalak and Hüsnü Tözeren

3.1 INTRODUCTION

Finite element methods are particularly suitable for bio-fluid problems because biological phenomena often involve non-linear governing equations, moving boundaries, and irregular boundary geometries. The chief category of biological flows for which finite element analyses have been given is the flow of blood. This includes high Reynolds number flow in large arteries and viscous flow of a suspension in very small blood vessels. In principle the finite element methods should also be useful for peristaltic flows such as occur in the intestines or ureter and for air flow in the lungs, for the movement of fluid in the canals of the ear, motion of cerebral-spinal fluid, for axonal flow of neural cells, for fluid flow in and out of the eye, for the deformation of the brain, and for the mechanics of swimming and flying. At the present time, of these possibilities there appear to be publications primarily in the field of blood flow and the review in this chapter will be restricted to this category.

Applications of finite element method in blood flow problems require different formulations depending on the Reynolds number of the flow, which means, for practical purposes, on the diameter of the blood vessel considered. In larger vessels, the particulate nature of blood may be ignored and blood may be considered as an incompressible and Newtonian fluid. However, due to high values of Reynolds number (100–10 000) the full Navier–Stokes equation must be considered. In the smallest blood vessels (capillaries) the acceleration terms of the fluid motion are negligible compared to unity, but blood may no longer be considered as a uniform Newtonian fluid. The discrete nature of the red blood cells and their mechanical properties must be taken into consideration to determine the velocities and pressures in blood.

There is an intermediate category of arterioles and venules from about 12 μm to 500 μm in diameter in which non-Newtonian effects are present, but analysis of the motion of each red blood cell is impractical. Finite element techniques could be used to determine velocity distributions in this range if a sufficiently general constitutive equation for blood were known and conditions at the boundary were better known. The blood cells tend to move radially inward leaving a narrow gap near the wall which has fewer blood cells in it (Goldsmith and Skalak, 1975). This leads to the Fahraeus–Lindqvist effect which is a reduction of apparent viscosity observed as the vessel diameter decreases. The constitutive and migration equations are not sufficiently well known in this range to allow accurate predictions by finite element methods.

3.2 CAPILLARY FLOW

Red blood cells are roughly disc-shaped with a diameter of about 8 μm and a thickness of about 2 μm. Capillary diameters range from about 4 to 10 μm, so in the capillaries the red blood cells more or less fill the lumen of the capillary and generally travel in single file. Further, the red blood cells are highly deformable and rarely maintain their disc shape during normal flow. The red blood cell consists of an elastic membrane filled with a haemoglobin solution which is a Newtonian fluid with a viscosity of about 6 centipoise. The suspending fluid of the red blood cells is the plasma which is a clear Newtonian fluid with a viscosity of about 1.2 centipoise at body temperature. The densities of the blood cells and the plasma are very nearly equal so that the flow may be regarded as a suspension of neutrally buoyant particles.

The Reynolds number of capillary blood flow is generally about 0.01 so that inertial terms for both the motion of the fluid and of the suspended particles may be neglected. Hence the motion of the blood plasma may be treated as a problem of an incompressible viscous flow at low Reynolds number. The Stokes equations of motion are then

$$\nabla p = \mu \nabla^2 \mathbf{u} \tag{3.1}$$

and the continuity equation is

$$\nabla \cdot \mathbf{u} = 0 \tag{3.2}$$

where $\mathbf{u}$ is the velocity and p is the pressure. The equations of motion of the particle, again at zero acceleration, must be satisfied together with the constitutive equations of the particle and continuity of the stresses at the surface of the particle. In general the particle will be deformed and moved by the stresses applied to it by the fluid. There are also integral conditions for each particle that must be satisfied which arise because the inertial and buoyant forces are considered negligible. These conditions are that the sum of the forces exerted on the particle and the sum of their moments must be equal to zero. This is usually known as the zero drag condition. In the case of a uniform speed of translation of the particles with a fixed deformed shape, the internal motion of the particle may be disregarded since the fluid internally will then exert, at most, a uniform pressure. If the particle is elastic, it is then in a state of quasi-equilibrium and the zero drag and zero moment conditions suffice to determine its motion.

The problem of viscous flow of the plasma may be treated by a number of finite element methods. The problem is exactly analogous to the solution of the displacement in an incompressible elastic solid and any of the methods suitable for this case also may be used for the plasma motion. For the fluid flow case the main choice is at first in the dependent variables. One possibility is to use stream functions if the problem is two-dimensional or axisymmetric. The other possibility is to use the fluid velocity and the pressure as dependent variables. In this case there are still several ways of proceeding (Zienkiewicz, 1977, p. 611). A variational principle can be formulated in which p enters as a Lagrangian multiplier. Other possibilities are to use the incompressibility condition Equation (3.2) with the use of a penalty function or in fact to treat the fluid as a compressible fluid of very small compressibility.

The question of the use of primitive variables versus stream function approaches has been discussed by Olson and Tuann (1978). They tentatively conclude that the stream function method is the least difficult and perhaps the most accurate for a given problem size. The use of the velocity and pressure as primary variables is advantageous in the case of suspension flow because the velocity boundary condition on the surface of the particles is most easily incorporated in this way. Further, a major saving is effected by the use of the variational principle in which the zero drag condition results as a natural boundary condition of the problem. In this approach it is not necessary to compute and integrate the stresses on the particle as they are automatically such that the integral is zero unless it is otherwise specified. The variational principle written for a suspension of rigid, neutrally buoyant particles in a viscous fluid is $\delta D_m = 0$, where

$$D_m = 2\mu \int_V e_{ij}^2 \,\mathrm{d}V - 2\int_V p\theta \,\mathrm{d}V - 2\int_{S_1} \gamma_i U_i \,\mathrm{d}S \tag{3.3}$$

In Equation (3.3), e_{ij} is the rate of deformation tensor, p is the pressure, θ is the rate of dilatation, $\theta = \nabla \cdot \mathbf{u}$, $\mathbf{u}$ is the velocity, and γ_i are components of the force per unit area on the surface S_1 which is any part of the boundary on which such stresses are specified. It is assumed that on the remainder of the boundary the velocity is prescribed. In adapting this variational principle to a finite element method, the procedure is straightforward, but caution must be observed to make the approximation of the pressure of lower degree than that of the velocity field. As pointed out by Olson and Tuann (1978), this restriction arises from the uncoupled nature of the continuity equation, i.e., it is an equation of constraint on the velocities rather than a full equation coupling velocities to the pressure.

Examples of the use of the above variational principle in conjunction with finite elements to solve the suspension flow of rigid particles placed axisymmetrically in a tube are given by Skalak *et al.*, (1972). Figure 3.1 shows a typical finite element grid used. The two components of velocity are unknowns at the corner and side

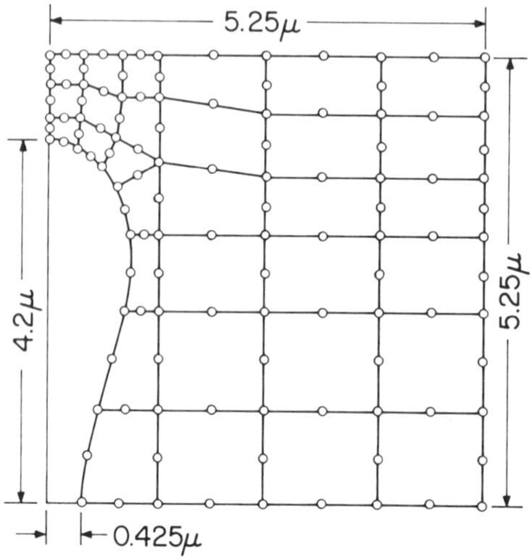

Figure 3.1 Finite element grid used for the capillary flow problem. The lower boundary is the axis of the circular tube and the upper boundary is the tube wall. Left- and right-hand boundaries are planes of symmetry. The section shown is the typical segment for an infinite line of particles located axisymmetrically and equally spaced (From Skalak *et al.*, 1972)

nodes. The elements are isoparametric curvilinear quadrilaterals with a quadratic approximation for the velocity components. The pressures are unknowns only at the corners with linear interpolation in the local coordinates. The section shown in Figure 3.1 is rotated about the axis of symmetry to generate the complete cylinder of flow and the unit is considered to be repeated indefinitely so that the problem solved is that of a line of equally spaced axisymmetric particles in a tube. Typical stream lines computed are shown in Figure 3.2. The flow between particles tends toward a Poiseuille flow. In Figure 3.3 the apparent viscosity for several diameter ratios λ and various hematocrits are shown. The diameter ratio is the cell diameter divided by the tube diameter. The hematocrit is the volumetric percentage of cells present.

The above methods have been extended to the case of elastic cells by Zarda (1975) and Zarda *et al.* (1977). The variational theorem indicated by Equation (3.3) can be extended to include cells which are elastic solids or elastic membranes filled with viscous fluid. Red blood cells are represented as closed elastic membranes filled with a viscous fluid. A red blood cell membrane is treated as a thin elastic shell and both the membrane strain energy and bending strain energy are included. The extended variational principle is $\delta D_n = 0$, where

$$D_n = D_m + 2\int_A T\overline{\lambda_1 \lambda_2}\,\mathrm{d}A + 2\int_A \dot{W}_m \mathrm{d}A + 2\int_A \dot{W}_b\,\mathrm{d}A \qquad (3.4)$$

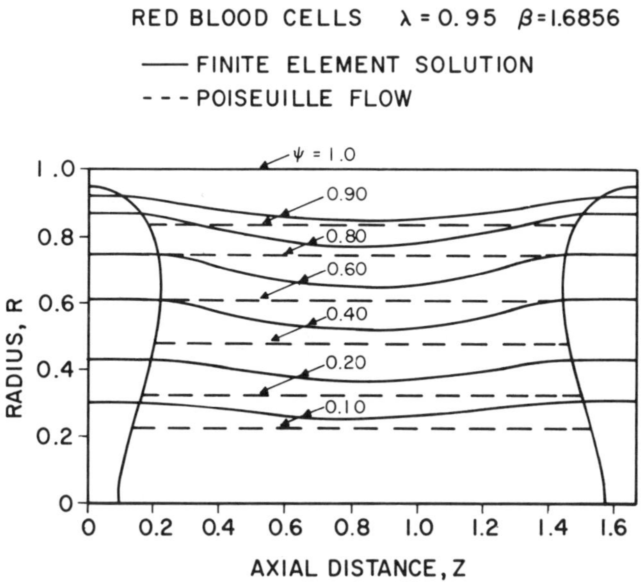

Figure 3.2 Streamlines (solid lines) computed from the velocity field solution for a finite element grid similar to Figure 3.1. The dotted lines are the comparable streamlines for Poiseuille flow. The dimensionless parameters λ and β are the particle radius and spacing divided by the tube radius (From Skalak *et al.*, 1972)

In Equation (3.4) T is an isotropic tension in the membrane which is similar to the pressure in incompressible fluids. The inclusion of this term produces motions in which the membrane cell area is a constant. The rate change of the surface area is measured by $\overline{\lambda_1 \lambda_2}$ where the dot indicates a time derivative and λ_1 and λ_2 are the principal extension ratios of the membrane. The term W_m is the strain energy of the membrane in deformational motions in the plane of the membrane. The term W_b is the bending strain energy. These energies are given by

$$W_m = \frac{B}{4}[\tfrac{1}{2}I_1^2 + I_1 - I_2] \tag{3.5}$$

$$W_b = \frac{D}{2}(K_1^2 + 2\nu K_1 K_2 + K_2^2) \tag{3.6}$$

In Equation (3.5), I_1 and I_2 are the invariants of the strain tensor and the coefficient B is an elastic shear modulus of the membrane (taken as $B = 0.5 \times 10^{-2}$ dyne/cm). The bending strain energy of Equation (3.6) is an assumed form in which the coefficient D is a bending stiffness (taken as $D = 10^{-12}$ dyne/cm) and the coefficient ν is similar to Poisson's ratio (taken as $\nu = 1/2$). The

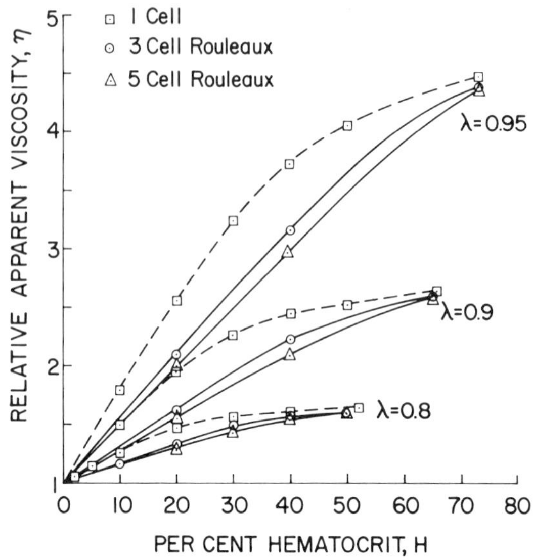

Figure 3.3 Relative apparent viscosity for a line of rigid red cells such as shown in Figures 3.1. and 3.2. The dimensionless parameter λ is the ratio of the cell diameter to the tube diameter. The hematocrit is varied by changing the spacing of the cells. The curves for rouleaux correspond to groups of three or five cells which adhere to each other (From Skalak *et al.*, 1972)

curvatures of the membrane are K_1 and K_2. Since the problem is axisymmetric, these are measured in the meridional and normal planes.

A sample result of a deformed cell and the finite element grid for the fluid flow is shown in Figure 3.4. The grid is deformed as the computation proceeds from a zero pressure drop to gradually increased pressure gradients. The nodes on the cell boundary serve to locate the membrane shell as well as serving as boundary nodes of the fluid. A shell element consisting of five nodes is used and the slopes at each end of the element are required to be equal to the slopes of the adjacent elements. At the axis of symmetry, the membrane is required to be normal to the axis. By running a range of pressure drops and initial diameter ratios, the variation of the apparent viscosity with pressure in tube diameter has been determined. Figure 3.5 shows some of the results. The parameter λ_i is the initial diameter ratio. Values of λ_i greater than 1 indicate that the cell is initially too large to fit into the capillary and must be deformed in order to pass through. In this case, the apparent viscosity approaches infinity as the pressure drop and velocity approach zero.

The above solutions use the nodal values of velocity and pressure as primary variables. The use of a stream function eliminates the need to include the pressure

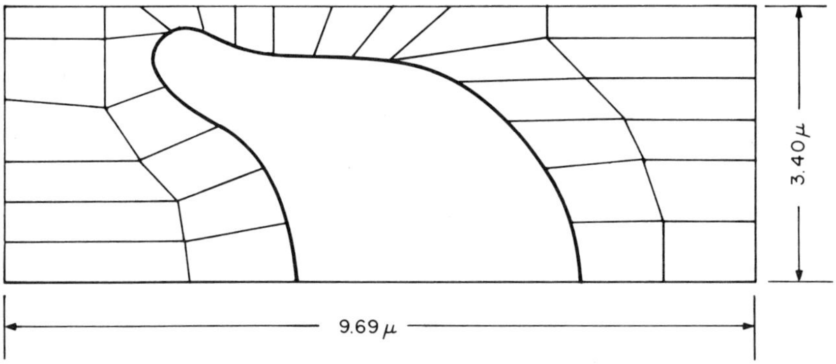

Figure 3.4 Finite element grid used for capillary flow problem taking deformability of the red blood cells into account. The lower boundary is the tube axis and the upper boundary is the tube wall. The section shown is assumed to be repeated to represent in infinite line of red blood cells located axisymmetrically and equally spaced

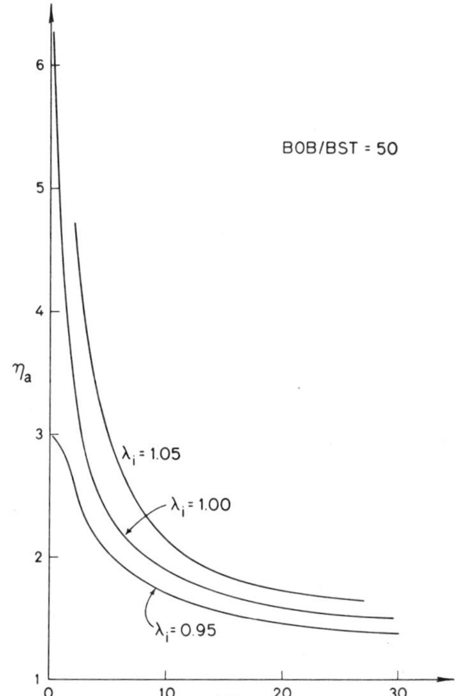

Figure 3.5 Relative apparent viscosity η_a, for a line of flexible red blood cells. The initial diameter ratio λ_i is the undeformed red blood cell diameter divided by the tube diameter. The hematocrit is 26% for all curves. The abscissa is the dimensionless pressure drop. The variations of η_a show the non-Newtonian behaviour of the suspension (From Skalak, 1980)

as a primary variable, but the partial differential equation on the stream function is of fourth order whereas the governing equations (3.1) and (3.2) are only of second order in terms of velocity and pressure. An example of the use of the stream function approach is given by Tong and Fung (1971). It can be seen that the boundary conditions, including the zero drag condition, are somewhat more difficult to apply than when velocity and pressure are used as primary variables.

Another study of capillary flow using the stream function approach has been given by Gupta *et al.* (1976). The flow is assumed to be axisymmetric and the blood cells are assumed to be rigid. Realistic deformed red blood cell shapes are used based on published experimental data. The results for the additional pressure drops are in qualitative agreement with experimental measurements.

The two-dimensional stream function approach has also been used by Vawter *et al.* (1974) to model the flow in a right-angle bifurcation at low Reynolds number. In addition, the effect of a constriction or sphincter at the entrance to the side branch is explored. The results show that the pressure drop in the side branch increases rapidly as the constricted width decreases.

3.3 BLOOD CELL DEFORMATION AND ADHESION

The finite element method has been useful in studying the deformations of red blood cells under the influence of osmotic and adhesive forces which produce large elastic deformations. When red blood cells are placed in a hypotonic medium, water passes through the cell membrane in response to the osmotic pressure. The cell swells and gradually takes on a spherical shape, after which it may lyse (burst). The same finite element approach as described above for capillary blood flow may be used in a quasi-static mode to solve for the successive shapes of the red blood cell during swelling (Zarda *et al.*, 1975). An unusual feature of the computation is that the pressure inside the cell is not known. The volume of the cell is prescribed and a Newton–Raphson type of iteration is performed to achieve the desired value. The internal pressure is, in effect, adjusted until the desired volume is achieved and the equilibrium equations are satisfied at each node. This is indicated in the output as zero velocity at each node.

A further use of the same basic finite element program is made for the problem of adhesion of one red blood cell to another. In this situation the red blood cells may form stacks which are called rouleaux. The analysis of such rouleaux is a problem of adhesion or fracture energy interacting with the elastic strain energy of the red blood cell membrane. It can be shown (Skalak *et al.*, 1978) that for an equilibrium position, the adhesive energy γ (dynes/cm) is given by

$$\gamma = \frac{\partial W}{\partial A_c} \tag{3.7}$$

where W is the strain energy of the cells involved which is a function of the area of contact, A_c, between the cells. The surface energy of the binding of the cells is γA_c.

The strain energy W is computed from the equilibrium shape for each different value of A_c. This is a finite element computation in which the constraint conditions of constant volume and constant surface area are enforced using a Newton–Raphson procedure as described above for the red cell sphering problem. A set of shapes computed by this method are shown in Figure 3.6. The vertical axis represents a plane of symmetry and the horizontal axis is the axis of rotational symmetry. It may be seen that the deformations are large and tend to a limiting shape which is a sector of a sphere.

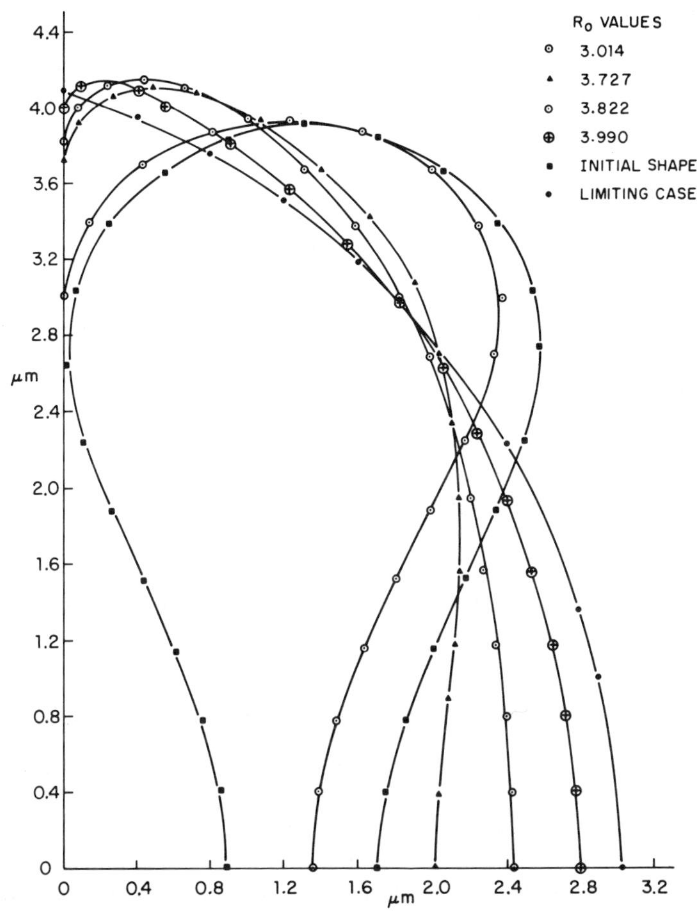

Figure 3.6 Shapes of red blood cell in a rouleaux computed by finite element representation of the axisymmetric cross-section by the nodes shown. The abscissa is the axis of rotational symmetry and the radial axis (ordinate) represents a plane of symmetry. The shapes shown correspond to one cell of a two cell rouleaux (From Skalak *et al.*, 1981)

Another type of blood cell problem which has been studied by finite element methods is the flow of white blood cells in straight and tapering capillary vessels. White blood cells are approximately spherical and are much stiffer than red blood cells. The white blood cells may be modelled as a viscoelastic solid (Bagge *et al.*, 1977; Schmid-Schonbein *et al.*, 1980). The flow of a white blood cell into a tapering capillary is a time-dependent problem due to the tapering of the tube as well as due to the viscoelastic properties of the cell. Further, the deformations of interest are large. This problem of large, axisymmetric deformation of a viscoelastic standard solid was treated by a finite element method by Tozeren (1977). An axisymmetric element of isoparametric quadrilateral cross-section is used with eight nodes and displacements and pressures as primary variables. The viscoelasticity of the cell requires knowledge of velocity field at each time step also. Finite differences are used to estimate the velocities at each time step. The problem is then a quasi-static computation at each time. A further complication of this problem is that the loading on the cell is applied by the suspending fluid (plasma). The Stokes equations for the plasma flow are simplified to the Reynolds equation of lubrication theory on the assumption that the thickness of the fluid film between the cell and the tube is small compared to the tube or particle diameter. The problem is one of a closely fitting particle which occupies most of the tube cross-section. The velocities and pressures in the lubrication layer are obtained by integration of Reynolds equation by a Runge–Kutta method. The pressures computed are applied to the cell and its deformation is then computed. Since this deformation affects the pressure distribution, an iterative scheme is used. The pressure and the cell deformation are alternately computed until both reach stationary (quasi-steady) values.

In summary, it may be said that for axisymmetric and two-dimensional flows, problems of interest in capillary blood flow have been formulated and some examples computed in satisfactory and illuminating manner by finite element methods. It remains to extend and apply the methods to realistic three-dimensional and non-axisymmetric cases.

3.4 PERISTALTIC FLOW

Peristaltic flow does not usually play a role in blood flow, although there may be some such effect in venous flow within actively working muscles. The principal peristaltic flows are in the intestines and the ureter. In peristaltic flows, the motion is driven by the active muscular motion of the walls, in contrast to arterial blood flow where the wall motion, at the time scale of the heart pulse, is completely passive. To model peristaltic flow, the wall motion is specified and the internal viscous flow is computed.

A paper by Stavitsky *et al.*, (1980) shows a method of computation of peristaltic flow by finite elements. The flow is treated as two dimensional and the flow is assumed to be a slow, viscous flow (Stokes flow). The formulation is in terms of the stream function ψ which satisfies the equation $\nabla^4\psi = 0$. An eighteen degree of freedom triangular element is used and solutions are derived for a single, asymmetric contraction. The deformations involved are very large and particularly illustrate the capability of the finite element method in this situation.

An earlier analysis of peristaltic pumping by a finite element method was given by Tong and Vawter (1972). They also used the Stokes equations (3.1) for the fluid but treated axisymmetric flow as a model of the ureter. The boundary motion was assumed to be periodic along the length of the tube and radial only. In a frame moving with the wave speed of the wall motion, the flow is steady. A stream function is used, but finite element approximations are written directly for various derivatives of the stream function. A triangular mesh is used and numerical results are shown to give good agreement with analytic solutions for long wavelengths and small amplitudes. An interesting feature of the flow is that some trapping under the crests of the boundary wave may occur even with only partial occlusion of the cross-section by the trough of the wave.

3.5 ARTERIAL BLOOD FLOW: STENOSIS AND BIFURCATIONS

The wave propagation and velocity distribution in arterial blood flow has probably received more attention in bio-fluid mechanics than any other topic. There are a number of reviews and books which give background and details (Skalak, 1966; McDonald, 1974; Caro *et al.*, 1978; Pedley, 1980). The propagation of the blood flow pulse is of medical interest because of the vital importance of the cardiovascular system and the high incidence of cardiovascular disease. Although the connection is still not fully known, it appears that the locations of initial incidence of atherosclerosis are affected by the fluid mechanical stresses associated with blood flow. Atherosclerotic lesions occur preferentially in regions of curvature and in the vicinity of bifurcations of the principal arteries. For this reason, the details of the flow patterns and shear stress distribution are of interest and research continues to discover how the fluid mechanical stresses interact with the many other factors involved, to produce the atherosclerotic plaques. These factors include lipid levels in the blood, blood pressure, smoking, obesity, stress, and many others.

In the arterial system, the non-linear viscous character of blood is not important because the shear stresses are relatively high. For arterial flow, the approximation of a Newtonian, incompressible fluid is quite accurate. The difficult parts of analyses of arterial blood flow lie in the irregular geometry of the arteries and the pulsatile nature of the flow. The Reynolds numbers are such that the full Navier–Stokes equations are needed for accurate representations of flow patterns. A great deal of work has been done on the wave propagation in blood flow both analytically using linearized theories, and numerically, by the method of characteristics. The analytical work usually needs to assume a straight circular tube and the method of characteristics is a one-dimensional approximation dealing only with mean velocity over any cross-section. These types of studies cannot provide any information on the details of flow patterns or stresses at bends and junctions. It is necessary to use finite element or finite difference techniques to provide full descriptions of the velocity and pressure distributions. The finite element method is particularly suitable because of the ease with which curved and branching boundaries may be represented. The pulsatility of the flow is a problem which is equally troublesome in any computational procedure and a fairly large number of time steps per cycle is needed to describe observed wave forms accurately.

Examples of the use of finite element methods for pulsatile flow in a straight circular artery and in a constricted or stenotic artery are given by Wille in Chapter 4 of this volume. The time derivatives are replaced by finite differences, but this still leaves the non-linear terms of the convective acceleration to be accommodated within the finite element spatial computation at each time step. This is accomplished by developing non-linear finite element equations in the incremental velocity and pressure changes since the previous time step. These non-linear equations are then solved by an iterative Newton–Raphson method. These computations are axisymmetric and represent the present state of the art. Similar three-dimensional computations have not yet been carried out. Another assumption of this analysis is that the walls of the artery are rigid, which precludes wave propagation and must introduce some error in the details of stress distribution on the walls. The error is probably not large, but merits exploration.

Another type of finite element analysis of pulsatile blood flow was given by Davids and Mani (1974). In this approach, equations of conservation of mass and momentum for each element are written by considering each element as a control volume. The convective terms are represented by momentum fluxes across the boundaries. The method is found suitable to incorporate the wall motion of the artery. The instantaneous radius is assumed to be a linear function of the pressure. This neglects the wall inertia, but is a reasonable first approximation to the arterial wall behaviour. The results show that the wall shear stresses are larger for the expanding tube and less for the contracting tube as compared to the rigid tube solutions.

The patterns of flow at branching points can be very complex and dependent on Reynolds number even for steady flow, as shown by experimental results (e.g. Karino *et al.*, 1979). Pulsatile flow through bifurcations is even more complex. There may be time-dependent zones of separated flow developed during each cycle. These details of flow are of interest when considering atherogenesis and there have been various computations made with this application in mind. These computations have been primarily for steady, laminar flow in two-dimensional geometries. Such a model may be a fairly good representation of the flow pattern on the plane of symmetry of a three-dimensional bifurcation of a cylindrical artery. But two-dimensional representation cannot reproduce the secondary flows which are often present in branching arteries.

An important advantage of the finite element method for simulation of arterial branch flows is that the irregular, natural geometry can be accurately represented.

For example, Gokhale *et al.*, (1978) developed a model of a section of canine aorta with two branches. The dimensions were obtained from the outline of a silicone rubber replica of the canine arteries. The arterial walls were considered to be rigid in the computations and the flow was assumed to be steady, Newtonian, and incompressible. The Navier–Stokes and continuity equations are discretized by the Galerkin form of finite element method. Nodal velocities and pressures are used as the primary variables. Isoparametric quadrilateral elements are first assigned and these are then subdivided into four triangles. Velocity is interpolated quadratically and pressure is approximated linearly over each triangular subelement. The convective terms of the acceleration lead to non-linear equations for the nodal values. Solutions for the nodal velocities and pressures are obtained by an iterative procedure. Shear rates at the walls are computed via the finite element approximation for several ratios of discharge through main and side branch arteries. For the particular aortic cast modelled, experimental data obtained by Lutz *et al.*, (1977) using an electrochemical technique allows estimation of the wall shear stress from chemical uptake rates. Good agreement is found between computed values of shear rates in the model and the measured shear rates on the mid-plane of the experimental casts. In the computations and the experiments it was found that the zones of separation and instabilities in the flow pattern were sensitive to small changes in flow rates and in the geometry of the wall. For these reasons, pulsatile flow and the actual three-dimensional geometry need to be incorporated before reliance can be placed on computed details of the flow.

Kandarpa., and Davids (1976) considered a similar, steady, two-dimensional laminar flow through a branching, straight-walled duct. The finite element procedure they used is unusual in that they integrate the governing equations for nodal velocities and pressures starting from the upstream end, where a parabolic velocity distribution is assumed. The equation of conservation of momentum is taken to be approximately satisfied in each element to find increments in velocities in the downstream direction. The velocities are assumed to vary linearly in the transverse direction and shear forces are approximated from the known velocity profile at the upstream end of each element. The results for the symmetric bifurcation flow show the degree of separation on the outer wall of the daughter vessel as a function of angle of bifurcation and Reynolds number.

A somewhat similar analysis has been given by Subbaraj and Patel (1980), in treating a straight-walled symmetrically branching duct. Cases of Reynolds numbers of 0 and 146 in a 45° bifurcation are solved by a method very similar to that described by Gokhale *et al.* (1978) mentioned above. Some other cases of zero Reynolds number flow through a side branch are given by Texon (1980).

3.6 CONCLUSION

The examples cited above indicate the utility of the finite element method in capillary blood flow, arterial haemodynamics, and peristaltic flow. The computational work to date has treated two-dimensional or axisymmetric flows primarily. To achieve more realistic results both in capillary and arterial blood flow, actual geometry and three-dimensional, unsteady flow computations are required. In capillary flow, the deformation of the red blood cells must be taken into account and for arterial flow, the vessel wall should be taken to be viscoelastic. Venous flow has not been treated by finite element methods. The veins are often subject to very large wall deformations and may collapse. These phenomena pose challenging numerical problems with large deformations.

Another complex and difficult area in which little numerical work has been done is the modelling of flow in small arterioles and venules where several particles may interact across the cross-section of a blood vessel. As a first approximation in this range, the blood may be considered as a non-Newtonian fluid (Zienkiewicz and Godbole, 1975; Godbole *et al.*, 1980). But the boundary conditions and dynamics of the hematocrit distribution across the tube are unknown and cannot be predicted by a homogeneous fluid theory. A finite element model of a two-phase flow (mixture theory) might be more appropriate.

The flow into microvessel branches taking the discrete nature of red and white blood cells into account is another complex problem of interest. It is important for the determination of the hematocrit and flow distribution in the capillary network. In conclusion, finite element methods have been useful in blood flow computations, but much more remains to be done.

ACKNOWLEDGMENT

This research was supported by U.S. National Institutes of Health Grant HL 16851.

REFERENCES

Bagge, U., Skalak, R., and Attefors, R. (1977) Granulocyte rheology, *Adv. Microcirc.*, **7,** 29–48.

Caro, C. G., Pedley, T. J., Schroter, R. C., and Seed, W. A. (1978) *The Mechanics of Circulation*, Oxford University Press, London.

Davids, N., and Mani, M. K. (1974) A finite element analysis of endothelial shear stress for pulsatile blood flow, *Biorheology*, **11,** 137–147.

Ghista, D. N., Kobayashi, A. S., and Davids, N. (1975) Analysis of some biomechanical structures and flows by computer-based finite element method, *Computers in Biology and Medicine*, **5** (1/2), 119–162.

Godbole, N., Nakazawa, S., and Zienkiewicz, O. C. (1980) Blood flow analysis by the finite element method, *Proceedings of the International Conference on Finite Elements in Biomechanics* (ed. B. R. Simon), University of Arizona, Feb. 18–20, 1980, 277–289.

Gokhale, V. V., Tanner, R. I., and Bischoff, K. B., (1978) Finite element solution of the Navier–Stokes equations for two-dimensional steady flow through a section of a canine aorta model, *J. Biomechanics*, **11,** 241–249.

Goldsmith, H. L., and Skalak, R. (1975) Hemodynamics, *Annual Review of Fluid Mechanics*, (eds. M. Van Dyke, W. G. Vincenti, and J. V. Wehausen), **7,** 213–247.

Gupta, B. B., Natarajan, R., and Seshadri, V. (1976) Analysis of blood flow in capillaries by finite element method, *Micro-vascular Research*, **12,** 91–100.

Kandarpa, K., and Davids, N. (1976) Analysis of fluid dynamic effects on atherogenesis at branching sites, *J. Biomechanics*, **9,** 735–741.

Karino, T., Dwong, H. H. M., and Goldsmith, H. L. (1979) Particle flow behavior in models of branching vessels, *Biorheology*, **16,** 231–247.

Lutz, R. J., Cannon, J. N., Bischoff, K. B., Dedrick, R. L., Stiles, R. K., and Fry, D. L. (1977) Wall shear stress distribution in a model canine artery during steady flow, *Circulation Res.*, **41,** 391–399.

McDonald, D. A. (1974) *Blood Flow in Arteries*, (2nd edn), Williams and Wilkens, Baltimore.

Nerem, R. M., and Cornhill, J. F. (eds.) (1978) The role of fluid mechanics in atherogenesis, *Proceedings* of a Symposium, Ohio State University, Columbus, Ohio, Aug. 24–25, 1978.

Olson, M. D., and Tuann, S. Y. (1978) Finite element solutions of primitive variables versus stream function Navier–Stokes equations, *Finite Elements in Fluids*, Vol. 3 (eds. R. H. Gallagher, O. C. Zienkiewicz, J. T. Oden, M. M. Cecchi, C. Taylor), John Wiley and Sons, New York, N. Y.

Pedley, T. J. (1980) *The Fluid Mechanics of Large Blood Vessels,* Cambridge University Press, London.

Schmid-Schönbein, G. W., Sung, K. L. P., Tözeren, H., Chien, S., and Skalak, R. (1980) The passive mechanical properties of human leukocytes (to be submitted for publication).

Skalak, R. (1966) Wave propagation in blood flow In: *Biomechanics Symposium* (ed. Y. C. Fung), Applied Mechanics Division, ASME, New York, N. Y.

Skalak, R. (1980) Physical and mathematical models blood flow: Theoretical analysis, In: *Erythrocyte Mechanics and Blood Flow* (eds. G. R. Cokelet, H. J. Meiselman, D. E. Brooks), A. L. Liss, Inc., New York, pp. 149–164.

Skalak, R., Chen, P. H., and Chien, S. (1972) Effect of hematocrit and rouleaux on apparent viscosity in capillaries, *Biorheology*, **9,** 67–82.

Skalak, R., Zarda, P. R., Jan, K–M, and Chien, S. (1978) Theory of rouleau formation, cardiovascular and pulmonary dynamics, *Journal of the Institute National de la Sante et de la Recherche Medical*, **71,** 200–308.

Skalak, R., Zarda, P. R., Jan, K–M, and Chien, S. (1981) Mechanics of rouleau formation, (Submitted to *Biophysical Journal*).

Stavitsky, D., Macagno, E., Christensen, J. (1980) Intestinal flow analysis using finite elements, *Proceedings of the International Conference on Finite Elements in Biomechanics* (ed. B. R. Simon) University of Orizina, Feb 18–20, 1980, pp. 223–240.

Subbaraj, K., and Patel, M. K. (1980) Finite element analysis of fluid flow at branching sites, *Proceedings of the International Conference on Finite Elements in Biomechanics* (ed. B. R. Simon), University of Arizona, Feb. 18–20, 1980, pp. 329–345.

Texon, M. (1980) *Hemodynamic Basis of Atherosclerosis*. Hemisphere Publishing Corp., and McGraw-Hill Book Col., New York N. Y.

Tong, P., and Fung, Y. C. (1971) Slow particulate viscous flow in channels and tubes – application in biomechanics, *J. Applied Mechanics, Trans. ASME,* **93,** 721–728.

Tong, P., and Vawter, D. (1972) An analysis of peristaltic pumping, *J. Applied Mechanics, Trans. ASME*, **94,** 857–862.

Tözeren, H. (1977) Flow of elastic and viscoelastic spheres in tubes, Ph. D. Thesis, Columbia University, New York, N. Y.

Vawter, D., Fung., Y. C., and Zweifach, B. W. (1974) Distribution of blood flow and pressure from a microvessel into a branch, *Microvascular Res.*, **8,** 44–52.

Wille, S. O. (1980) Pressure and flow patterns in the vicinity of an arterial stenosis (Chapter 4 in this volume).

Zarda, P. R., Chien, S., and Skalak, R. (1975) Elastic deformations of red blood cells, *J. Biomechanics*, **10**, 211–221.

Zarda, P. R., Chien, S., and Skalak, R. (1977) Interaction of a viscous incompressible fluid with an elastic body, *Computational Methods for Fluid-Structure Interaction Problems*, (eds. T. Belytschko and T. L. Geers), *ASME-AMD*, **26**, 65–82.

Zienkiewicz, O. C. (1977) *The Finite Element Method* (3rd edn), McGraw-Hill Book Co., New York, N. Y.

Zienkiewicz, O. C., and Godbole, P. N. (1975) Viscous, incompressible flow with special reference to non-Newtonian (plastic) fluids, In: *Finite Elements in Fluids*, Vol. I, *Viscous Flow alnd Hydrodynamics* (eds. R. H. Gallagher, J. T. Oden, C. Taylor, O. C. Zienkiewicz), John Wiley and Sons, New York, N. Y., pp. 25–55.

Finite Elements in Biomechanics
Edited by R. H. Gallagher, B. R. Simon, P. C. Johnson, and J. F. Gross

Chapter 4

Finite Element Simulations of the Pulsatile Blood Flow Patterns in Arterial Abnormalities

S. O. Wille

4.1. INTRODUCTION

The characteristic localization of early atherosclerotic lesions has led many workers to believe that haemodynamic factors play an important role in the pathogenesis of atherosclerosis. The preferable sites of arterogenesis are, according to Fry (1972), wall segments in the arterial tree which are exposed to high shear stresses causing destruction of the endothelium. Caro *et al.*, (1971) suggest that the early atherosclerotic lesions are coincident with sites in the arterial tree with low flow which thereby have a lowering of the diffusion at the interface between the vessel wall and the blood. The contradiction between these two hypotheses implies that both the precise localization of early atherosclerosis and the haemodynamic flow conditions within the vessel have to be investigated more closely. Measurement of velocity profiles in blood vessels has been carried out by means of pulsed ultrasound doppler flowmeters (Peronneau *et al.*, 1974). However, the space resolution of this instrument is poor, especially along the wall.

Experimental investigations which have been carried out on physical models have mainly been concerned with measuring flow velocities by a laser doppler flowmeter (Durst *et al.*, 1976), visualization of the streamlines either by contrast dye (Roach, 1972), or hydrogen bubbles (Azuma and Fukushima, 1976). The large amount of data which has to be collected in order to map velocity profiles by the laser doppler flowmeter, especially for pulsatile flow, has complicated these measurements. The visualization of streamline patterns by contrast-dye injection into the moving fluid has been successful for steady flow. When pulsatile flow is studied, the problem of dye mixing in the fluid complicates these investigations. Visualization of streamlines by hydrogen bubbles has been more successful as the bubbles dissolve in the fluid after a short time.

In view of the limitations of the experimental methods, mathematical models play a useful role in the analysis of flow and pressure patterns and may give useful additional information, particularly where the geometries are complicated. The finite element method is an especially effective approach to mathematical modelling.

The finite element formulation and the solution of the dynamic axisymmetric Navier – Stokes equations is described for the subject problem. Triangular linear elements are used for interpolating pressure and triangular quadratic elements for velocities. The non-linear set of differential equations are solved by the Newton – Raphson method. The time derivatives are replaced by finite differences using the one-step θ-method. This allows modification of the time step scheme during the Newton – Raphson iteration procedure. During the first Newton – Raphson iterations, the more stable Euler backward scheme is applied. In the last iterations the time step is changed to the trapezoidal scheme to achieve greater accuracy.

Two models of arterial abnormalities are simulated. The first one is an arterial stenosis with a constriction area of 75 %. The other model describes the pulsatile pressure and flow patterns in an arterial aneurysm. The solutions of the Navier–Stokes equations are calculated at 24 points in time during the cardiac cycle. In this paper a selected number of pressure and flow patterns is presented in terms of velocity vectors, streamlines and isobars. The calculated velocity field in the vicinity of the stenosis shows that a vortex is developed at the wall distal to the stenosis as the velocity decreases from the peak systolic value. At the site of the vortex a local pressure minimum is found due to the conversion of pressure to kinetic energy. When the flow is reversed, the reversal is first occurring along the wall, thus forcing the vortex towards the centre of the tube. As the reverse flow velocity increases, a vortex is also developed at the proximal site of the stenosis. The pressure and flow patterns in the aneurysm model demonstrate a vortex formation during most of the cycle. The pressure within the aneurysm is nearly constant. At the downstream end of the expansion, high pressure gradients are found.

4.2 THE DYNAMIC NAVIER–STOKES EQUATIONS

The equations governing axisymmetric, time-dependent flow are the Navier–Stokes equations combined with the equations of continuity.

$$\frac{1}{\rho}\frac{\partial p}{\partial x}+\frac{\partial u}{\partial t}-\mu\left(\frac{1}{r}\frac{\partial u}{\partial r}+\frac{\partial^2 u}{\partial x^2}+\frac{\partial^2 u}{\partial r^2}\right)+u\frac{\partial u}{\partial x}+v\frac{\partial u}{\partial r}=0 \tag{4.1}$$

$$\frac{1}{\rho}\frac{\partial p}{\partial r}+\frac{\partial v}{\partial t}-\mu\left(\frac{1}{r}\frac{\partial v}{\partial r}-\frac{v}{r^2}+\frac{\partial^2 v}{\partial x^2}+\frac{\partial^2 v}{\partial r^2}\right)+u\frac{\partial v}{\partial x}+v\frac{\partial v}{\partial r}=0 \tag{4.2}$$

$$\frac{v}{r}+\frac{\partial u}{\partial x}+\frac{\partial v}{\partial r}=0 \tag{4.3}$$

where u is velocity in the x (longitudinal) direction, v is velocity in the r (radial) direction, p is pressure, ρ is density of the fluid, and μ the kinematic viscosity. The flow is assumed to be laminar and Newtonian and the fluid homogeneous and incompressible. By using these equations the geometrical shape of the boundary must be symmetric around the main axis of flow.

4.3 THE FINITE ELEMENT FORMULATION

The velocities are approximated with quadratic polynomials and the pressure is approximated by a linear polynomial. Let u_i and p_i denote the nodal values of velocities and pressure on an isoparametric triangular element with interpolation polynomials N_i and L_i respectively. The velocities and pressure at an arbitrary point on the element are then given by

$$u = \Sigma N_i u_i$$
$$v = \Sigma N_i v_i$$
$$p = \Sigma L_i p_i$$

Among the different finite element formulations the one described by Taylor and Hood (1973) is chosen. With this approach the finite element equations are formed by the Galerkin process and the direct coupling method is used to obtain their solution. The viscosity terms are integrated by parts to reduce the required order of inter-element continuity. The time derivatives for the velocities are approximated by finite differences. Considering the equation

$$\frac{\partial \phi}{\partial t} = f$$

the approximation is defined by

$$\frac{\phi - \overline{\phi}}{\Delta t} = (1-\theta) f + \theta \overline{f} \qquad 0 < \theta < \tfrac{1}{2}$$

The bars in the formula denote the values obtained at the previous time iteration step. For $\theta = 0$ the expression becomes Euler's backwards formula, and for $\theta = \frac{1}{2}$ the scheme becomes the trapezoidal rule. The Euler backward formula is the most stable of these schemes but suffers from less accuracy (Lambert, 1973).

In the present work the backward difference scheme is also tried, which is given by

$$(\phi - 4\overline{\phi} + 3\overline{\overline{\phi}}) = \tfrac{1}{3} \Delta t f$$

where the double bar indicates the value two time steps backwards. This scheme is claimed (Lambert, 1973) to give the same accuracy as the trapezoidal rule and to have equal stability to the Euler backward scheme. However, this method has been applied in the present work without success. It gave divergent solutions where the Euler backward scheme converged.

The set of non-linear equations obtained by the previously described procedure is solved by an iterative Newton–Raphson method. The final system of equations to be solved may then be expressed on matrix form as

$$
\begin{bmatrix}
\dfrac{\partial F_{1i}}{\partial u_j} & \dfrac{\partial F_{1i}}{\partial v_j} & \dfrac{\partial F_{1i}}{\partial p_j} \\[2ex]
\dfrac{\partial F_{2i}}{\partial u_j} & \dfrac{\partial F_{2i}}{\partial v_j} & \dfrac{\partial F_{2i}}{\partial p_j} \\[2ex]
\dfrac{\partial F_{3i}}{\partial u_j} & \dfrac{\partial F_{3i}}{\partial v_j} & \dfrac{\partial F_{3i}}{\partial p_j}
\end{bmatrix}
\begin{bmatrix} \Delta u_j \\[2ex] \Delta v_j \\[2ex] \Delta p_j \end{bmatrix}
= -
\begin{bmatrix} F_{1i} \\[2ex] F_{2i} \\[2ex] F_{3i} \end{bmatrix}
$$

where:

$$
\begin{aligned}
F_{1i} = \iint_e \Bigg[& N_i \Sigma N_k (u_k - \bar{u}_k) + \Bigg[\mu \Bigg(-\frac{N_i}{r} \Sigma \frac{\partial N_k}{\partial r} u_k + \frac{\partial N_i}{\partial x} \Sigma \frac{\partial N_k}{\partial x} u_k \\
& + \frac{\partial N_i}{\partial r} \Sigma \frac{\partial N_k}{\partial r} u_k \Bigg) + N_i \Bigg[(\Sigma N_k u_k) \Sigma \frac{\partial N_k}{\partial x} u_k + (\Sigma N_k v_k) \Sigma \frac{\partial N_k}{\partial r} u_k \\
& + \frac{1}{\rho} \Sigma \frac{\partial L_k}{\partial x} p_k \Bigg]\Bigg] (1-\theta) \Delta t + \Bigg[\mu \Bigg(-\frac{N_i}{r} \Sigma \frac{\partial N_k}{\partial r} \bar{u}_k + \frac{\partial N_i}{\partial x} \Sigma \frac{\partial N_k}{\partial x} \bar{u}_k \\
& + \frac{\partial N_i}{\partial r} \Sigma \frac{\partial N_k}{\partial r} \bar{u}_k \Bigg) + N_i \Bigg[(\Sigma N_k \bar{u}_k) \Sigma \frac{\partial N_k}{\partial x} \bar{u}_k \\
& + (\Sigma N_k \bar{v}_k) \Sigma \frac{\partial N_k}{\partial r} \bar{u}_k + \frac{1}{\rho} \Sigma \frac{\partial L_k}{\partial x} \bar{p}_k \Bigg]\Bigg] \theta \Delta t \Bigg] \mathrm{d}x \, \mathrm{d}r \\
& - \int_s \mu N_i \Sigma \frac{\partial N_k}{\partial n} u_k (1-\theta) \Delta t \, \mathrm{d}s - \int_s \mu N_i \Sigma \frac{\partial N_k}{\partial n} \bar{u}_k \theta \Delta t \, \mathrm{d}s
\end{aligned}
\tag{4.4}
$$

$$
\begin{aligned}
F_{2i} = \iint_e \Bigg[& N_i N_k (v_k - \bar{v}_k) + \Bigg[\mu \Bigg(-\frac{N_i}{r} \Sigma \frac{\partial N_k}{\partial r} v_k + \frac{N_i}{r^2} \Sigma N_k v_k + \frac{\partial N_i}{\partial x} \Sigma \frac{\partial N_k}{\partial x} v_k \\
& + \frac{\partial N_i}{\partial r} \Sigma \frac{\partial N_k}{\partial r} v_k \Bigg) + N_i \Bigg[(\Sigma N_k u_k') \Sigma \frac{\partial N_k}{\partial x} v_k + (\Sigma N_k v_k) \Sigma \frac{\partial N_k}{\partial r} v_k \\
& + \frac{1}{\rho} \Sigma \frac{\partial L_k}{\partial r} p_k \Bigg]\Bigg] (1-\theta) \Delta t + \Bigg[\mu \Bigg(-\frac{N_i}{r} \Sigma \frac{\partial N_k}{\partial r} \bar{v}_k + \frac{N_i}{r^2} \Sigma N_k \bar{v}_k \\
& + \frac{\partial N_i}{\partial x} \Sigma \frac{\partial N_k}{\partial x} \bar{v}_k + \frac{\partial N_i}{\partial r} \Sigma \frac{\partial N_k}{\partial r} \bar{v}_k \Bigg) + N_i \Bigg[(\Sigma N_k \bar{u}_k) \Sigma \frac{\partial N_k}{\partial x} \bar{v}_k \\
& + (\Sigma N_k \bar{v}_k) \Sigma \frac{\partial N_k}{\partial r} \bar{v}_k + \frac{1}{\rho} \Sigma \frac{\partial L_k}{\partial r} \bar{p}_k \Bigg]\Bigg] \theta \Delta t \Bigg] \mathrm{d}x \, \mathrm{d}r
\end{aligned}
$$

$$-\int_s \mu N_i \Sigma \frac{\partial N_k}{\partial n} v_k (1-\theta)\Delta t \, ds - \int_s \mu N_i \Sigma \frac{\partial N_k}{\partial n} \bar{v}_k \theta \Delta t \, ds \tag{4.5}$$

$$F_{3i} = \iint_e L_i \left(\frac{1}{r} \Sigma N_k v_k + \Sigma \frac{\partial N_k}{\partial x} u_k + \Sigma \frac{\partial N_k}{\partial r} v_k \right) dx \, dr \tag{4.6}$$

$$\begin{aligned} \frac{\partial F_{1i}}{\partial u_j} = \iint_e \Bigg[N_i N_j + \Bigg[\mu \left(-\frac{N_i}{r}\frac{\partial N_j}{\partial r} + \frac{\partial N_i}{\partial x}\frac{\partial N_j}{\partial x} + \frac{\partial N_i}{\partial r}\frac{\partial N_j}{\partial r} \right) \\ + N_i \left(N_j \Sigma \frac{\partial N_k}{\partial x} u_k + \frac{\partial N_j}{\partial x} \Sigma N_k u_k + \frac{\partial N_j}{\partial r} \Sigma N_k v_k \right) \Bigg] \\ \times (1-\theta)\Delta t \Bigg] dx\,dr - \int_s \mu N_i \frac{\partial N_j}{\partial n}(1-\theta)\Delta t \, ds \end{aligned} \tag{4.7}$$

$$\frac{\partial F_{1i}}{\partial v_j} = \iint_e N_i N_j \Sigma \frac{\partial N_k}{\partial r} u_k (1-\theta)\Delta t \, dx\,dr \tag{4.8}$$

$$\frac{\partial F_{1i}}{\partial p_j} = \iint_e \frac{N_i \partial L_j}{\rho \, \partial x} (1-\theta)\Delta t \, dx\,dr \tag{4.9}$$

$$\frac{\partial F_{2i}}{\partial u_j} = \iint_e N_i N_j \Sigma \frac{\partial N_k}{\partial x} v_k (1-\theta)\Delta t \, dx\,dr \tag{4.10}$$

$$\begin{aligned} \frac{\partial F_{2i}}{\partial v_j} = \iint_e \Bigg[N_i N_j + \Bigg[\mu \left(-\frac{N_i}{r}\frac{\partial N_j}{\partial r} + \frac{N_i}{r^2} N_j + \frac{\partial N_i}{\partial x}\frac{\partial N_j}{\partial x} + \frac{\partial N_i}{\partial r}\frac{\partial N_j}{\partial r} \right) \\ + N_i \left(N_j \Sigma \frac{\partial N_k}{\partial r} v_k + \frac{\partial N_j}{\partial x} \Sigma N_k u_k + \frac{\partial N_j}{\partial r} \Sigma N_k v_k \right) \Bigg] \\ \times (1-\theta)\Delta t \Bigg] dx\,dr - \int_s \mu N_i \frac{\partial N_j}{\partial n}(1-\theta)\Delta t \, ds \end{aligned} \tag{4.11}$$

$$\frac{\partial F_{2i}}{\partial p_j} = \iint_e \frac{N_i}{\rho} \frac{\partial L_j}{\partial r} (1-\theta)\,\Delta t \, dx\,dr \tag{4.12}$$

$$\frac{\partial F_{3i}}{\partial u_j} = \iint_e L_i \frac{\partial N_j}{\partial x} dx\,dr \tag{4.13}$$

$$\frac{\partial F_{3i}}{\partial v_j} = \iint_e L_i \left(\frac{N_j}{r} + \frac{\partial N_j}{\partial r} \right) dx\,dr \tag{4.14}$$

$$\frac{\partial F_{3i}}{\partial p_j} = 0 \tag{4.15}$$

Using the Newton–Raphson scheme to solve a set of non-linear equations the correction to the solution at the previous iteration is obtained.

$$u = \Delta u + \bar{u}$$
$$v = \Delta v + \bar{v}$$
$$p = \Delta p + \bar{p}$$

where Δu, Δv, and Δp are the corrections and u, v, and p are the solutions within a time step. When the Newton–Raphson method has converged for a Δt with the corresponding boundary conditions, the solution for the next time step can be calculated. In the equation system $\bar{u}$, $\bar{v}$, and $\bar{p}$ are the solutions at the previous time step. The convergence of the Newton–Raphson scheme is considered by evaluating:

$$\varepsilon_v = \frac{\max(\Delta u^2 + \Delta v^2)}{\max(u^2 + v^2)}$$

$$\varepsilon_p = \frac{|\Delta p|}{|p|}$$

Usually it is possible to obtain a solution for which ε_v and ε_p are approximately 10^{-5}. The equation solver used is especially designed for block-diagonal matrices and described elsewhere (Wille, 1980).

4.4 PULSATILE FLOW IN A STRAIGHT CIRCULAR TUBE

The finite element model of the straight circular tube with boundary conditions is shown in Figure 4.1. The boundary conditions are zero velocities along the wall and zero normal derivatives of the velocities at both the inlet and outlet. Along the axis of symmetry the normal velocity and the normal derivative of the tangential

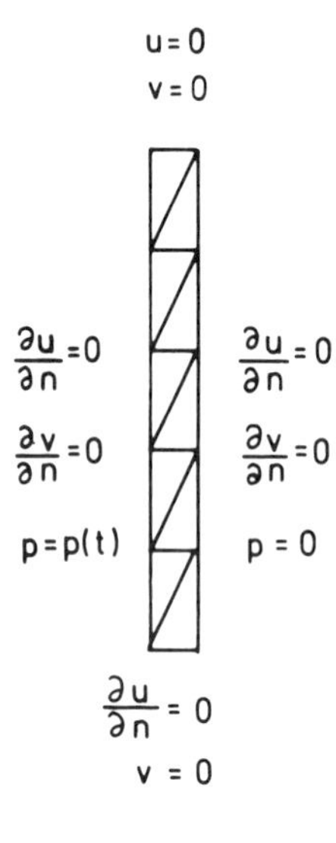

Figure 4.1 The finite element model of the straight circular tube. The boundary conditions $u = 0$, $v = 0$, correspond to the wall; $\partial u/\partial n = 0$, $v = 0$ correspond to the axis of symmetry; $\partial u/\partial n = 0$, $\partial v/\partial n = 0$, $p = 0$ correspond to the outlet; and $\partial u/\partial n = 0$, $\partial v/\partial n = 0$, $p = p(t)$ correspond to the inlet. $p(t)$ is the inlet pressure with similar characteristics as physiological pressures found in arteries and the time course of $p(t)$ is found in Figure 4.2a

velocity are zero. The flow in the tube is driven by the pressure gradient shown in Figure 4.2a. The shape of this pressure gradient is chosen to obtain a velocity waveform with characteristics similar to physiological velocity pulses.

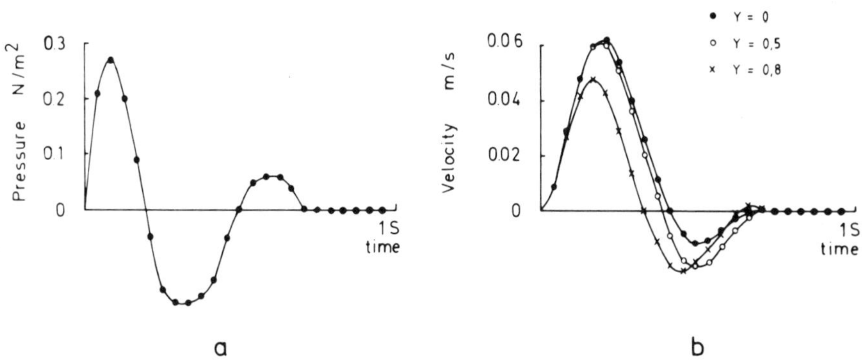

Figure 4.2(a) The pressure gradient which drives the flow in the straight circular tube as function of time. (b) The resulting pulse velocity profiles at different distances from the centre axis as function of time. The pulse profile at $y = 0$ is at the axis of symmetry of the tube, the profile at $y = 0.5$ shows the velocities at a distance equal to half the radius away from the centre axis, and pulse profile at $y = 0.8$ shows the profile 0.8 times the radius away from the centre axis

The radius of the tube is $r = 5 \times 10^{-3}$ m and the kinematic viscosity is $\mu = 4 \times 10^{-6}\,\mathrm{m^2\,s^{-1}}$. The maximum mean velocity during the cycle is $\bar{u} = 4.7 \times 10^{-2}\,\mathrm{m\,s^{-1}}$, thus giving a maximum Reynolds number of 120. By taking the angular frequency as 2π, the Womersley α becomes 6.27.

In Figure 4.2b the response of the pressure gradient, the time–velocity waveforms are shown for different distances from the axis of symmetry. The details of the velocity profiles at 16 instances in the cycle, each 1/24 second or 15 degrees apart, are shown in Figure 4.3. It can be seen that in the first part of the cardiac cycle the velocity profiles are almost flat, except for a thin boundary layer at the wall. When flow reverses direction, the lamina at the wall reverses first while there is still forward flow in the central core of the fluid. In the end of the cycle the velocities decay and become approximately zero.

4.5 PULSATILE FLOW THROUGH A STENOSIS

The finite element model of the stenosis with boundary conditions is shown in Figure 4.4. The boundary conditions at the wall are zero velocities. At the outlet the normal derivatives of the velocities are zero and along the axis of symmetry the normal velocity and the normal derivative of the tangential velocity are zero. At the inlet the tangential velocity is zero, whilst the normal velocities are those of the velocity profiles obtained for pulsatile flow in a straight tube. These velocity

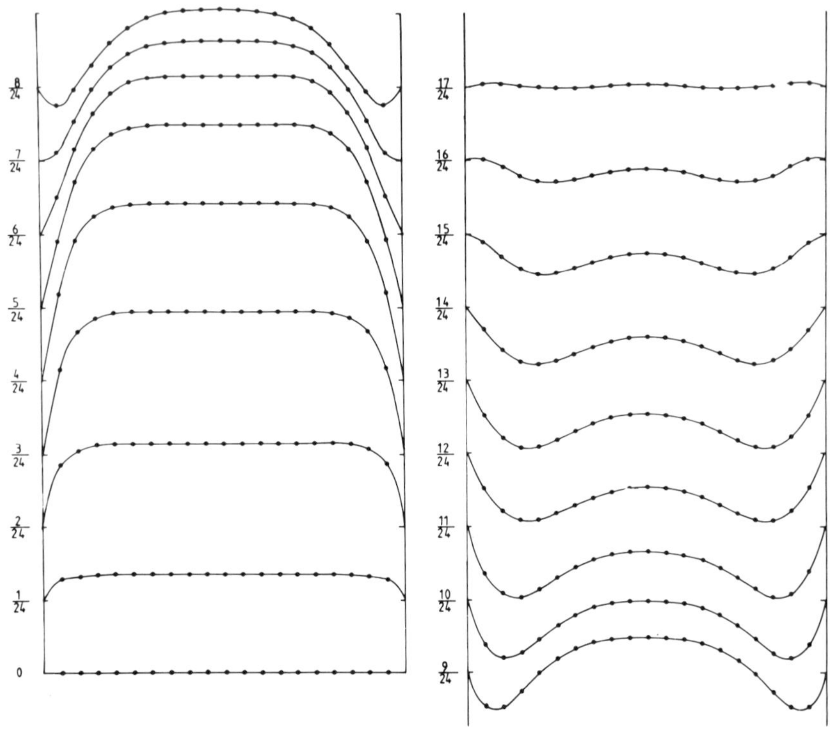

Figure 4.3 The velocity profiles in a straight circular tube as a response to the pressure gradient shown in Figure 4.2a. The length of the cardiac cycle is 1 s and the velocity profiles are shown at intervals of 1/24 s. The velocity profiles between 18/24 s and 24/24 s are flat with zero velocities and they are not shown in the figure

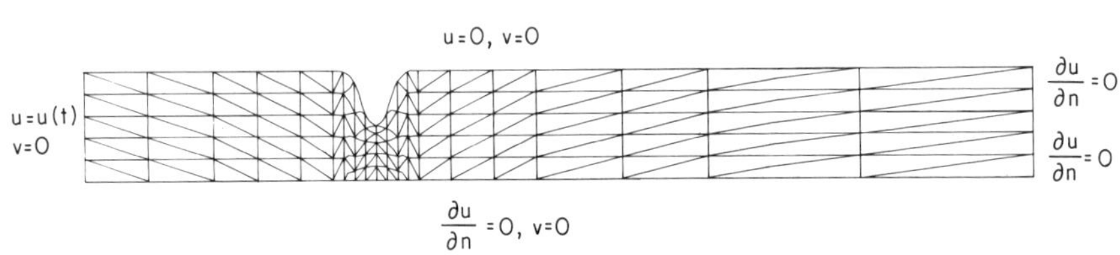

Figure 4.4 The finite element model of the stenosed axisymmetric tube. The boundary condition $u = 0, v = 0$ correspond to the wall; $\partial u/\partial n = 0, v = 0$ correspond to the symmetry axis; $\partial u/\partial n = 0, \partial v/\partial n = 0$ correspond to the outlet; and $v = 0, u = u(t)$ correspond to the inlet. $u(t)$ is given in Figure 4.3

profiles are used as boundary conditions at the inlet at each instant in time where the pressure and velocity patterns in the stenosis model are calculated. The radius at the inlet of the stenosis model is $r = 5 \times 10^{-3}$ m and the kinematic viscosity $\mu = 4 \times 10^{-6}\ \mathrm{m^2\,s^{-1}}$. The maximum constriction of the model is 75 % of the inlet area. With the velocity profiles in Figure 4.3 as boundary conditions at the inlet, the pressure gradient across the stenosis model which is obtained is shown in

Figure 4.5. Figure 4.6–4.11 show the velocity vectors, streamlines and pressure patterns at each instant in time.

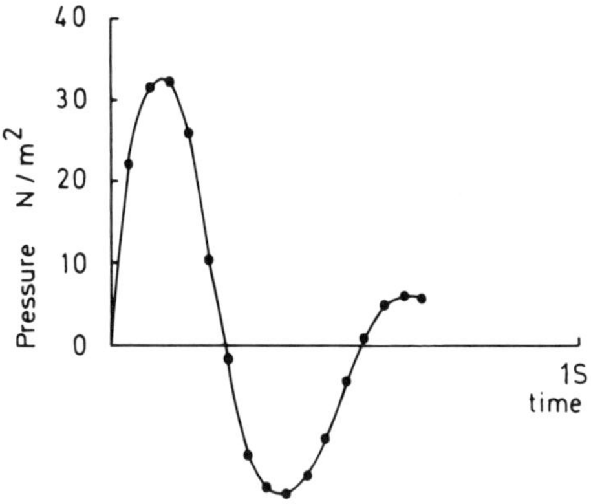

Figure 4.5 The pressure gradient across the model as function of time

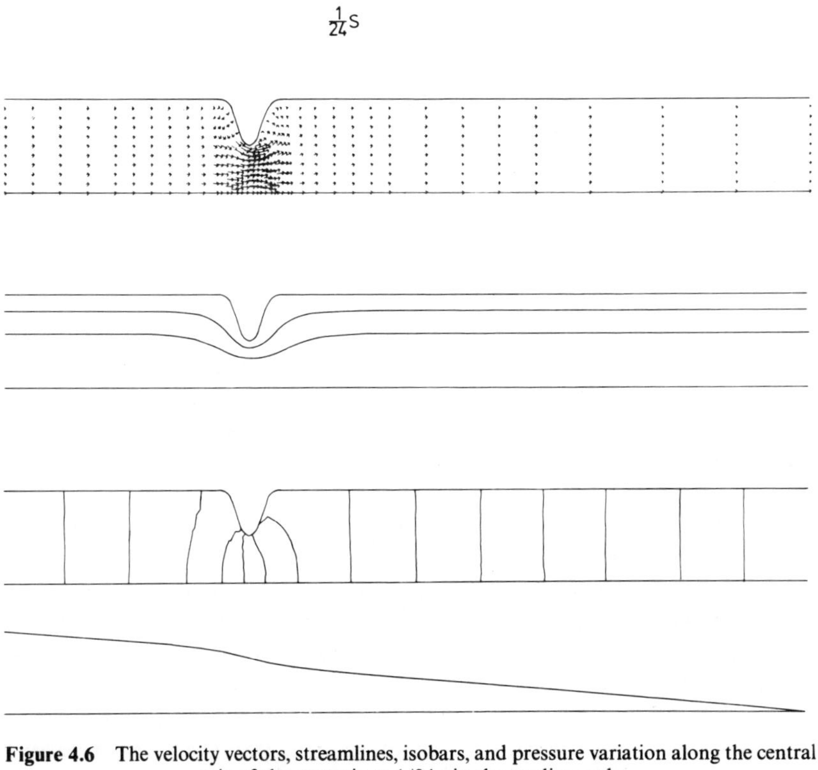

Figure 4.6 The velocity vectors, streamlines, isobars, and pressure variation along the central axis of the stenosis at 1/24 s in the cardiac cycle

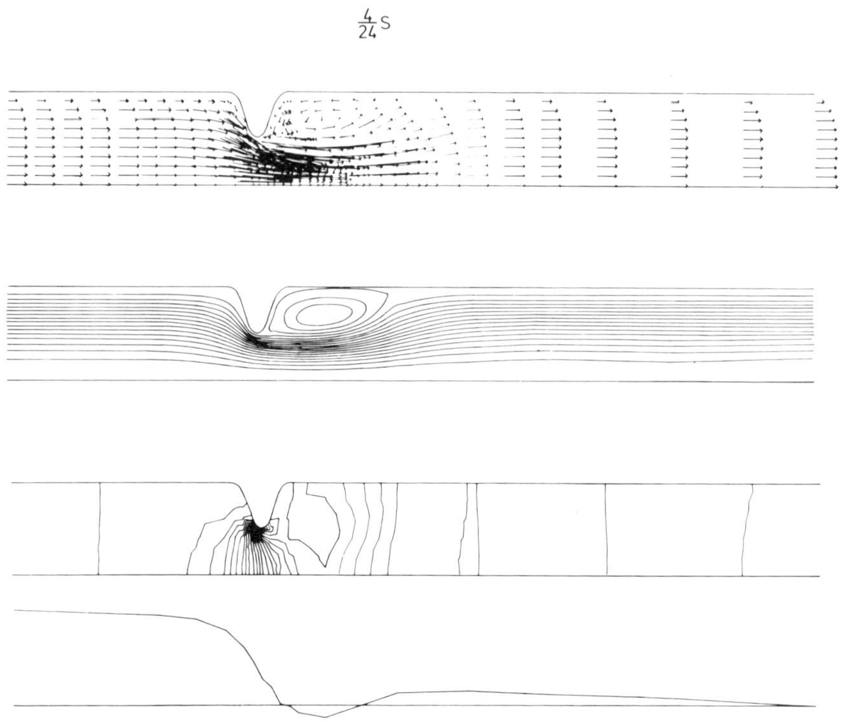

Figure 4.7 The velocity vectors, streamlines, isobars, and pressure variation along the central axis of the stenosis at 4/24 s in the cardiac cycle

In the early part of systole, 1/24 s (Figure 4.6), the flow is accelerated and no separation is seen downstream of the constriction. The pressure gradient is slightly greater at the stenosis than in the upstream and downstream part of the model. Later in the cycle at 4/24 s (Figure 4.7), a vortex formation has developed downstream of the stenosis, and the relative pressure drop across the stenosis has increased. The size of the vortex is increased during the following time steps (7/24 s, Figure 4.8) and a local pressure minimum develops at the level of the vortex. The influence of the reversed flow downstream of the stenosis at 10/24 s is seen in Figure 4.9, where the fluid lamina along the wall are moving upstream through the stenosis and thus forcing the vortex towards the axis. At 13/24 s (Figure 4.10) the velocity vector plot shows that another vortex has developed upstreams of the stenosis. At 16/24 s (Figure 4.11) there are only two vortices with low velocities left, one upstream and one downstream of the stenosis. The calculations were not carried out for subsequent points in time because the numerical methods produced divergent solutions.

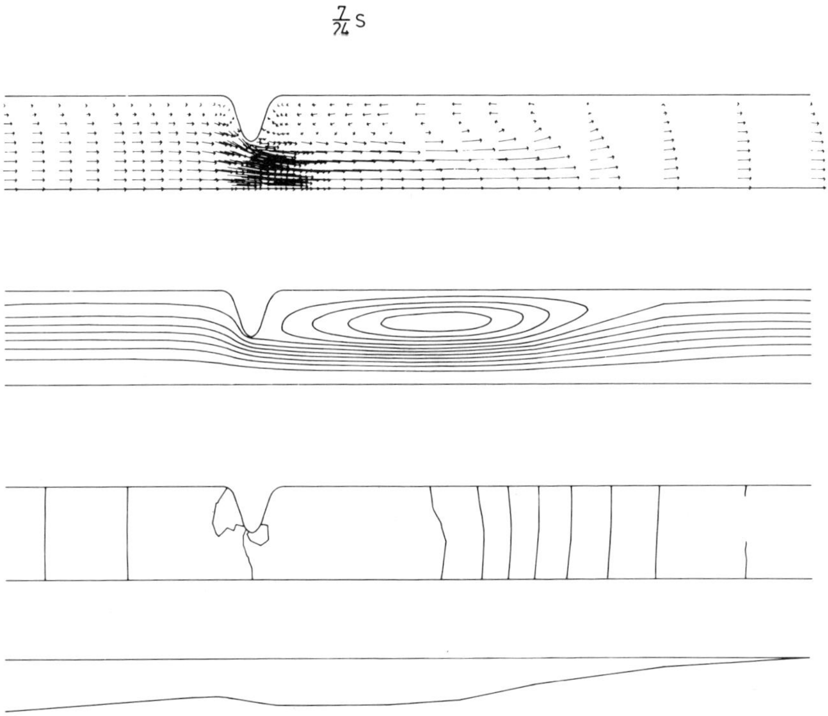

Figure 4.8 The velocity vectors, streamlines, isobars, and pressure variation along the central axis of the stenosis at 7/24 s in the cardiac cycle

In the model of the arterial stenosis a vortex formation has been shown to appear downstream of the constriction in the systolic part and upstream of the constriction in the diastolic part of the cycle. As the flow is reversed in diastole, the vortex is forced towards the centre of the vessel whilst the reversed velocities are occuring in the area close to the wall. At the level of the vortex a low-pressure zone is developed.

In the present simulation the walls of the stenosed model are supposed stiff. This is not the case in actual blood vessels, which have viscoelastic properties with varying diameter during the cardiac cycle. However, the relative change in diameter is in the order of 10 % (McDonald, 1974) and it is therefore believed that the size of the error introduced by assuming a fixed diameter is small.

The velocity profiles used as boundary conditions for the stenosis at the inlet are the corresponding velocity profiles in a long straight tube. However, the feedback from the stenosis may influence the velocity profiles at the inlet. The correct velocity profiles at the inlet are probably slightly different from those in a straight tube. Inlet velocity profiles are not available from experimental or

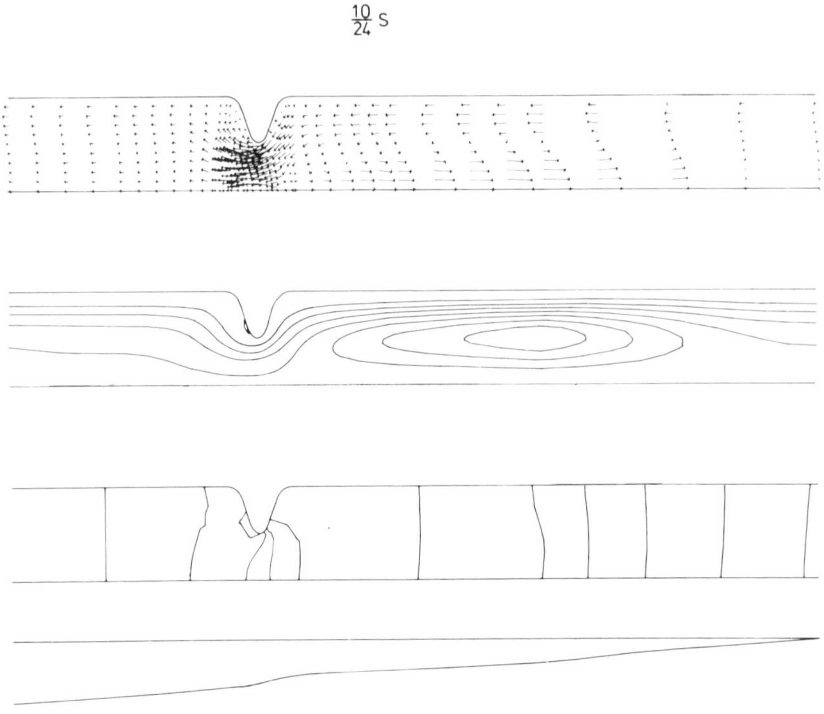

Figure 4.9 The velocity vectors, streamlines, isobars, and pressure variation along the central axis of the stenosis at 10/24 s in the cardiac cycle

theoretical work. The results reported here should not be affected by small deviations that may be present in the inlet velocity profile due to feedback.

With the present method it is not possible to obtain convergent solutions towards the end of the cardiac cycle. The fact that the downstream vortex grows out through the outlet of the tube, thus violating the outlet boundary conditions, may be the reason for divergent solution.

In the last part of the cardiac cycle, where it was not possible to obtain convergent solutions, we believe that the velocities will decay, and that the effects of the small movements in late diastole will have negligible influence on the next cycle. At least, if we use the last obtained solution (16/24 s) as the solution at the end of the diastole, the effects of the two vortices disappeared after two time steps.

4.6 PULSATILE FLOW THROUGH AN ANEURYSM

The boundary conditions for the aneurysm model are similar to those of the stenosis model. The finite element model of the aneurysm is shown in Figure 4.12.

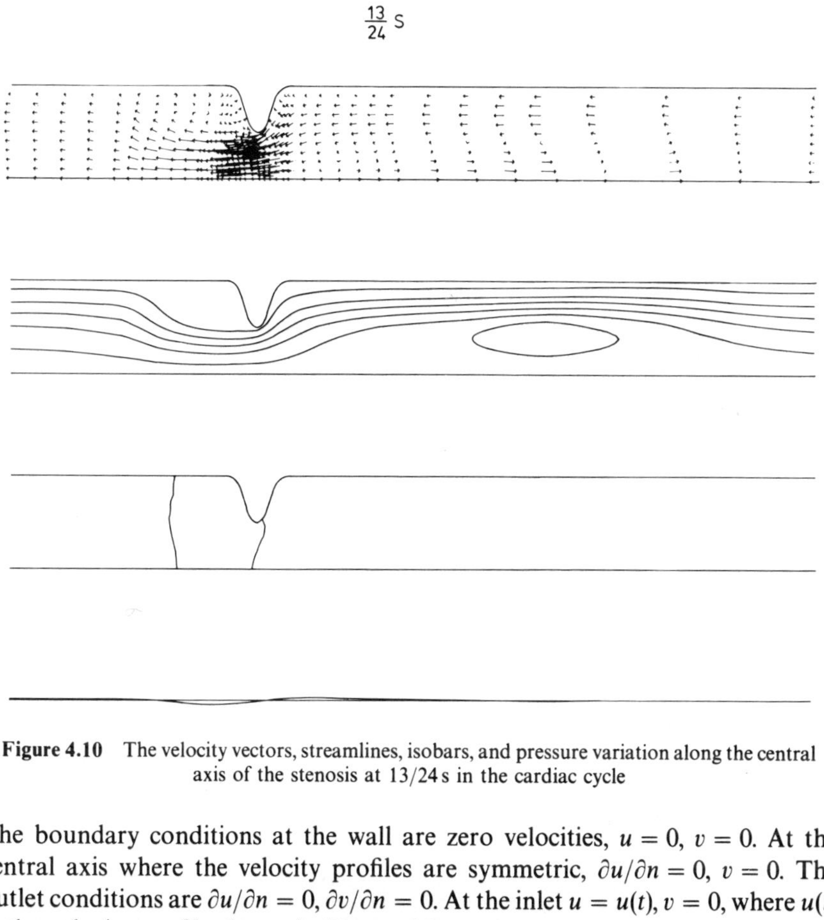

Figure 4.10 The velocity vectors, streamlines, isobars, and pressure variation along the central axis of the stenosis at 13/24 s in the cardiac cycle

The boundary conditions at the wall are zero velocities, $u = 0$, $v = 0$. At the central axis where the velocity profiles are symmetric, $\partial u/\partial n = 0$, $v = 0$. The outlet conditions are $\partial u/\partial n = 0$, $\partial v/\partial n = 0$. At the inlet $u = u(t)$, $v = 0$, where $u(t)$ is the velocity profile shown in Figure 4.3.

The radius of the aneurysm at the inlet and outlet vessel is $r = 2.5 \times 10^{-3}$ m. The radius of the aneurysm itself is 7.5×10^{-3} m which is three times the radius at the inlet. The cross-sectional area of the aneurysm is thus nine times the area at the inlet. The kinematic viscosity of the fluid is $\mu = 4 \times 10^{-6}\,\mathrm{m^2\,s^{-1}}$. The maximum mean velocity is $\bar{u} = 4.7 \times 10^{-2}\,\mathrm{m\,s^{-1}}$. The flow in the aneurysm is characterized by the Reynolds number which is $Re = 60$ and the Womersley constant $\alpha = 3.14$. The values of both the Reynolds number and the Womersley constant are within the physiological range of blood flow.

The simulations of pulsatile flow in the aneurysm are shown in Figures 4.14–4.19. Figure 4.13 shows the pressure gradient across the aneurysm model as function of time as response to the velocity profiles in Figure 4.3 applied to the inlet. The pressure and flow patterns in the early part of systole, only 1/24 s after the initialization of a cardiac cycle, is seen in Figure 4.14.

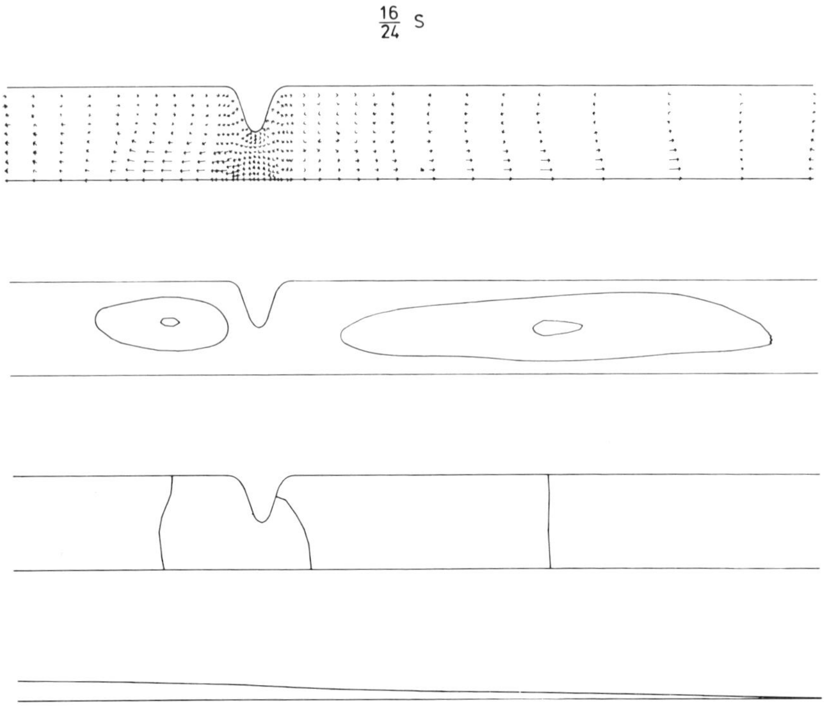

Figure 4.11 The velocity vectors, streamlines, isobars, and pressure variation along the central axis of the stenosis at 16/24 s in the cardiac cycle

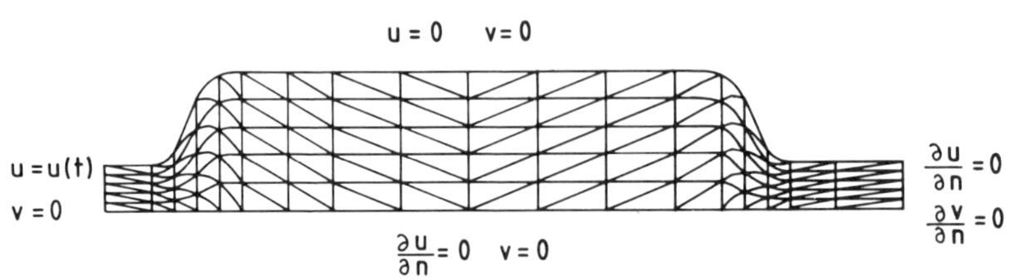

Figure 4.12 The finite element model of the axisymmetric aneurysm. The boundary conditions $u = 0, v = 0$ correspond to the wall; $\partial u/\partial n = 0, v = 0$ correspond to the axis of symmetry; $\partial u/\partial n = 0, \partial v/\partial n = 0$ correspond to the outlet; and $u = u(t), v = 0$ correspond to the inlet. The inlet velocity profiles, $u(t)$, are those obtained in the straight circular tube and they are shown in Figure 4.3

The velocity vector diagram demonstrates low velocities inside the aneurysm, but there is a small vortex formation at the downstream end. This vortex is more clearly apparent in the streamline diagram. The presence of this vortex is due to the vortex in the end diastole of the previous cycle which did not decay entirely

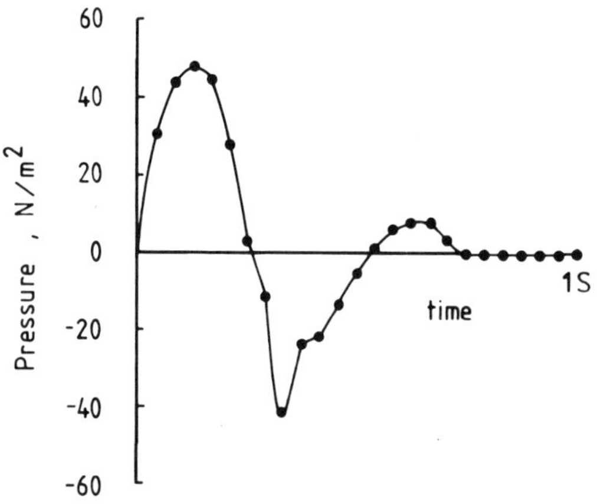

Figure 4.13 The pressure gradient in the aneurysm as a function of time for the successive velocity profiles shown in Figure 4.3 used as boundary conditions

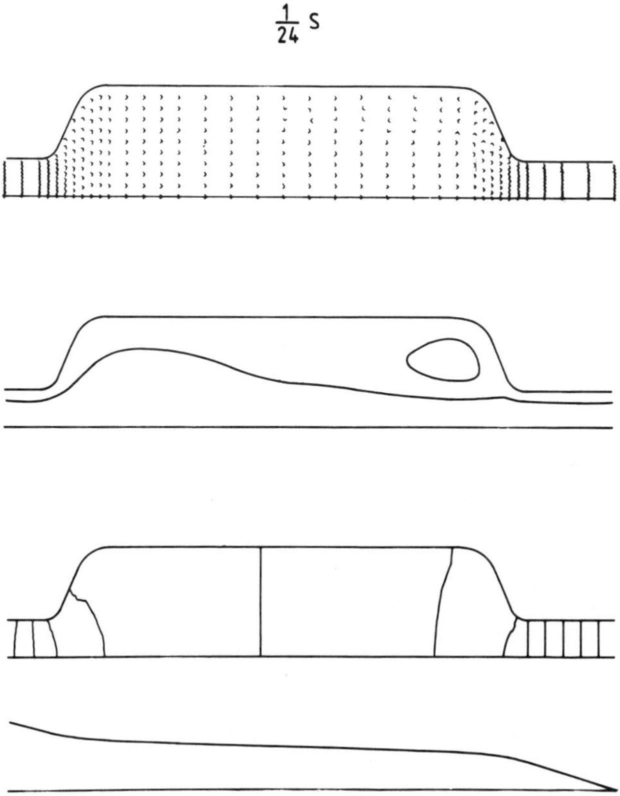

Figure 4.14 The velocity vectors, streamlines, isobars, and pressure variation along the central axis of the aneurysm at 1/24s in the cardiac cycle

before the beginning of the proceeding cycle. The pressure at 1/24 s is seen to be fairly constant within the aneurysm, whilst there is a greater pressure drop in the inlet and outlet vessel.

In Figure 4.15 at 4/24 s the downstream vortex has disappeared, whilst a vortex upstream of the aneurysm has been developed. At the downstream part of the aneurysm where the outlet vessel begins, a high pressure gradient has developed. This pressure gradient will tend to expand the aneurysm further downstream and will probably cause weakening of the vessel wall.

$\frac{4}{24}$ s

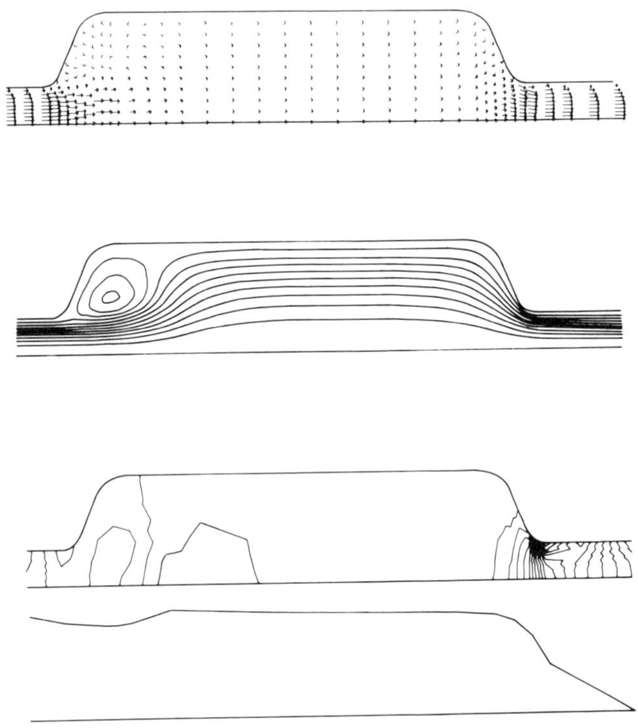

Figure 4.15 The velocity vectors, streamlines, isobars, and pressure variation along the central axis of the aneurysm at 4/24 s in the cardiac cycle

At 7/24 s (Figure 4.16) a negative pressure gradient is developed in the direction of flow. However, the kinetic energy of the fluid is still sufficient to provide a forward flow through the aneurysm.

Figure 4.17 shows that at 10/24 s the net flow in the aneurysm is reversed. The

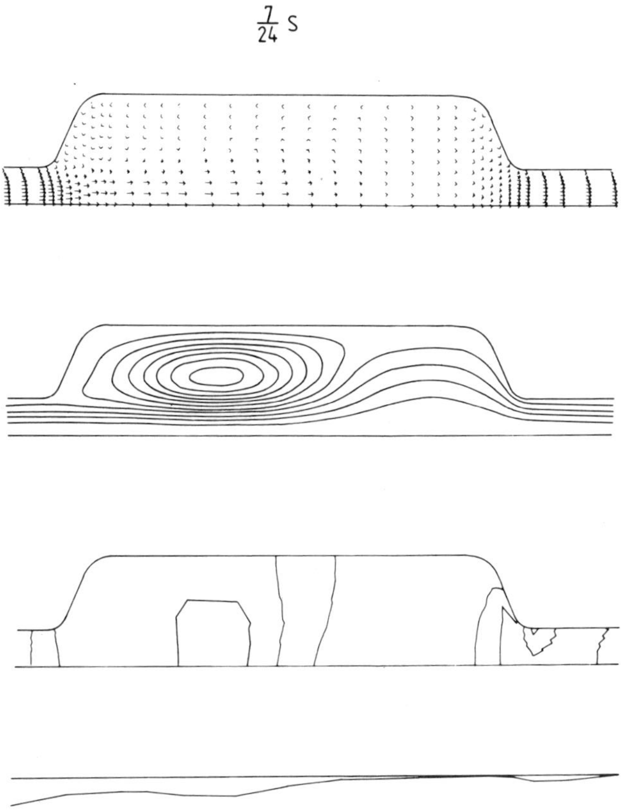

Figure 4.16 The velocity vectors, streamlines, isobars, and pressure variation along the central axis of the aneurysm at 7/24 s in the cardiac cycle

reversed flow is greater along the wall and the backward flowing fluid comes from the lamina between the vortex and the wall. The vortex is thus forced towards the central axis of the aneurysm.

At 13/24s (Figure 4.18) another very small vortex at the downstream wall of the aneurysm has developed. This vortex is due to the high velocities in the reversed flow at the outlet and it is of the same nature as the vortex shown in Figure 4.15 (4/24 s) in the accelerating phase of the systole. The two vorticies inside the aneurysm are rotating in different directions, the largest vortex anticlockwise and the smaller one clockwise.

In Figure 4.19 at 16/24 s the fluid lamina along the wall are changing direction and again there is forward flow. However, there is no net transportation of fluid through the aneurysm. From this point in time the flow in the model decreases during the rest of the cardiac cycle.

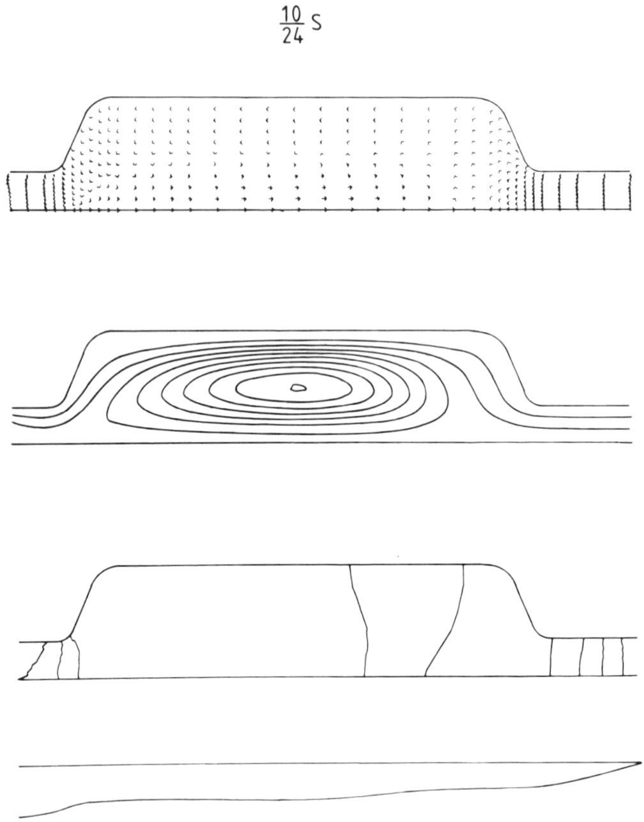

Figure 4.17 The velocity vectors, streamlines, isobars, and pressure variation along the central axis of the aneurysm at 10/24 s in the cardiac cycle

The development of aneurysms in the arterial tree is most frequently found in the abdominal aorta as a result of a general weakening of the wall. The expansion of the vessel is gradually filled with a thrombus formation. Although changes in the mechanical and chemical properties of the arterial wall may be responsible for the initialization of aneurysms, the haemodynamics within the involved arterial segment are believed to be of considerable importance in the formation of the thrombus and growth of the aneurysm.

The flow patterns in the aneurysm model demonstrate clearly that a vortex is present during most of the cardiac cycle. The intensity and size of the vortex, however, vary but reach a maximum in the late systolic phase. This vortex represents a stagnant zone in the flow through the aneurysm, and it is likely that the low velocities would cause blood clotting which will form a thrombus. In

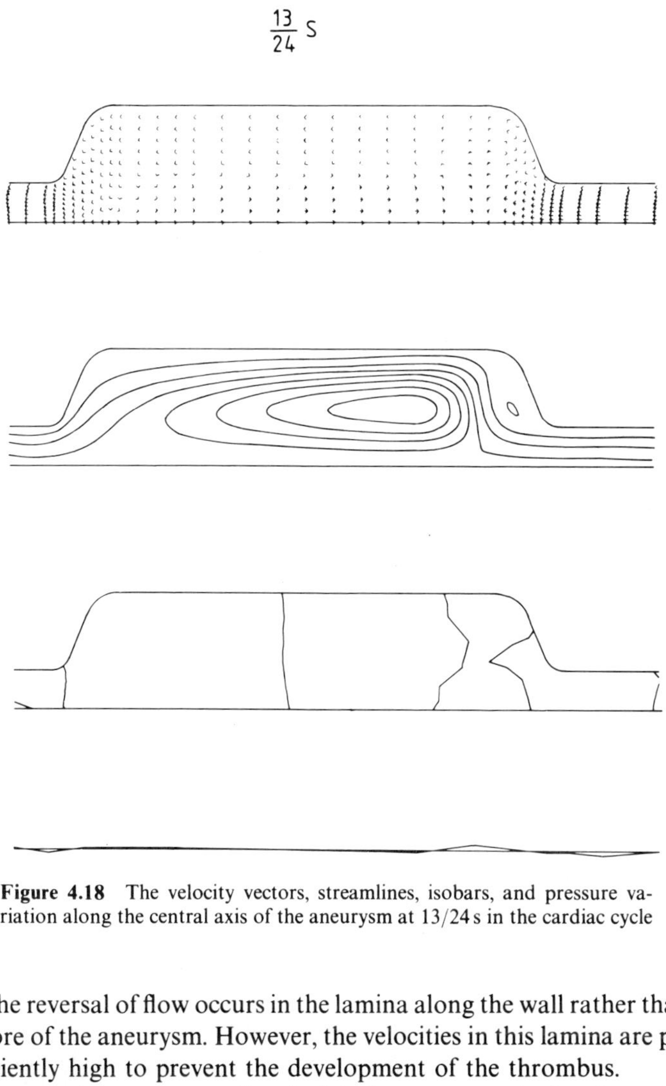

Figure 4.18 The velocity vectors, streamlines, isobars, and pressure variation along the central axis of the aneurysm at 13/24 s in the cardiac cycle

diastole the reversal of flow occurs in the lamina along the wall rather than in the central core of the aneurysm. However, the velocities in this lamina are probably not sufficiently high to prevent the development of the thrombus.

There is a small pressure gradient inside the aneurysm during the entire cycle. However, a local pressure minimum can be observed at the vortex centre. In the radial direction the pressure is equally constant. The small pressure gradient inside the aneurysm for pulsatile flow is in agreement with the results of steady flow (Wille, 1980).

The important and high pressure gradient is found downstream at the outlet vessel. This pressure gradient will tend to expand the already existing aneurysm further downstream. The phenomenon of a downstream-growing aneurysm has been clinically observed. The high shear forces exerted on the wall associated with

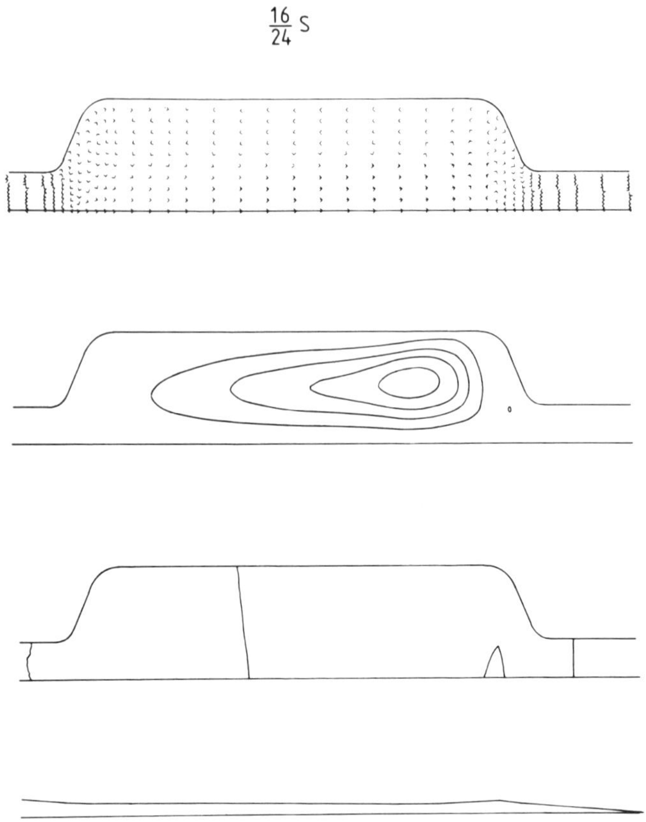

Figure 4.19 The velocity vectors, streamlines, isobars, and pressure variation along the central axis of the aneurysm at 16/24 s in the cardiac cycle

the pressure gradient are likely to damage the endothelium layer and increase the possibility of a rupture.

In diastole a secondary vortex is developed in the downstream part of the aneurysm. This vortex is small and probably of less importance to the pathology.

The present aneurysm model has the advantage of being axisymmetric, and thus the number of non-linear equations in the system have been reduced considerably. In future work three-dimensional models will be designed, so other geometries such as bifurcations and bends in the arterial system can be analysed.

ACKNOWLEDGEMENT

The author is grateful to Pål Bergan, Olav Dahl, and Lars Wallφe for valuable discussions and encouragement.

REFERENCES

Azuma, T., and Fukushima, T. (1976) Flow patterns in stenotic blood vessel models, *Biorheology*, **13**, 337–355.

Caro, C. G., Fitz-Gerald, J. M., and Schroter, R. C. (1971) Atheroma and arterial wall shear. Observations, correlations and proposal of a shear dependent mass-transfer mechanism for atherogenesis, *Proc. Roy. Soc. Lond.* (B), **177**, 109–159.

Durst, F., Melling, A., and Whitelaw, J. H. (1976) *Principles and Practice of Laser-Doppler Anemometry*, Academic Press, London.

Fry, D. L. (1972) Localizing factors in arteriosclerosis, In: *Atherogenesis and Coronary Heart Disease* (ed. W. Likoff *et al.*), Grune and Stratton Inc., New York, pp. 85–104.

Lambert, J. D. (1973) *Computational Methods in Ordinary Differential Equations*, John Wiley & Sons, London.

McDonald, D. A. (1974) *Blood Flow in Arteries*, The Camelot Publishing Co., Baldwin Park, California.

Péronneau, P., Hinglais, J., Xhaard, M., Delouche, P., and Phillipo, J. (1974) The effects of curvature and stenosis on pulsatile flow in vivo and vitro, *Cardiovascular Application of Ultrasound* (ed. R. S. Renemann), North Holland Publishing Co., Amsterdam, pp. 203–215.

Roach, M. R. (1972) Poststenotic Dilatation in Arteries. *Cardiovascular fluid dynamics* (ed. Bergel), Academic Press, New York, pp. 111–139.

Taylor, C., and Hood, P. (1973) A numerical solution of the Navier–Stokes equation using the finite element technique, *Computers and Fluids J.* **1**, 73–100.

Wille, S. O. (1980) A finite element solution of the Navier–Stokes equations for two dimension and axis-symmetric flow, *Modelling, Identification and Control*, **2**, 105–117.

Finite Elements in Biomechanics
Edited by R. H. Gallagher, B. R. Simon, P. C. Johnson, and J. F. Gross

Chapter 5

Finite Element Analyses in Soft Tissue Mechanics

George C. Lee and Nien T. Tseng

5.1 INTRODUCTION

In recent years there has been a growing interest in using finite element analysis for the modelling of the behaviour of biological systems. Soft biological tissues have been considered as passive elastic continua in various distinct types. For example, certain soft tissues may be regarded as incompressible continua while others are treated as compressible. In addition, for a given soft tissue, assumptions of material homogeneity and anisotropy are first assessed based on available experimental evidence before specific analytical solutions can be sought. In most cases, however, the basic approach in finite elements remains more or less the same.

To discuss finite element analysis in soft-tissue mechanics, it is useful to consider it within a broader perspective of biomechanics, by including finite element analysis as a component. The details of finite element analysis such as the question of convergence will not be elaborated in this chapter. There are many recent publications that deal with various aspects of finite element approaches for specific tissues. Some of these as examples are included in the list of references at the end of the chapter.

Finite element analysis in soft-tissue mechanics must be viewed somewhat differently from its application in traditional structural mechanics research. One of the many characteristics is that the 'normal responses' of soft tissue cannot be uniquely defined. There is always a range in the magnitude of pressure or deformation of the tissue that is considered to be normal. This contributes to the difficulty in assessing accuracy in analytical solution procedures. A second distinct feature in soft-tissue mechanics is the simultaneous consideration of mechanical, electrical, and biochemical responses, particularly at the microscopic level. Finite element formulations in this regard must be considerably broadened, from those that are typically used in structural mechanics research. Additionally, the experimental data available on soft tissue are generally limited. Even for those tissues where data are available, there are often more questions than answers in the conditions and limitations of the data. In most cases the difficulties of experimental observations overwhelm the desirability of the form of data for theoretical purposes. This has considerably elevated the complexity of analytical solutions which are dictated by what is given. In finite element analysis the same is true. On the other hand, it may be said that finite element analyses are the most powerful tools available to obtain many quantitative descriptions of the behaviour of soft tissue that could not be realized otherwise. Finite element analysis in man-made systems is often the means to a satisfactory solution for a problem. In biological systems its role is to provide a quantitative basis for a renewed study of the same problem in an evolving fashion.

5.2 SOFT TISSUE

Soft tissue may be considered in several basic categories: soft connective tissues (lung tissue, skin, blood vessels, ligaments, tendons, mesentery, and other membranes), muscles, organs, and the brain. Other than the brain, most soft tissues are composites of several mechanically and biochemically different constituents. The mechanics of the soft tissue, therefore, largely depend upon the individual responses and the quantitative description of the proportion and disposition of these basic constituents.

The most fundamental type of basic component in soft tissue are the elastic fibres. They are the major force-bearing components of the tissue. Elastic fibres may be considered as passive elements. The most notable are the elastins (Gotte, 1977) and collagens (Elden, 1976). The mechanical characteristics and the disposition of these fibres are quite different. They form as bundles in the tendon, as plane matrices in membranes, or as spatial frameworks embedded in proteins and other biological solids in the lung. They typically possess the following characteristics when considered as a continuum.

5.2.1 Large deformation

The force–deformation relationship of various soft tissues are always non-linear with extension ratios ranging between 10% and over 100% under different loading conditions. In typical two-dimensional modelling of membranes, skin, or blood vessels, material incompressibility is generally assumed. In certain three-dimensional models such as the lung parenchyma, principles of compressible continua have been used. In most cases the material has been assumed to be homogeneous and initially isotropic. Some attempt to include directional response has been made in the study of two-dimensional structures such as the skin, by assuming transverse isotropy.

5.2.2 Hysteresis

Soft tissue contains a large fluid component. Also the deformational responses of the tissue are dependent on the transport, electrical, and biochemical characteristics of its cellular and molecular constituents. These factors are believed to be responsible for the loading path-dependent and rate-dependent nature of soft tissue. However, very few responses have been quantified. In typical continuum studies, elastic modelling is generally used. The rate-dependent behaviour has been characterized by viscoelastic models by many investigators. The path-dependent characteristic (hysteresis) has been a major difficulty in soft-tissue modelling because of its non-conservative nature. In this area a few investigators have attempted inelastic or quasi-elastic models. There is a definite need to establish broad mechanics principles beyond the normal limits of non-linear elasticity theory in order to better characterize soft biological tissues.

5.2.3. The active feedback system

Most soft-tissue responses are more or less controlled and coordinated. For convenience, the passive mechanical properties of elastic fibres and biological membranes have been modelled separately. The significance of force transmission and the interactive nature of the active muscles with the passive soft-tissue components has not been well established. This subject will not be considered here.

5.3 SOFT TISSUE MECHANICS

Mechanics of soft tissue contains a broad spectrum of problems ranging from the standard structural analysis type to the integrated study of mechanical and biochemical properties. In other words, in certain problems the assumption of a hyperelastic continuum can reasonably be made while in others, the non-conservative characteristics must be properly taken into consideration.

There is very little which can be said about systematic development of solid mechanics models of non-conservative systems although it is clearly a long-term challenge in biomechanics. Within the framework of a passive, conservative system a large number of studies have been carried out (using both compressible and incompressible formulations). In any case, it is essential to have clearly defined strain energy density functions or stress–strain relationships of the material. In the following, some challenging aspects of soft-tissue mechanics will be outlined before a discussion of finite element analysis application is made.

5.3.1 Mechanical properties of the material

There have been two separate types of experimental approaches in seeking experimental data for the purpose of formulating the material properties of soft tissue: those that attempted to measure *in vivo* properties and those that dealt with experiments on excised specimens. Obviously, excised specimens are considerably easier for experimental measurements but the question always remains regarding the validity of the resulting constitutive laws to predict the behaviour of the intact system. In soft-tissue mechanics, proper experimental measurement of mechanical properties remains the most important task. More often than not, a complete set of desirable experimental data can never be obtained. Imaginative techniques of experimental measurements should be attempted within a carefully thought-out theoretical framework. It may be also noted that some discussion has been made in the literature regarding suitable choices of the functional form for the material law of soft tissue. This is of relatively less importance compared with the problem of the accuracy of experimental results on which curve fitting is made.

5.3.2 The free body and the boundaries

Most mechanical modelling requires the identification of a physical domain on which the problem is defined and formulated. In soft biological tissues the physical model (particularly the boundary conditions) is not easily identified. Often the boundary conditions are part of the desired solution to the problem. In many soft-tissue mechanics studies, the material constitutive relationships are formulated based on data of excised specimens where a stress-free state is assumed to exist. Typically, this condition provides the reference state for the stresses or deformations of a physical model. However, tissue behaviour characterized on this basis cannot be readily correlated with *in vivo* conditions where the reference states (within the structure and along its boundary) are generally not completely known. One of the most difficult and interesting soft-tissue mechanics problems is to find a reference state (including the boundary conditions) within the tissue so that results can be meaningfully related to the intact system.

5.3.3 Microscopic–macroscopic interaction

Perhaps the underlying importance of studying soft-tissue behaviour lies in the understanding of the interactive nature of the macroscopic behaviour with the constituents at the microscopic level in the material. At the present time major efforts in soft-tissue studies are concentrated on gross behaviour. This is understandable because such experimental data are comparably easier to obtain and certain continuum mechanics principles can be readily used to characterize the mechanical properties. However, the major direction of life science research has been and is heading toward successively smaller dimensions where many branches of science and engineering disciplines converge and interplay closely. Therefore, in macroscopic level modelling the implications of microscopic level phenomena are important objectives to be kept in mind.

5.4 FINITE ELEMENT ANALYSIS IN LUNG TISSUE MECHANICS

The role of finite element analysis in soft-tissue mechanics may be viewed from two separate aspects. For problems where the physical model and boundary conditions of the tissue can be reasonably defined and where a set of constitutive relationships are given, finite element analysis can be carried out by using the principles of non-linear elasticity or viscoelasticity, similar to other solid mechanics applications. The non-linear material behaviour is typically linarized in a step-by-step fashion and solutions are sought either incrementally and/or iteratively. For geometric non-linearities, the strain–displacement relationships described by Lagrangian coordinates (either in total or in an updated fashion) are generally preferred. Depending on the non-linear nature of the material and the available boundary and loading conditions, finite elements can be formulated on the basis of potential energy or complementary energy. In any case, for a given type of tissue the available information and the desired solution will generally dictate the formulation of a special finite element procedure. However, the basic principles and approaches involved remain more or less the same.

The other type of finite element application is more involved and complex. To illustrate this type of application of finite element analysis in soft-tissue mechanics, research on lung tissues conducted by the author and others will be described briefly here.* Examples from other research areas are outlined above .

5.4.1 Some objectives of lung tissue mechanics

One important problem area in lung tissue mechanics is to quantify the deformation and pressure gradients in the lung and along its boundaries (chestwall) due to the combination of body forces and externally imposed forces or kinematic conditions. The regional differences of stresses and deformations in the lung have significant implications in the understanding of the pattern of gas exchange variation and efficiency within the lung. It is obvious that in order to model the lung by any analytical approach (including the finite element method), it is first necessary to know the constitutive relationships of the material. Therefore, the first step was the determination of the strain energy density function for the lung parenchyma.

5.4.2 Characteristics of lung parenchyma

The mechanical characteristics of lung parenchyma have been studied by many investigators, both experimentally and analytically. Some typical references are given at the end of this chapter.

* A few selected references are given at the end of this chapter. No attempt is made to be inclusive.

The large deformations and non-linear nature of lung tissue arise from the composite action of elastin and collagen fibres. This is illustrated in Figure 5.1. Elastins are extremely extensible fibrous proteins. They consist mainly of polypeptide chains and are usually tangled and sparsely cross-linked. The modulus of elasticity is low (approximately 3 to 8 × 10^6 dynes/cm^2). Elastin fibres may be extended to above 100 % elongation before failure. The collagen fibres are relatively inextensible (approximately 10 % elongation at rupture). The mechanical properties of aggregates of collagen depend on their spatial disposition, density, and relationship with other tissue. At high stresses the length–force relationship of collagen is approximately linear with a modulus of elasticity estimated at 1 × 10^6 dynes/cm^2. The composite response of elastin and collagen fibres may be described as follows. The collagens are primarily responsible for the increasing stiffness of the lung tissue at higher stress levels, while the elastins are in action in the entire stress range of the lung tissue. At low lung volumes, collagen fibres are presumed to unwind by undergoing rigid body motions. The characteristics of elastins, collagens and their combinations are described in Figure 5.1.

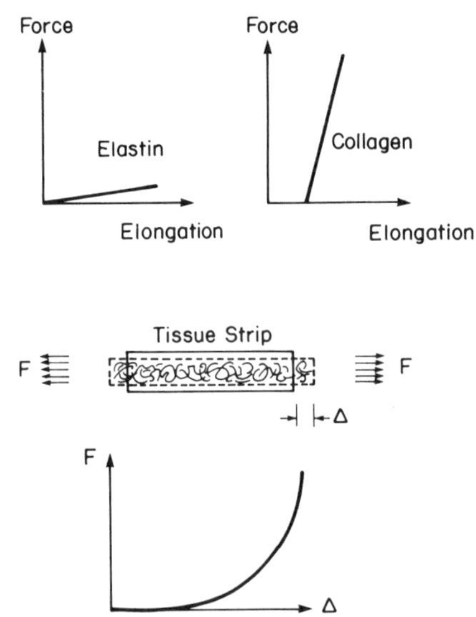

Figure 5.1 Schematic illustration of the analytical model of the force–elongation relationship of soft tissue strips based on the combined properties of elastin and collagen fibres

In a normal human lung there are approximately 3–4 × 10^8 alveoli (with 200 μm average diameter). The alveoli are the smallest respiratory units and gas exchange takes place through the alveolar walls. In a normal lung, the total gas exchange surface of all alveolar walls may be as large as 30–40 m^2. The pressure–volume relationship of the lung contains a large hysteresis loop because of the special characteristics of the lining layer of the alveolar walls (pulmonary surfactant) and the surface tension of the air–fluid interface. This is illustrated in Figure 5.2. This hysteresis loop is considerably reduced if the lung is ventilated by physiological solution (shown in Figure 5.2). The present state-of-the-art in modelling of lung parenchyma is to consider only the deflation or the inflation curve, by using hyperelasticity where a strain energy density function of the material is assumed to exist).

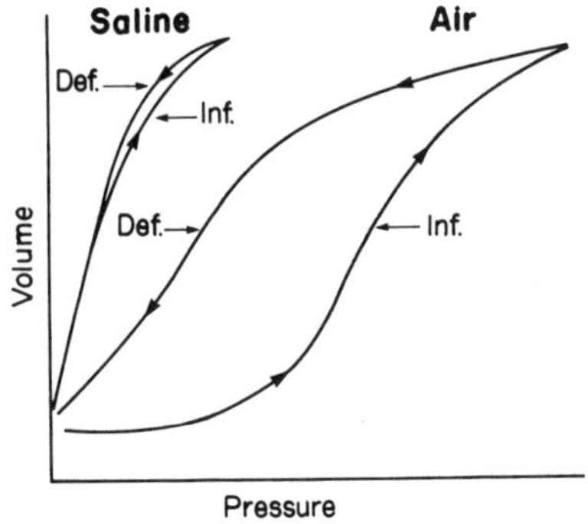

Figure 5.2 Typical pressure vs. volume relationships of the lung. It is shown that the P–V curve for an air-ventilated lung has a large hysteresis when compared with that for the saline-ventilated lung

Many microscopic and macroscopic experimental observations have provided the basis for the assumption that the lung (90% of which is alveoli) may be modelled as a homogeneous and initially isotropic continuum.

5.4.3 The constitutive relationship

The pressure–volume (P–V) relationship of the lung as shown in Figure 5.2 is the only well-established experimental evidence available. However, the P–V curves represent only the relationship between the negative hydrostatic pressure (uniformly distributed tension) and the volume strain (single degree of freedom model). This relationship is insufficient to characterize the stress–strain properties of a non-linear, compressible continuum undergoing large deformations. Additional three-dimensional experimental data are necessary for the formulation of a strain energy density function.

For a homogeneous and initially isotropic compressible continuum, the strain energy density function $W(\varepsilon_{ij})$ may be written as

$$W = W(I_1, I_2, I_3) \tag{5.1}$$

where ε_{ij} is the Lagrangian strain tensor, and

$$I_1 = \varepsilon_{kk}, \quad I_2 = \frac{1}{2}\varepsilon_{ij}\varepsilon_{ji}, \text{ and}$$

$$I_3 = \frac{1}{3}\varepsilon_{ij}\varepsilon_{jk}\varepsilon_{ki} \text{ are the strain invariants.}$$

Alternatively, Equation (5.1) may be written in the series form of the strain invariants,

$$\begin{aligned} W(I_1, I_2, I_3) &= W_0 + W_1 + W_2 + W_3 + W_4 + \dots \\ &= W_0 + A_0 I_1 + (A_1 I_1^2 + A_2 I_2) + (A_3 I_1^3 + A_4 I_1 I_2 + A_3 I_3) \\ &\quad + (A_6 I_1^4 + A_7 I_1^2 I_2 + A_8 I_1 I_3 + A_9 I_2^2) + \dots \end{aligned} \tag{5.2}$$

In Equation (5.2) W_0 is a constant which does not enter the constitutive equation and can be set to zero. W_1 consists of terms linearly related to the strain tensor (I_1 only), W_2 consists of terms that are quadratic of strain tensor or linear combinations of I_1^2 and I_2, etc. The constitutive law has the form

$$t_{ij} = \frac{\partial W}{\partial \varepsilon_{ij}} = \frac{\partial W}{\partial I_1}\frac{\partial I_1}{\partial \varepsilon_{ij}} + \frac{\partial W}{\partial I_2}\frac{\partial I_2}{\partial \varepsilon_{ij}} + \frac{\partial W}{\partial I_3}\frac{\partial I_3}{\partial \varepsilon_{ij}} \tag{5.3}$$

Depending upon the available experimental data, Equation (5.2) may be used to generate a strain energy density function. It may be noted that for an isotropic linear elastic material without initial stress, $W = W_2 = A_1 I_1^2 + A_2 I_2$, and the constitutive equation, Equation (5.3), is given by

$$t_{ij} = \frac{\partial W_2}{\partial \varepsilon_{ij}} = 2A_1 I_1 \delta_{ij} + A_2 \varepsilon_{ij} = \lambda I_1 \delta_{ij} + 2\mu\varepsilon_{ij} = \lambda\varepsilon_{kk}\delta_{ij} + 2\mu\varepsilon_{ij} \tag{5.4}$$

where

$$A_1 = \frac{\lambda}{2} \quad \text{and} \quad A_2 = +2\mu$$

Several attempts to formulate the strain energy density function have been reported in the literature, either based on Equation (5.2) or on some other functional forms. (See the list of references at the end of the chapter.) Here Equation (5.2) was used with a combination of three-dimensional experimental data (Hoppin, Lee, and Dawson, 1977) and an analytical model denoted as the 'averaged alveolus' (Frankus and Lee, 1974). The latter model was based on P–V curves and a finite element computational procedure, and is illustrated in Figure 5.3. The strain energy density function of the lung parenchyma was formulated using Equation (5.2) up to the fourth-order terms in the strain tensor ε_{ij}. The coefficients of Equation (5.2) are

$$\begin{array}{lll} A_0 = -0.633 & A_1 = 0.316 & A_2 = 5.181 \\ A_3 = 0.122 & A_4 = -5.457 & A_5 = 4.769 \\ A_6 = -0.187 & A_7 = 2.239 & A_8 = 1.184 \\ A_9 = -0.934 & & \end{array}$$

5.4.4 Finite element formulation

Finite element formulation begins with the consideration of equilibrium between the virtual work of external forces and the inner product of the conjugate stress

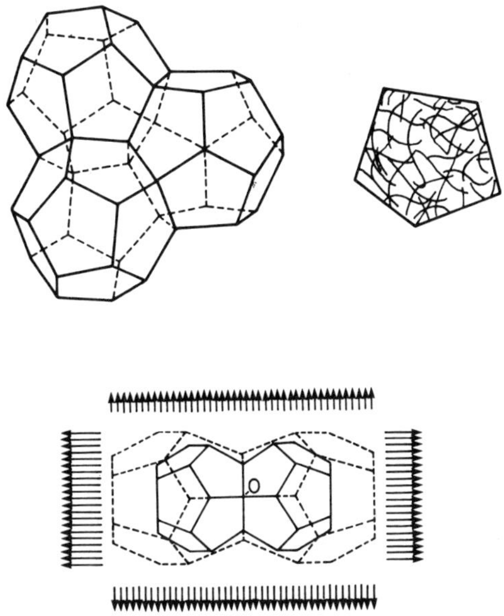

Figure 5.3 Finite element computation of the mechanical properties of the lung parenchyma based on the 'averaged alveolus'. The tissue walls of the dodecahedral model are made up of randomly oriented elastic fibres. Their properties are derived from the *P*–*V* curve of the lung, and substantiated by the approach illustrated in Figure 5.1

tensor, t_{ij} (referred to the deformed configuration per unit of undeformed area) and the virtual Lagrangian strain tensor, $\delta\varepsilon_{ij}$. The virtual work equation may be written as

$$\int_A \delta\bar{u}\,.\,\bar{t}\mathrm{d}A + \int_V \delta\bar{u}\,.\,\bar{f}\mathrm{d}V = \int_{A_0} \delta\bar{u}\,.\,F\,.\,\bar{t}_0\mathrm{d}A + \int_{V_0} \rho_0\delta\bar{u}\,.\,\bar{b}\mathrm{d}V = \int_{V_0} t_{ij}\delta\varepsilon_{ij}\,\mathrm{d}V \tag{5.5}$$

where

$\bar{u}$ is the displacement vector;
det $\mathrm{F} = \rho_0/\rho$ in which ρ_0 and ρ are respectively the mass density in the undeformed and deformed configurations;
F is the deformation gradient tensor;
$\bar{b} = f/\rho$ is the body force per unit mass in the deformed configuration;
$\bar{t}_0 = T\cdot N$ in which $T = [t_{ij}]$ is the Piola–Kirchhoff conjugate stress tensor and N is the unit normal vector in the undeformed configuration;
A is the surface area in the deformed configuration which has prescribed surface tractions;

A_0 is the corresponding surface area in the undeformed configuration which has prescribed surface tractions;

V and V_0 are the volume in the deformed and undeformed states, respectively.

In order to accommodate both the material and the geometrical non-linearities, the strain–displacement relationships are given by

$$\varepsilon_{ij} = \frac{1}{2}\left(\frac{\partial u^i}{\partial x_j} + \frac{\partial u^j}{\partial x_i} + \frac{\partial u^k}{\partial x_i}\frac{\partial u^k}{\partial x_j}\right) \tag{5.6}$$

The finite element formulation (Lin and Lee, 1978) using constant strain tetrahedron elements for the lung will be briefly described in the following.

Let $u^i(x_1, x_2, x_3)$ denote the displacement components. The elemental displacement field of the tetrahedron can be interpolated by

$$u^i = a^i_r x^r + d^i \qquad (i, r = 1, 2, 3) \tag{5.7}$$

where d^i represents the components of rigid-body translation. For the nth node of the element, the nodal displacements are

$$u^i_n = a^i_r x^r_n + d_i \qquad (n = 1, 2, 3, 4) \tag{5.8}$$

The coefficients a^i_r, d_i are obtained from Equations (5.7) and (5.8). The displacement field is then related to the nodal displacements by

$$\{u^i\} = [u^i_n][B^n_r \mid c^n]\left\{\begin{array}{c} x^r \\ \hline 1 \end{array}\right\} \qquad (n = 1, 2, 3, 4;\ i, r = 1, 2, 3) \tag{5.9}$$

in which

$$[B^n_r] = -\frac{1}{D}\begin{bmatrix} \begin{vmatrix} x^2_2 - x^2_4 & x^3_2 - x^3_4 \\ x^2_3 - x^2_4 & x^3_2 - x^3_4 \end{vmatrix} & -\begin{vmatrix} x^1_2 - x^1_4 & x^3_2 - x^3_4 \\ x^1_3 - x^1_4 & x^3_3 - x^3_4 \end{vmatrix} & \begin{vmatrix} x^1_2 - x^1_4 & x^2_2 - x^2_4 \\ x^1_3 - x^1_4 & x^2_3 - x^2_4 \end{vmatrix} \\ -\begin{vmatrix} x^2_3 - x^2_1 & x^3_3 - x^3_1 \\ x^2_4 - x^3_1 & x^3_4 - x^3_1 \end{vmatrix} & \begin{vmatrix} x^1_3 - x^1_1 & x^3_3 - x^3_1 \\ x^1_4 - x^1_1 & x^3_4 - x^3_1 \end{vmatrix} & -\begin{vmatrix} x^1_3 - x^1_1 & x^2_3 - x^2_1 \\ x^1_4 - x^1_1 & x^2_4 - x^2_1 \end{vmatrix} \\ \begin{vmatrix} x^2_2 - x^2_1 & x^3_2 - x^3_1 \\ x^2_4 - x^2_1 & x^3_4 - x^3_1 \end{vmatrix} & -\begin{vmatrix} x^1_2 - x^1_1 & x^3_2 - x^3_1 \\ x^1_4 - x^1_1 & x^3_4 - x^3_1 \end{vmatrix} & \begin{vmatrix} x^1_2 - x^1_1 & x^2_2 - x^2_1 \\ x^1_4 - x^1_1 & x^2_4 - x^2_1 \end{vmatrix} \\ -\begin{vmatrix} x^2_2 - x^2_1 & x^3_2 - x^3_1 \\ x^2_3 - x^2_1 & x^3_3 - x^3_1 \end{vmatrix} & \begin{vmatrix} x^1_2 - x^1_1 & x^3_2 - x^3_1 \\ x^1_3 - x^1_1 & x^3_3 - x^3_1 \end{vmatrix} & -\begin{vmatrix} x^1_2 - x^1_1 & x^2_2 - x^2_1 \\ x^1_3 - x^1_1 & x^2_3 - x^2_1 \end{vmatrix} \end{bmatrix}$$

and

$$D = \begin{bmatrix} x^1_1 & x^2_1 & x^3_1 & 1 \\ x^1_2 & x^2_2 & x^3_2 & 1 \\ x^1_3 & x^2_3 & x^3_3 & 1 \\ x^1_4 & x^2_4 & x^3_4 & 1 \end{bmatrix}$$

In $[B^n_r]$ the first subscript n denotes the row index and the second subscript r denotes the column index.

Also, in Equation (5.9),

$$[C^n] = \frac{1}{D} \begin{bmatrix} -\begin{vmatrix} x_2^1 & x_2^2 & x_2^3 \\ x_3^1 & x_3^2 & x_3^3 \\ x_4^1 & x_4^2 & x_4^3 \end{vmatrix} \\ \begin{vmatrix} x_1^1 & x_1^2 & x_1^3 \\ x_3^1 & x_3^2 & x_3^3 \\ x_4^1 & x_4^2 & x_4^3 \end{vmatrix} \\ -\begin{vmatrix} x_1^1 & x_1^2 & x_1^3 \\ x_2^1 & x_2^2 & x_2^3 \\ x_4^1 & x_4^2 & x_4^3 \end{vmatrix} \\ \begin{vmatrix} x_1^1 & x_1^2 & x_1^3 \\ x_2^1 & x_2^2 & x_2^3 \\ x_3^1 & x_3^2 & x_3^3 \end{vmatrix} \end{bmatrix}$$

The displacement gradients, $\partial u^i/\partial x_j$ are derived from Equation (5.9) as

$$\left[\frac{\partial u^i}{\partial x_j}\right] = [u_n^i][B_j^n] \qquad (n = 1, 2, 3, 4;\ i, j = 1, 2, 3) \tag{5.10}$$

By substituting Equation (5.1) into Equation (5.6), the strain–nodal displacement relationship may be written as

$$\varepsilon_{ij} = \frac{1}{2}(u_n^i B_j^n + u_n^j B_i^n + u_n^k B_i^n u_m^k B_j^m) \qquad (m, n = 1, 2, 3, 4;\ i, j, k = 1, 2, 3) \tag{5.11}$$

The nodal force–displacement relationship and the tangent stiffness matrix may be obtained from Equation (5.5) by considering an element V_0 in the reference state and volume V in the deformed state.

The left-hand side of Equation (5.5) may be written as

$$\int_{A_0} \delta\bar{u} \,.\, F \,.\, \bar{t}_0 \mathrm{d}A + \int_{V_0} \rho_0 \delta\bar{u} \,.\, \bar{b} \mathrm{d}V = f_n^k \delta u_n^k$$

in which f_n^k is the equivalent nodal force in the kth direction of node n resulted from distributed body and/or surface force, and δu_n^k is a system of virtual nodal displacement.

If the element is a constant stress or constant strain element, the right-hand side of Equation (5.5) may be written as

$$\int_{V_0} t_{ij}\delta\varepsilon_{ij}\mathrm{d}V = V_0 t_{ij}\delta\varepsilon_{ij} = V_0 t_{ij}\frac{\partial\varepsilon_{ij}}{\partial u_n^k}\delta u_n^k$$

Therefore Equation (5.5) reduces to

$$f_n^k \delta u_n^k = V_0 t_{ij} \frac{\partial \varepsilon_{ij}}{\partial u_n^k} \delta u_n^k \tag{5.12}$$

Since δu_n^k is arbitrary, non-trivial solution of Equation (5.12) yields

$$f_n^k = V_0 t_{ij} \frac{\partial \varepsilon_{ij}}{\partial u_n^k} \tag{5.13}$$

Introducting Equation (5.11) into Equation (5.13), the nodal force– displacement relationship is given by

$$f_n^k = V_0 t_{ij} (B_n^i \delta_{jk} + B_n^i B_m^j u_m^k) \qquad (m, n = 1, 2, 3, 4;\ i, j, k = 1, 2, 3) \tag{5.41}$$

where n = nodal number, k = direction of force, and t_{ij} are the stress tensor components.

The local tangent stiffness $ESK(nk, mt)$ at nodal force f_n^k and nodal displacement u_m^t is defined by

$$ESK(nk, mt) = \frac{\partial f_n^k}{\partial u_m^t} \qquad (m, n = 1, 2, 3, 4;\ k, t = 1, 2, 3) \tag{5.15}$$

By introducting Equation (5.14) into Equation (5.15), the tangent stiffness matrix can be obtained. This 12×12 matrix is given by

$$.ESK(nk, mt) = V_0 \left[t_{ij} B_{mi} B_{nj} \delta_{kt} + \left(\frac{\partial^2 I_\alpha}{\partial \varepsilon_{ii} \partial \varepsilon_{rs}} \frac{\partial W}{\partial I_\alpha} + \frac{\partial I_\alpha}{\partial \varepsilon_{ij}} \frac{\partial I_\beta}{\partial \varepsilon_{rs}} \frac{\partial^2 W}{\partial I_\alpha \partial I_\beta} \right) \right.$$

$$\left. \times (B_{ni} B_{mr} \delta_{jk} \delta_{st} + B_{li} B_{mj} B_{mr} u_l^k \delta_{st} + B_{ni} B_{lr} B_{ms} u_l^t \delta_{jk} + B_{li} B_{nj} B_{pr} B_{ms} u_l^k u_p^t) \right]; \tag{5.16}$$

$$l, m, n, p = 1, 2, 3, 4;\ i, j, k, r, s, t, \alpha, \beta = 1, 2, 3 \ldots$$

The global stiffness matrix $[SK_{ij}]$ can be assembled based on the element stiffness of Equation (5.16) according to

$$[SK_{ij}] = \sum ESK(nk, mt) \tag{5.17}$$

This constant strain tetrahedron element was used to determine the stresses and strains in the lung due to the action of gravitational force and boundary movements. In that formulation, the human lung geometry was modelled for the purpose of substantiating experimental data. It was assumed to be an ellipsoid which is symmetrical about the plane through the lateral axes (see Figure 5.4). Numerical solutions for this three-dimensional model were obtained using an iterative scheme.

Subsequent to the tetrahedron element formulation, a simplified finite element model was developed by the authors by assuming the lung to be an axisymmetric paraboloid. The primary purpose of this simplified study was to investigate the

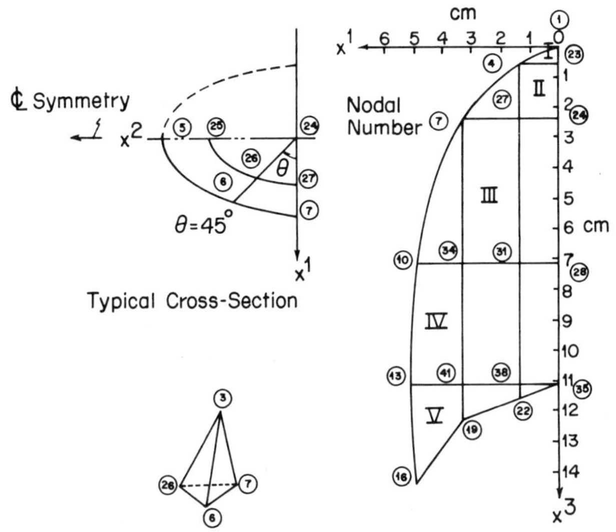

Figure 5.4 Illustration of typical finite element modelling of the lung by using tetrahedron elements. The lung is examined in five horizontally divided regions with Region I containing the Apex and Region V containing the base. For 1/4 of the lung, 41 nodes and 86 elements were used

effects of the heart cavity and interlobar fissures on the gravity-induced deformational behaviour of the lung. The analysis was only applicable to an upright posture. Axisymmetric isoparametric elements were used with the above virtual displacement principle. Numerical solutions were obtained by using a combined incremental-iterative scheme.

5.4.5 Finite element solution and illustrative results

There are several unique features involved in finite element analysis of the lung which will be discussed before examining the results. The smallest possible configuration of the chestwall cavity (bounded by rib cage and diaphragm) is greater than the collapsed lung volume. Therefore the tissue network of the *in vivo* lung is subjected to tension. Direct measurements of these tensile stresses cannot be made. A lack of knowledge of the initial stress state in the model posed the most difficult and challenging question in quantitative analysis of soft biological tissues. In the authors' studies the initial state of the lung is based on the assumption that the stress and strain in the *in vivo* lung are uniformly distributed at the total lung capacity (TLC) with a pressure of approximately 25 cm H_2O in the absence of gravity. With this assumption, a reference state of the lung is obtained by deflating the lung uniformly to zero stress and strain.

The boundary conditions for the *in vivo* lung are complicated. The lung is physically attached to the trachea at the hila and bounded by the rib cage and the diaphragm without fibrous connection. The lung is surrounded by the visceral pleura. The outer surface of the visceral pleura is smooth and slides against the inside membrane surface of the chestwall. The sliding is facilitated by a thin layer of interstitial fluid within the pleural space. This attachment is due to the negative pleural pressure. The same slipping characteristics are also present between the fissures of the lobes so that the lobes are more or less free to move against each other along the fissure planes. In human lungs there are five lobes, two in the left lung and three in the right lung. The left and right sides of the chest cavity are separated by the mediastinum, in which lie the heart, great vessels, and the oesophagus.

In general, there are three types of boundary conditions; Dirichlet, Neumann, and mixed. The boundary conditions for the lung model are the mixed type. The displacement of each point on the pleural surface is allowed to move freely along the inner surface of the chestwall if tangential force exists at that point. However, in the equilibrium configuration, the resultant force (or pleural pressure) must be perpendicular to the pleural surface, which is prescribed (see Figure 5.5).

There are several types of nodal points in the finite element model (shown in Figure 5.5):

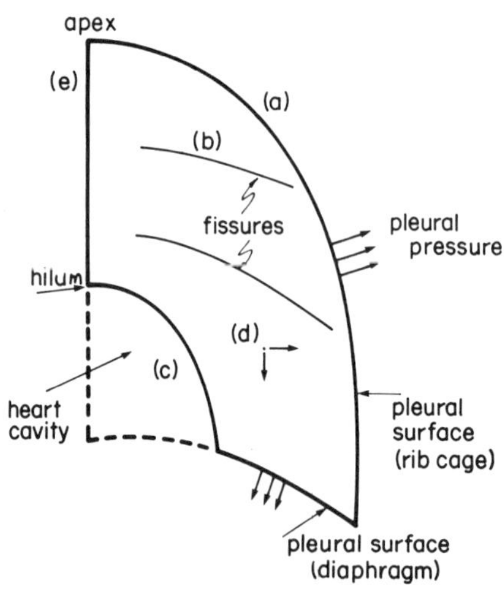

Figure 5.5 Lung model containing five different types of nodes as indicated by (a) through (e)

(a) Nodes on the outer surface, which are allowed to move along the prescribed boundary until there is no tangential force.
(b) Nodes on the fissure planes which are allowed to slide along the prescribed planes until there is no tangential force.

(c) Nodes on the prescribed surface of the heart cavity which are allowed to move along this curved surface only.
(d) Internal nodes which have the freedom to move in the three-dimensional space.
(e) Nodes on the symmetrical axis can only move along that axis.

Any element containing nodes that are restricted to move along some prescribed surface needs coordinate transformation such that the freedoms are perpendicular and tangential to the curved boundary. The stiffness in the perpendicular direction at this point was artificially set to a very large value so that motion in that direction was prevented. The new element stiffness matrix $[ESK]'$ is obtained from $[T]^{\mathrm{T}}[ESK][T]$, where $[T]$ is the transformation matrix.

Calculations for the lung model included the following steps:

(a) Define a configuration of the lung for which stresses and strains in the lung are sought.
(b) Inflate the lung from its reference state by uniform expansion to a volume.
(c) Change the outer boundary shape and introduce gravitational force at the nodes incrementally until the desired final equilibrium configuration is reached.

As a stressed continuum, a given lung volume with different outer boundary shapes, will map different equilibrium configurations both kinetically and kinematically. This will result in different distributions of regional responses in the lung. Therefore, the prescription of boundary shapes at different lung volumes has an important influence on the results.

In the following, typical finite element results from the lung models at the total lung capacity (TLC), the functional residual capacity (FRC), and the residual volume (RV) are given. These volumes are the standard measures used by pulmonary physiologists.

Figure 5.6 shows the finite element solutions reported by Liu and Lee (1978) and West and Matthews (1972) compared with experimental observations reported by Milic-Emili *et al.* (1966) for regional lung volumes. The regional distribution of lung volumes due to gravity is plotted against the total lung volume. If the lung is weightless and is uniformly distended, the relationship should be a straight line with a unit slope. The five horizontally divided regions deform differently due to gravity. The solid lines represent finite element results of Liu and Lee which were based on large displacement formulation. They correlate well with experimental results (dashed line in Figure 5.6). The dash-dot lines of Figure 5.6 are finite element results of West and Matthews which were based on small displacement formulation. The model deviation from the measured data, particularly at low lung volumes, is rather profound.

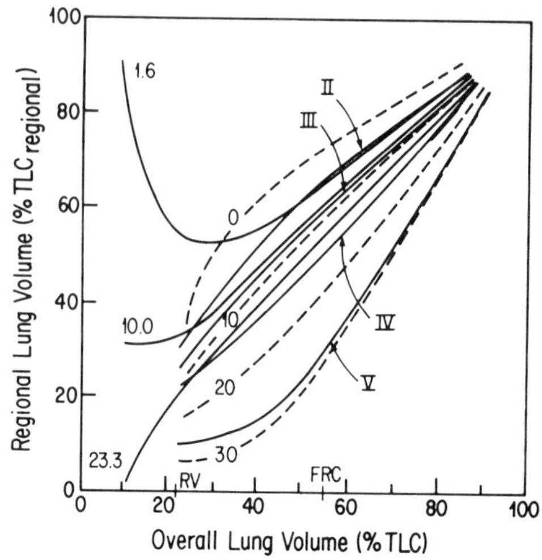

Figure 5.6 The relationship between the regional lung volume and the overall lung volume in the head-up posture. For uniform expansion without gravity there should be one 45° straight line. With gravity, the lung deforms non-uniformly, expressed by different regions. The large strain tetrahedron finite element solutions (Liu and Lee, 1978) compare extremely well with experimental observations reported by Milic-Emili *et al.* (1966). An earlier finite element solution based on small displacement theory West and Matthews (1972) did not predict the regional behaviour at low lung volumes

In Figure 5.7 are some preliminary finite element results of Lee, Tai, and Tseng (1978). These are based on the axisymmetrical model using an isoparametric element formulation. The purpose of the study was to investigate the effects of the heart cavity and fissures on the deformational characteristics in the lung. The results indicate that these structural characteristics reduce the non-uniformity of regional deformation due to gravitational forces.

Figure 5.8 shows recent results obtained by the authors on the stress distribution in the lung along a vertical section in an isovolume manoeuvre. The middle curve of Figure 5.8 represents uniform expansion. It would be a horizontal straight line in the absence of gravitational force. There is a large difference in the magnitude of the stress near the apex of the lung between diaphragmatic breathing (dots in Figure 5.8) and ribcage breathing (crosses in Figure 5.8).

5.4.6 Accuracy of results

One of the questions often asked about large-deformation finite element analysis is what is the level of accuracy of the results? In many soft-tissue studies it is

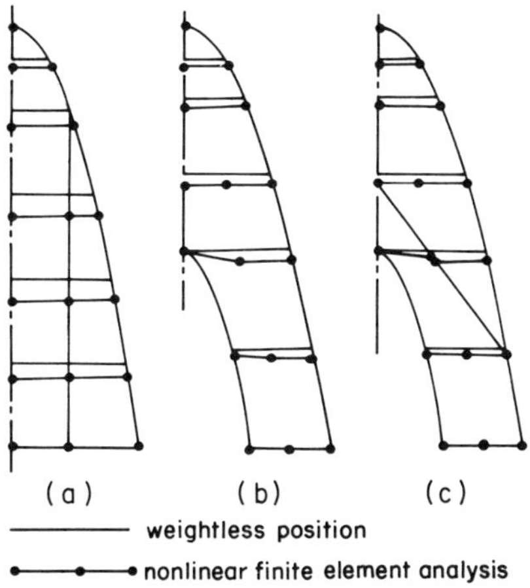

Figure 5.7 Lung tissue displacements from its uniform weightless expansion state at functional residual capacity as predicted by non-linear finite element analysis are shown for (a) lung as a solid mass of paraboloid, (b) lungs with mediastinal cavity and hilar support constraint, and (c) lungs with fissure, mediastinal cavity, and hilar support. The reduction of distortional displacements may be seen in the order of (a) through (c). This qualitatively illustrates the importance of these structural characteristics of the lung in reducing the non-uniformity of regional deformation due to gravitational distortion. The major function of fissures as seen in this simple example is to relieve partially the shear distortion of the parenchyma (Lee, Tai, and Tseng, 1978)

important that some qualitative and quantitative assessments of the results be made. For this reason a separate study on the local compliance of the lung parenchyma was made by Lee, Tseng, and Tai (1979). From the viewpoint of pulmonary physiologists, local compliance represents a measure of the relative degree of flexibility of a local region in the lung with respect to the overall flexibility of the entire lung (assumed to be a homogeneous and initially an isotropic elastic compressible continuum). The lung is often modelled as a spherical shape as shown in Figure 5.9. Three cases were considered. The first includes non-linear material and large strain, for which only the finite element solutions are available. The second case is for linear material and large strain, for which an analytical solution and the finite element solution are obtained. The third case is a typical linear material and small strain, for which standard elasticity solutions are available.

The simplest model for the problem of local compliance is that of a spherical domain with radius r_2. This is shown in Figure 5.9 where an inner spherical space

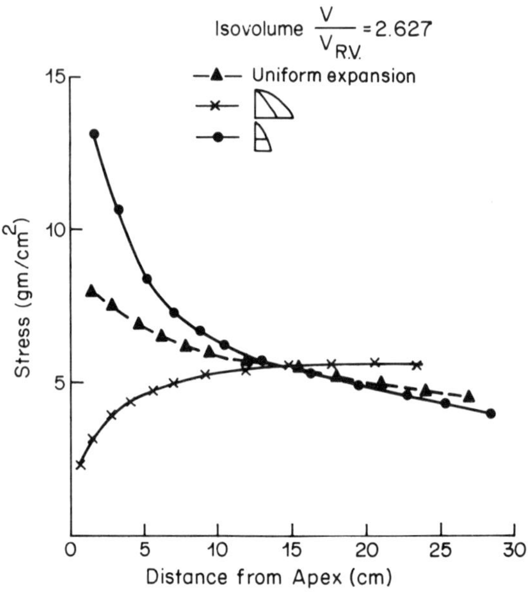

Figure 5.8 Finite element solutions for the pressure distributions in isovolume manoeuvres based on the axisymmetrical lung model

of radius r_1, subjected to pressure P_i, is considered to be the local region. Here r_2 is assumed to be much greater than r_1. At the outer boundary a transpulmonary pressure P_L is assumed. For uniform expansion, $P_L = P_i$. The local compliance of the inner domain with respect to the total domain may be expressed by the ratio $r_i/(r_1)_{\text{initial}}$ as a function of P_i and P_L.

Based on the formulation of Green and Adkins (1970) for a linear elastic material the equation of equilibrium may be written as

$$4(\lambda+2\mu)\left[\rho^3\left(\frac{\mathrm{d}r}{\mathrm{d}\rho}\right)^3+\frac{3}{2}\rho^4\left(\frac{\mathrm{d}r}{\mathrm{d}\rho}\right)^2\frac{\mathrm{d}^2r}{\mathrm{d}\rho^2}\right]-4(\lambda+2\mu)\left[\rho^2r\left(\frac{\mathrm{d}r}{\mathrm{d}\rho}\right)^2+2r^3\right]$$
$$+2\lambda\left[2\rho^2r^2\left(\frac{\mathrm{d}^2r}{\mathrm{d}\rho^2}\right)+4\rho^2r\left(\frac{\mathrm{d}r}{\mathrm{d}\rho}\right)^2\right]+8\mu\left[r\rho^2\left(\frac{\mathrm{d}r}{\mathrm{d}\rho}\right)^2+r^3\right]=0 \qquad (5.18)$$

where λ and μ are the Lame's constants, and $r = r(\rho)$ where ρ is the independent variable in the radial direction.

It may be shown that the solution of Equation (5.18) has the form

$$r(\rho) = c_1\rho + c_2\rho^A \qquad (5.19)$$

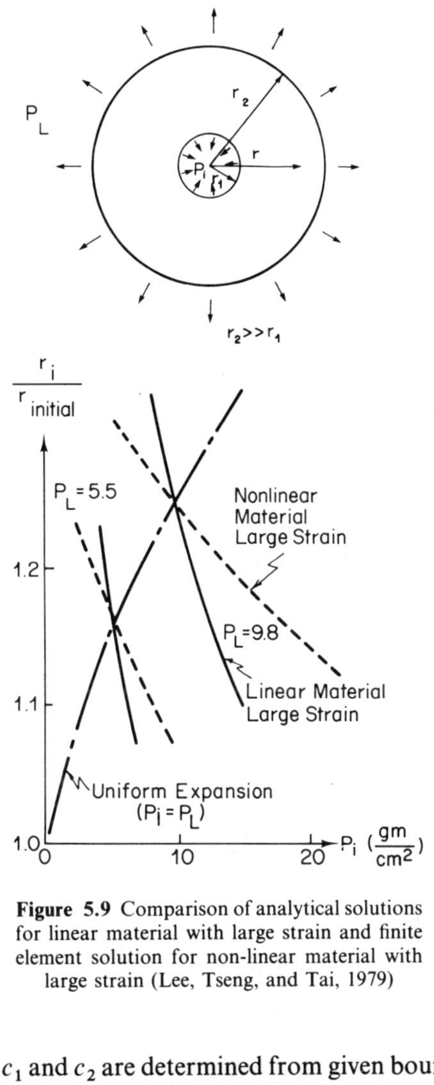

Figure 5.9 Comparison of analytical solutions for linear material with large strain and finite element solution for non-linear material with large strain (Lee, Tseng, and Tai, 1979)

where the constants c_1 and c_2 are determined from given boundary conditions. A value for A can be determined using experimental data on Poisson's ratio as a function of P_L and a typical pressure–volume curve for the lung (uniform expansion relationship between the transpulmonary pressure and the extension ratio).

The stresses and deformation in the spherical domain may be readily determined using Equation (5.19). A typical result for a measure of local compliance is also given in Figure 5.9 where the radius change of the inner sphere is plotted against the change of the pressure P_i for two different values of the transpulmonary pressure P_L. The dash-dotted line in Figure 5.9 represents the uniform expansion curve ($P_i = P_L$). The solid lines are solutions to Equation (5.18) and are almost the same as the finite element solutions (not shown in the figure). The dotted lines are finite element solutions obtained using the non-linear material law and the finite deformation approach.

Estimation of the local compliance using the infinitesimal strain approach amounts to using the slope of the tangent of the dotted curve. The finite-deformation solution of Equation (5.18) is non-linear, as expected. The values of the slopes lie between the constant value of infinitesimal strain solution and those values based on non-linear material properties and finite deformation theory.

5.5 SUMMARY

Finite element analysis has a significant role in soft-tissue mechanics, particularly when it is coordinated with experimental research. One important new emphasis in finite element modelling is to consider its applications in very small microscopic dimensions where mechanical, electrical, and biochemical responses must be simultaneously considered.

An example of finite element analysis of lung tissue conducted by the authors suggests that it is the only approach by which some quantitative estimation of the lung deformation can be obtained. An important role of finite element analysis in soft-tissue mechanics in the immediate future is to enable us to obtain quantitative input to facilitate continued experimental and numerical research efforts.

ACKNOWLEDGMENT

The research conducted by the authors reported in this chapter was supported by NIH Grant HL-14495 of the National Heart, Lung, and Blood Institute.

REFERENCES

Elden, H. R. (1976) Physical properties of collagen fibers, *Int. Rev. Conn. Tissue Res.*, **7,** 283.

Frankus, A., and Lee, G. C. (1974) A theory for distortion studies of lung parenchyma based on alveolar membrane properties, *J. Biomech.*, **7,** 101–107.

Fung, Y. C. (1973) Biorheology of soft tissue, *Biorheology*, **10,** 139–155.

Fung, Y. C. (1974) A theory of elasticity of the lung, ASME *J. Appl. Mech.*, **41,** 8–14.

Fung, Y. C., Tong, P., and Patitucci, P. (1978) Stress and strain on the lung, *J. Engng. Mech.* ASCE, **104, (EM1)** 201–224.

Gotte, L. (1977) Recent observations on the structure and composition of elastin, *Adv. in Exp. Med. and Biology*, **27,** 105.

Green, A. E., and J. E. Adkins (1970) *Large Elastic Deformations* (2nd edn), Clarendon Press, Oxford pp. 8–10.

Hoppin, F. G., Lee, G. C., and Dawson, S. V. (1975) Properties of lung parenchyma in distortion, *J. Appl. Physiol.*, **39,** 742–751.

Hoppin, F. G., and Hildebrandt, J. (1977) Mechanical properties of the lung, in: *Bioengineering Aspects of the Lung* (ed. J. B. West) Marcel Dekker, New York, 1977, Ch. 2, pp. 83–162.
Karakaplan, A. D., Bieniek, M. P., and Skalak, R. (1980) A mathematical model of lung parenchyma, *J. Biomech. Engng.*, Transact. ASME, **102,** 124–136.
Lanir, Y., and Fung, Y. C. (1974) Two-dimensional mechanical properties of rabbit skin. I. Experimental system. II. Experimental results, *J. Biomech.*, **7,** 29–34, and **7,** 171–182.
Lanir, Y. (1979) A structural theory for the homogeneous biaxial stress–strain relationships in flat collagenous tissues, *J. Biomech.*, **12,** 423.
Lee, G. C., and Frankus, A. (1975) Elasticity properties of lung parenchyma derived from experimental distortion data, *Biophysical J.*, **15,** 481–493.
Lee, G. C., Tai, R. C., and Tseng, N-T. (1978) Finite deformation modeling of regional behavior in lungs, *Proc. 1st Internat. Conf. on Mechanics in Medicine and Biology*, Aachen, Germany.
Lee, G. C., Tseng, N-T., and Tai, R. C. (1979) Local compliance of lungs predicted by infinitesimal and large strain approaches, ASCE *Proc. 3rd Engng. Mechanics Conf.*, Austin, Texas, pp. 688–691.
Liu, J-T., and Lee, G. C. (1978) Static finite deformation analysis of the lung, *J. Engng. Mech.* Div. ASCE, EM1, pp. 225–238.
Matthews, F. L., and J. B. West (1972) Finite element displacement analysis of a lung, *J. Biomech.*, **5,** 591–600.
Mead, J., Takishima, T., and Leith, D. (1970) Stress distribution in lungs: a model of pulmonary elasticity, *J. Appl. Physiol.*, **28,** 596–608.
Milic-Emili, J., *et al.* (1966) Regional distribution of inspired gas in the lung, *J. Appl. Physiol.* **21,** (3) 749–759.
Oden, J. T. (1972) *Finite Elements of Nonlinear Continua*, McGraw-Hill Book Co. Inc., New York.
Pao, Y. C., Chevalier, P. A., Rodarte, J. R., and Harris, L. D. (1978) Finite element analysis of the strain variations in excised lobe of canine lung, *J. Biomech.*, **11,** 91–100.
Soong, T. T., and Huang, W. N. (1973) A stochastic model for biological tissue elasticity in simple elongation, *J. Biomech.*, **6,** 451–458.
Suhihara, T., Martin, C. J., and Hildebrandt, J. (1971) Length–tension properties of alveolar wall in man, *J. Appl. Physiol.*, **30,** 874–878.
Vawter, D. L., Fung, Y. C., and West, J. B. (1978) Elasticity of excised dog lung parenchyma, *J. Appl. Physiol.*, **45,** 261–269.
Vawter, D. L., Fung, Y. C., and West, J. B. (1979) Constitutive equation of lung tissue elasticity, *J. Biomech. Engng.*, Trans. ASME, **101,** 38–45.
Weibel, E. R. (1963) *Morphometry of the Human Lung*, Academic Press, New York.
West, J. B., and Matthews, F. L. (1972) Stresses, strains, and surface pressures in the lung caused by its weight, *J. Appl. Physiol.*, **32,** 332–345.
West, J. B. (1977) *Bioengineering Aspect of the Lung*, Marcel Dekker, New York.

Finite Elements in Biomechanics
Edited by R. H. Gallagher, B. R. Simon, P. C. Johnson, and J. F. Gross

Chapter 6

Deformation of the Lung: The Role of Interfacial Forces

Donald L. Vawter and Wilfred H. Shields

6.1 INTRODUCTION

There is considerable interest in the physiological community in the regional differences in alveolar expansion (influencing ventilation perfusion ratios), perivascular and peribronchial pressures, distribution of pleural pressure, distortion of the lung due to its own weight, effect of diaphragm movement on regional volume expansion, and the effects of surface tension. The boundary value problems for these macroscopic problems are quite complex because of material non-linearities, finite deformation, and complicated loading. Only in rare cases can an analytic solution be found.

The finite element method is an attractive approach for solving such complicated boundary value problems. The present model is applicable when the elasticity can be characterized by a strain energy function and when the energy required to create new surface can be calculated.

There have been other presentations of finite element analyses of macroscopic lung deformation in the literature. These analyses do not consider the effect of surface tension explicitly.

The first finite element analysis was presented by West and Matthews (1972). This important paper was instrumental in showing the importance of gravity on the deformation of the lung and was probably the catalyst for more refined analyses. Because of limited knowledge on the material behaviour of the lung, an elastic model based on an extension of uniaxial stress–strain behaviour was assumed. From the viewpoint of continuum mechanics such an extension to finite elasticity is inconsistent and hence the quantitative results are subject to question. In particular the analysis assumed the tissue behaviour was isotropic even when distorted, which is not at all consistent with experimental observations. A rather coarse grid was used and demonstration of a convergent solution was not presented. The analysis was based on purely elastic behaviour although in determining the material properties, air-filled pressure volume behaviour was used. The effects of surface tension are implicit in the analysis, although assessing interfacial forces by merely increasing tissue stiffness is questionable.

Recently Liu and Lee (1978) presented a three-dimensional analysis of the finite deformation of the lung under gravity. The analysis deals with non-linear elasticity and finite deformation consistently. The major drawback is that the effects of surface tension are not considered explicitly. The analysis was based on tetrahedral elements and a total of 40 nodes were used. No mention is made of the effect of grid size on convergence.

In the following sections a finite element model will be developed in which finite elasticity is handled consistently and in which, for the first time, the effects of surface tension are treated explicitly. The model is applicable to three-dimensional problems; however, numerical results will be presented only for an axisymmetric problem. More complete details of the formulation may be found elsewhere, (Vawter, 1980 a, b).

6.2 VARIATIONAL FORMULATION

The principle of stationary potential energy requires that in an equilibrium configuration the total potential energy of a body must be at a stationary value with respect to all kinematically admissible deformations. For an elastic system the total potential energy, Π_E, is given (Pian and Tong, 1971) by

$$\Pi_E = \int_{V_0} (W' - \bar{F}_i u_i)\,\mathrm{d}V_0 - \int_{S_\sigma} \bar{T}_i u_i\,\mathrm{d}S \tag{6.1}$$

where W' is the strain energy per unit volume, $\bar{F}_i$ are the prescribed body forces, u_i are the cartesian displacements, $\bar{T}_i$ are prescribed boundary tractions, V_0 is the initial volume, and S_σ is the surface on which tractions are prescribed. In this paper the summation convention is used.

For problems involving surface tension, the energy associated with creating new surfaces must be added. If we denote the surface energy density by E then the total potential energy, Π, is given by

$$\Pi = \Pi_E + \int_{V_0} E\,\mathrm{d}V_0 = \int_{V_0} (W - \bar{F}_i u_i)\,\mathrm{d}V_0 - \int_{S_\sigma} \bar{T}_i u_i\,\mathrm{d}S \tag{6.2}$$

where $W = W' + E$. The relationship between E and the surface tension, γ, will be discussed subsequently. In particular, we will show that E is a function of the surface tension and the strain invariants. There are many methods for obtaining the stationary value of Π and we have chosen the incremental approach outlined by Pian and Tong (1971).

6.3 VARIATIONAL PRINCIPLE FOR INCREMENTAL PROBLEMS

If a body is in equilibrium with initial displacements $\mathbf{u}_0$ and loading p_0, and a small increment of load, Δp, is applied resulting in displacements $\Delta\mathbf{u}$ (where we have assumed proportional loading), the change in potential energy, $\Delta\Pi$, can be written as

$$\begin{aligned}\Delta\Pi &= \Pi(\mathbf{u}_0 + \Delta\mathbf{u}, p_0 + \Delta p) - \Pi(\mathbf{u}_0, p_0)\\ &= \Delta p\, F_1(\mathbf{u}_0, p_0) + F_2(\mathbf{u}_0, \Delta\mathbf{u}, p_0, \Delta p) + F_3(\mathbf{u}_0, \Delta\mathbf{u}, p_0, \Delta p)\end{aligned} \tag{6.3}$$

where F_1, F_2, F_3 are functionals: F_1 is independent of Δp and $\Delta\mathbf{u}$; F_2 contains the second-order terms in Δp and $\Delta\mathbf{u}$; and F_3 contains all higher-order terms. The above is a general functional relationship which will be expanded in detail subsequently. The new equilibrium configuration is determined by the stationary value of $\Delta\Pi$. For small Δp the first-order approximation of $\Delta\mathbf{u}$ is determined by the stationary value of F_2 alone, that is,

$$\delta F_2(\mathbf{u}_0, \Delta\mathbf{u}, p_0, \Delta p) = 0 \tag{6.4}$$

F_2 can be written conveniently in terms of W (Pian and Tong, 1971) as

$$\begin{aligned}F_2 = \int_{V_0} \Bigg[&(1/2)(E_{ijkl})_0 \Delta e_{ij} \Delta e_{kl} + (1/2) S_{ij} \frac{\partial \Delta u_k}{\partial a_i}\frac{\partial \Delta u_k}{\partial a_j}\\ &- \Delta\overline{F}_i \Delta u_i \Bigg] \mathrm{d}V_0 - \int_{S_\sigma} \Delta\overline{T}_i \Delta u_i \,\mathrm{d}S\end{aligned} \tag{6.5}$$

where

$$e_{ij} = (1/2)\left[\frac{\partial u_i}{\partial a_j} + \frac{\partial u_j}{\partial a_i} + \frac{\partial u_k}{\partial a_i}\frac{\partial u_k}{\partial a_j}\right]\Bigg|_{\mathbf{u}=\mathbf{u}_0} \qquad \text{(Green strains)}$$

$$(E_{ijkl})_0 = \frac{\partial^2 W}{\partial e_{ij}\partial e_{kl}}\bigg|_{\mathbf{u}=\mathbf{u}_0}; \qquad S_{ij} = \frac{\partial W}{\partial e_{ij}}\bigg|_{\mathbf{u}=\mathbf{u}_0}$$

$$\Delta e_{ij} = (1/2)\left[\frac{\partial \Delta u_i}{\partial a_j} + \frac{\partial \Delta u_j}{\partial a_i} + \frac{\partial (u_k)}{\partial a_i}\bigg|_0 \frac{\partial \Delta u_k}{\partial a_j} + \left(\frac{\partial u_k}{\partial a_j}\right)\bigg|_0 \frac{\partial \Delta u_k}{\partial a_i}\right]$$

and a_i are the undeformed Cartesian coordinates. Notice that for an elastic analysis S_{ij} is the Kirchhoff stress. However, the physical interpretation of S_{ij} is not simple when surface tension is included.

6.4 FINITE ELEMENT FORMULATION

At this point no approximations have been made about the displacements $\Delta\mathbf{u}$. In the finite element approach we divide the body into a finite number of non-overlapping elements and make assumptions about $\Delta\mathbf{u}$ within an element. In particular, we assume the displacements within an element are related uniquely to the displacements at the corners of an element by known interpolation functions. If we denote the corner displacements by $\Delta\mathbf{q}$ we have within an element

$$\Delta\mathbf{u} = \mathbf{A}\,\Delta\mathbf{q} \tag{6.6}$$

where the matrix $\mathbf{A}$ is a function only of the Cartesian coordinates. Once the form of $\mathbf{A}$ is chosen the strains $\Delta\mathbf{e}_{ij}$ can easily be related to the nodal displacements $\Delta\mathbf{q}$ and Equation (6.5) can be written more compactly as

$$F_2 = (1/2) \sum_{\text{all elements}} \Delta\mathbf{q}^{\mathrm{T}}\mathbf{K}\,\Delta\mathbf{q} - \mathbf{F}\,\Delta\mathbf{q} \tag{6.7}$$

where $\mathbf{K}$ and $\mathbf{F}$ are functions of the initial stress, strain, and geometry, but not functions of $\Delta\mathbf{q}$. The nodal displacements, $\Delta\mathbf{q}$, for an element are a subset of the nodal displacements of the overall body, $\Delta\mathbf{q}^*$, and Equation (6.7) can be rewritten as

$$F_2 = \frac{1}{2}\Delta\mathbf{q}^{*\mathrm{T}}\mathbf{K}_G\Delta\mathbf{q}^* - \mathbf{F}_G\Delta\mathbf{q}^* \tag{6.8}$$

where $\mathbf{K}_G$ and $\mathbf{F}_G$ are termed the global stiffness and force matrices respectively. The elements of $\Delta\mathbf{q}^*$ are independent and the stationary value of F_2 is given by

$$\mathbf{K}_G\Delta\mathbf{q}^* = \mathbf{F}_G \tag{6.9}$$

This system of linear equations can be solved for the incremental displacements $\Delta\mathbf{q}^*$ giving a new equilibrium state. Another small increment in Δp may then be applied and the process repeated until the total load has been applied.

6.5 STRAIN ENERGY OF THE LUNG PARENCHYMA

We will use the strain energy density function proposed by Fung, Tong, and Pattitncci (1978), namely

$$W' - \frac{C}{2D_A}\exp\left[\alpha I_1^2 + \beta I_2\right] \tag{6.10}$$

where C, α, β are material constants determined from experiments, D_A is the typical alveolar diameter when unstressed, and I_1 and I_2 are the well-known invariants of the strains defined in Equations (6.5),

$$I_1 = e_{11} + e_{22} + e_{33}$$
$$I_2 = e_{11}e_{22} + e_{22}e_{33} + e_{33}e_{11} - e_{12}^2 - e_{23}^2 - e_{31}^2$$

6.6 SURFACE TENSION BEHAVIOUR

It is well known that interfacial forces are significant and must be included when analysing the mechanical behaviour of the lung. In order to undersand the complexity of the problem consider Figure 6.1, which is a histological preparation showing the lung parenchyma. Although some shrinkage of the tissue undoubtedly occurred during processing, and hence the preparation may

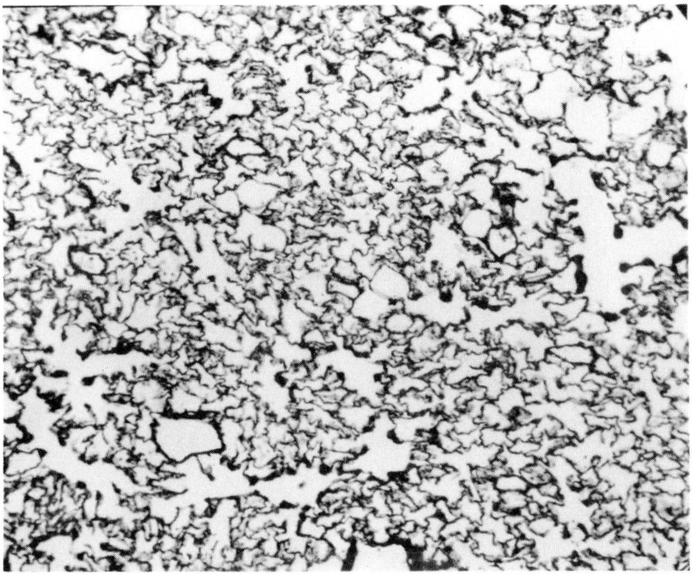

Figure 6.1 Histological Section. 10 μ section magnified 60 ×. Standard haemotoxin and eosin staining

not precisely depict the *in vivo* configuration, the basic structure of the lung parenchyma is shown. The sources of elastic support are the individual alveolar walls. Each wall, except on the lung surface, is surrounded by air on both sides. The alveolar walls themselves are coated by a surface-active film called lung surfactant. The forces developed because of this interface are termed interfacial forces. Because of these forces, energy is required to create new surface within the lung. The rate of change of this energy with respect to surface area is the surface tension, γ.

The importance of surface tension can be seen in Figure 6.2 where pressure–volume curves for both air-filled and saline-filled lungs are presented. The air–surfactant interface is eliminated for a saline-filled lung, dramatically altering the pressure–volume relationship. In the lung, surface tension varies with surface area. The exact variation is unknown because the change in surface area as the lung changes volume cannot be measured directly.

Estimates of the variation have been made by 'washing' the surfactant layer out of the lung and determining the surface tension of the extract using the so-called Langmuir–Wilhelmy surface tension balance method (Brown *et al.*, 1959). There are many uncertainties associated with this approach, primarily because there is no guarantee that the extract is representative of the surface layer which coats the alveoli. Also, there is no assurance that the experimental configuration of the extract is similar to the configuration occurring in the *in vivo* lung. In particular,

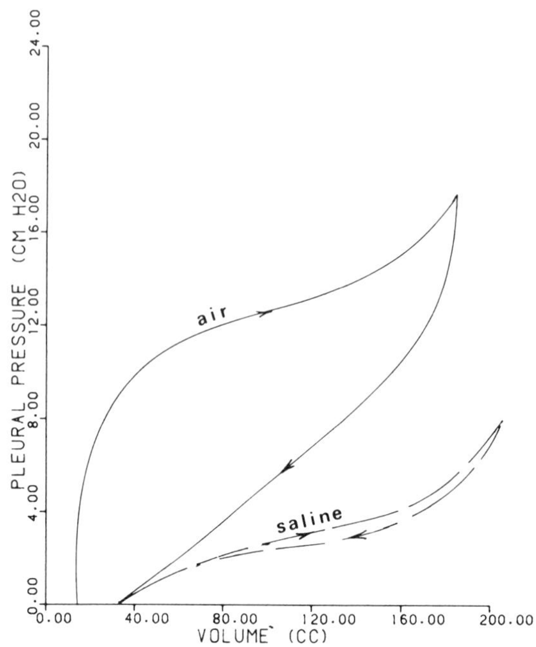

Figure 6.2 Air- and saline-filled pressure–volume curves. Notice the large hysteresis in the surface tension curve (After Shields, 1979)

the large changes in surface area which occur in the lung extract test may not occur in the intact lung during normal breathing.

Another estimate for the variation in surface tension with area has been presented by Flicker and Lee (1974). They estimated surface area from microscopic observations of the dimensional changes occurring in the alveoli just beneath the pleural surface. Combining these observations with those of the pressure–volume curves of air-filled lungs, they were able to estimate the variation of surface tension with lung volume. The major hypothesis of their analysis, which is untested, is that the alveoli far from the surface and the subpleural alveoli behave similarly.

The results of the lung wash and subpleural findings are presented in Figure 6.3. D_A is the average alveolar diameter in the stress-free state. The abscissa is the normalized surface area which will be defined and discussed subsequently. Notice that there are wide variations in the curves and that the surface tension asymptotically approaches a maximum for the extract models but does not in the subpleural model.

Because of the discrepancies among the experimental findings, we have incorporated various models for the surface tension into this finite element

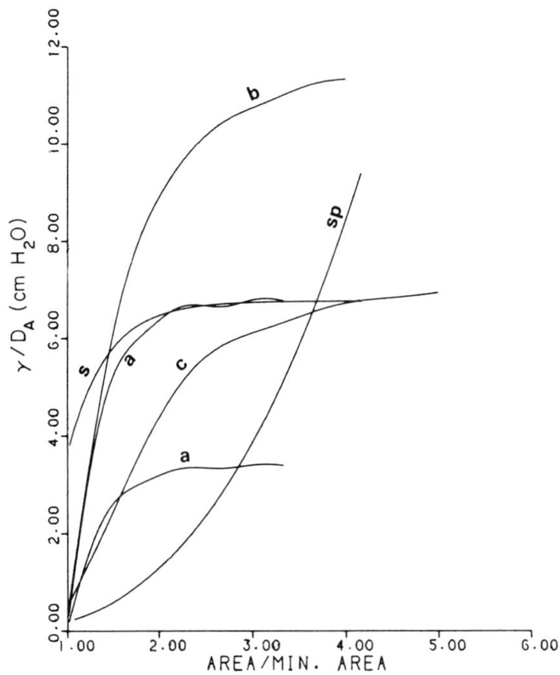

Figure 6.3 Surface tension vs. surface area. Curves a, b, c, and s are from lung wash experiments. Curve sp is from subpleural observations. Constants for the models are found in Table 6.1. Curve a′ has twice the alveolar diameter of curve a

analysis. We have chosen to incorporate five models of surface tension. In three of the models the surface tension is expressed as a Fourier sine series of the surface area after Fung *et al.* (1978). These curves are based on the lung extract curves of Clements (1962), Slama *et al.* (1973), and Scarpelli (1965). We also have used a power law relationship based on the observations of subpleural alveoli (Fliker and Lee, 1974; Vawter and Lee, 1976), and an exponential relationship based on the lung wash curves of Scarpelli (1968).

The relationships for the surface tension, γ, are: *Fourier Series Model*:

$$\gamma = \gamma_{\min} + (\gamma_{\max} - \gamma_{\min}) \left[\frac{A-1}{\eta} + \sum_{n=1}^{5} C_n \sin\left(n\pi \frac{(A-1)}{\eta} \right) \right] \tag{6.11}$$

where $A^2 = \frac{4}{3}(I_1 + I_2) - 1$ and the values of the constants, $\gamma_{\min}$, $\gamma_{\max}$, η, C_1 to C_5 are given in Table 6.1. A is an approximation for the normalized surface area. For details see Fung *et al.* (1978).

Subpleural model:

$$\gamma = C_1 A^{C_2} \tag{6.12}$$

Table 6.1 Constants for surface tension equations

Curve	γ_{max}/D_A (cm H_2O)	γ_{min}/D_A (cm H_2O)	η	C_1	C_2	C_3	C_4	C_5
a	6.78	0.17	2.33	0.519	0.165	0.066	0.031	0.020
b	11.36	0.339	3.00	0.425	0.108	0.380	0.013	0.007
c	6.95	0.509	4.00	0.386	0.067	0.005	−0.013	−0.010
s	–	–	–	6.78*	4.83	2.35	–	–
sp	–	–	–	0.1966*	2.71	–	–	–

* C_1/D_A (cm H_2O), $D_A = 60.1\ \mu m$

with C_1 and C_2 given inTable 6.1.
Exponential model:

$$\gamma = C_1\,(1 - C_2 \exp(-C_3 A)) \tag{6.13}$$

with C_1, C_2, and C_3 given in Table 6.1.

In the finite element formulation the quantity of interest is not the surface tension but the surface energy density E which is related (see Vawter, 1980a) to γ by:

$$E = \frac{12}{D_A}\int_1^A \gamma \, dA + C_0$$

where C_0 is an arbitrary constant chosen such that $E\,(A = 1) = 0$. For the surface tension models presented above E is given as follows.
Fourier series model:

$$E = \frac{12}{D_A}\Bigg[\gamma_{min}(A-1) - (\gamma_{max} - \gamma_{min})\,2\eta\,(A-1)^2 - (\gamma_{max} - \gamma_{min}) \sum_{\eta=1}^{5} \frac{C_n}{n\pi}\eta \times \cos\left\{\frac{n\pi(A-1)}{\eta} - 1\right\}\Bigg] \tag{6.14}$$

Subpleural model:

$$E = \frac{12C_1}{D_A(1+C_2)}\,[A^{(1+C_2)} - 1] \tag{6.15}$$

Exponential model:

$$E = \frac{12C_1}{D_A}\left[A + \frac{C_2}{C_3}\exp(-C_3 A) - 1\right] \tag{6.16}$$

Constant surface tension, γ,:

$$E = \frac{12C_1}{D_A}(A-1) \tag{6.17}$$

6.7 AXISYMMETRIC FORMULATION

In order to study both the deformation of the lung under its own weight and the stress concentrations around large airways and blood vessels, we have developed a computer code based on the above formulation. We have solved only the axisymetric problem although the approach is valid for three-dimensional problems. The displacements are radial and axial and there are four components of strain (radial, axial, circumferential, and shear in the plane formed by the radial and axial axes).

We chose quadrilateral elements and used a bilinear interpolation function. For such an element Δq is 8×1 consisting of the values of the radial and axial displacements at the corners of the quadrilateral. This leads to an 8×8 element stiffness matrix ($\mathbf{K}$) and a global stiffness ($\mathbf{K}_G$) matrix of size $2n \times 2n$, where n is the number of nodes of the entire body.

The computer code is straightforward once the form of the interpolation function is chosen. Spatial integrations are performed numerically and the solution of Equation (6.9) is accomplished by a standard triple-factor method. Care is taken to account for the sparseness of $\mathbf{K}_G$. For a 10×16 mesh, one would need approximately 140 000 words to store every term in $\mathbf{K}_G$. By taking into account the sparseness we have reduced the storage requirement to just over 7000 words.

6.8 NUMERICAL RESULTS FOR A LUNG LOADED BY ITS OWN WEIGHT

In order to show the effectiveness of the approach, demonstrate convergence, and emphasize the necessity of characterizing interfacial tension effects explicitly, let us consider the problem of a lung loaded by gravity only. We shall consider the lung to be a body of revolution as shown in Figure 6.4. The geometry is chosen to closely follow that of the 'narrow' lung of the West and Matthews (1972) analysis. The height and basal diameter at FRC (functional residual capacity) are 25 cm and 16.74 cm respectively. The specific gravity of the lung at FRC is assumed to be 0.23, following West. We have shown elsewhere that the density of the lung in its stress-free state is 0.365 (Vawter, 1977). Using these values gives an FRC volume 1.58 times the stress-free volume. West and Matthews used a value of 1.825 in their analysis. The expanded shape of the chest cavity is assumed to be geometrically similar to its stress-free shape and the lung is assumed to fill the chest cavity at all times. We have assumed that because of the fluid interface there are no shear forces acting on the pleural surface. The effect of hilar support and diaphragm shape will be discussed subsequently.

The results can be conveniently divided into five areas:

1. Convergence of the finite element approximation.
2. The effect of surface tension and the necessity of including surface energy explicitly.

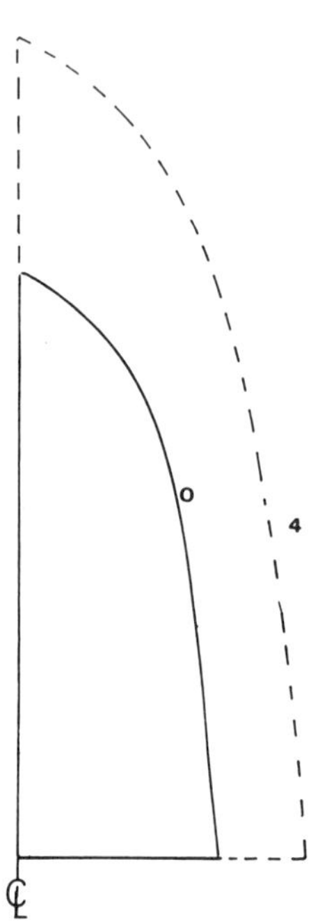

Figure 6.4 Geometric model of the lung's exterior surface

3. The sensitivity of the results to the choice of surface tension model.
4. The role of the diaphragm.
5. The role of hilar support.

6.8.1 Convergence of the solution

Although an analytic solution does not exist for the lung loaded by gravitational forces, convergence may be checked. In the formulation the absence of shear at the pleural surface was assumed. This is a natural boundary condition and is only satisfied approximately by the numerical approach. Thus the absence of shear in the numerical results is indicative of a convergent solution. We have calculated the average shear at the surface as a percentage of the normal stress (pressure at the pleural surface) for a variety of mesh sizes and present the results in Table 6.2. Note that with only 12 elements the shear is less than 3% of the pressure and falls to less than 1% for a 160-element mesh. Table 6.2 also gives the maximum pressure at the pleural surface for various mesh sizes. We have assumed that the fine grid is 'exact' when calculating pressure errors. Based on these results, we used a 10×16 mesh and 5 loading increments for the remainder of the results.

6.8.2 Necessity of including surface tension explicitly

In order to understand the effect of surface tension, let us consider the following case with the elastic constants in Equation (6.10) are given by $C/D_A = 2.45\,\text{kPa}$

Table 6.2 Convergence of solution
A. Mesh

Mesh size	Elements	Degrees of freedom	Shear (%)	Pressure error (%)
3 × 4	12	40	−2.77	−11.2
5 × 8	40	108	−1.59	−4.14
8 × 12	96	234	−1.08	−1.51
10 × 16	160	374	−0.81	–

B. Increments

Increments	Maximum pressure	Error (%)
1	9.661	−2.88
2	9.741	−2.07
3	9.810	−1.38
5	9.883	−0.64
10	9.947	–

(25 cm H_2O), $\alpha = 0.183$, $\beta = -0.291$. For a justification of these values see Vawter (1980a). In the following results the lung has been uniformly expanded to a volume 2.74 times its stress-free volume and then gravitational and interfacial forces are applied. For the geometry chosen, this expansion corresponds to a gas volume of 5.50 litres. The gas volume at the stress-free state is 1.47 litres. The gas volume ratio is different (3.74 vs. 2.74) from the lung volume ratio (gas + tissue) because of the presence of incompressible lung parenchyma. It can easily be shown the relationship between the gas volume ratio (GVR) and the lung volume ratio (LVR) is given by:

$$\mathrm{GVR} = (\mathrm{LVR} - \mathrm{sg})/(1 - \mathrm{sg}) \tag{6.18}$$

where sg is the specific gravity of the lung (gas + tissue) at the stress-free state. For our model sg = 0.365. In the absence of interfacial and gravitational forces the gas volume ratio would be constant throughout the lung. However, when expansion is not uniform regional differences in the gas volume ratio occur. In Figure 6.5 we plot the regional gas volume ratio (RGVR) against distance down the lung. RGVR is defined as the local gas volume ratio multiplied by 100 and divided by the overall gas volume ratio. If the volume expansion were uniform, RGVR would be 100 throughout the lung. As we see in Figure 6.5 the apex of the lung is more highly expanded than the base. Curve E is the result when there is no surface tension, while for curve ST the surface tension is given by Equation (6.12). The implications of using Equation (6.12) will be discussed in a later section.

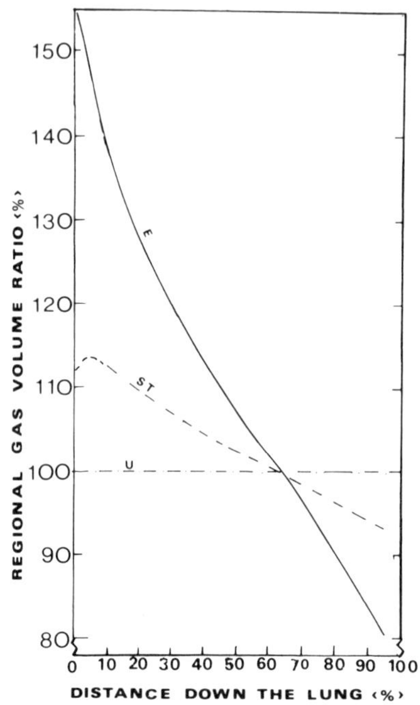

Figure 6.5 Regional expansion of lung. Curve E is for no surface tension, curve ST is for normal surface tension, curve U is for uniform expansion

Implicit in the presentation is the assumption that regional expansion is constant across the lung for a given position down the lung. This assumption will later be shown to be valid for a lung without hilar support and with a flat diaphragm.

The effect of surface tension is to cause a more uniform expansion of the lung. The base becomes more highly expanded while the apex is less overexpanded. It should be noted that in the midlung region the regional gas volume ratio is nearly the same with or without surface tension present. This phenomenon will be exploited below.

In Figure 6.6 we plot the pressure (normal component of the Eulerian stress) which must be present on the pleural surface to hold the lung in the configuration prescribed above (i.e. a volume expansion of 2.74, no shear on the surface, and no distortion in the absence of gravity and surface tension). As surface tension increases, the overall stress level increases as expected but the apex-to-base pressure difference decreases. The pressure will be shown to be very sensitive to the choice of surface tension model (i.e. Equations (6.11), (6.12), (6.13), give significantly different numerical results).

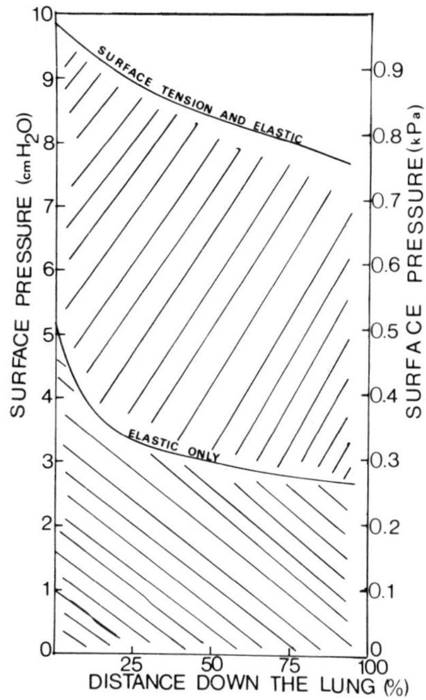

Figure 6.6 Distribution of pleural pressure. Pressure distribution required to keep the lung in the expanded shape of Figure 6.1. Notice the large effect of surface tension

The curves in Figures 6.5 and 6.6 were plotted when the expression for the surface tension is given by Equation (6.12). One may ask whether the same volume and pressure distribution can be attained by eliminating surface tension and changing the material constants determining tissue elasticity. This is, of course, what is done when surface tension is included only implicitly. Although C/D_A, α, or β may be changed we have studied only the effects of changing C/D_A. In Figure 6.7 the air volume distribution is shown in the absence of surface tension for values of C/D_A from 2.45 to 39.2 kPa (25 to 400 cm H_2O). Although the elastic curves can approximate the previous curve in the middle and lower parts of the lung, they cannot do so near the apex.

The role of surface tension in determining apical volume is clarified in Figure 6.8. For this curve the original values of the elastic constants ($C/D_A = 2.45$ kPa) were used but different values were used for the surface tension (held constant by using Equation (6.17)). As the surface tension increases the apical regional volume falls dramatically and actually becomes less than that of the base. The elastic analysis shows no such tendency again showing the importance of incorporating interfacial effects explicitly.

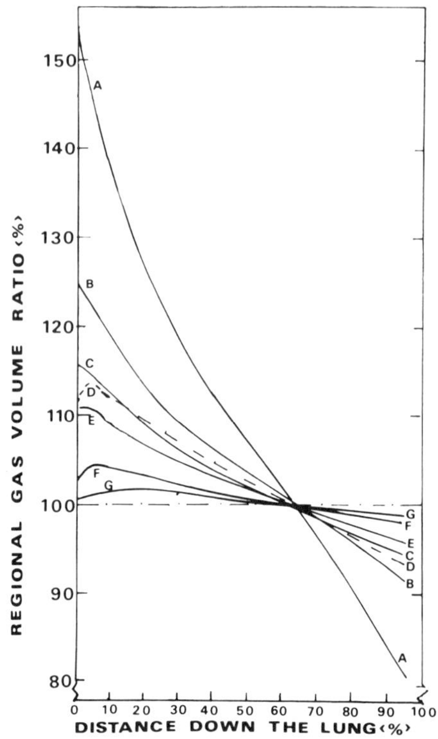

Figure 6.7 Effect of stiffness on regional expansion. Values of C/D_A for A, B, C, E, F, G, are 2.45, 4.90, 7.35, 9,60, 19.20, 38.4 kPa (25, 50, 75, 100, 200, 400, cm H_2O). Surface tension is zero for the above curves. Curve D is repeated from Figure 6.2 and has normal surface tension

This is further illustrated in Figure 6.9, in which we plot the pleural pressure for the midlung (a distance about 57 % down the lung) against C/D_A and γ. This linear relationship is to be expected since a very special point was chosen in the lung. Figures 6.7 and 6.8 show that the volume expansion in the midlung region is insensitive to changes in the elastic or surface tension parameters. In states of equal strain a linear relationship would be expected between the material constants (C/D_A and γ) and the pressure. By choosing appropriate scaling the curves nearly coincide. This gives a correspondence between C/D_A and γ. A value of C/D_A may be chosen which will give the same pressure in the midlung as the corresponding value of γ. Although these pressures are the same, the apical expansion is not.

This scaling information was used here to illustrate the difference between elastic and surface tension effects. In Figure 6.10 apical gas volume ratio against

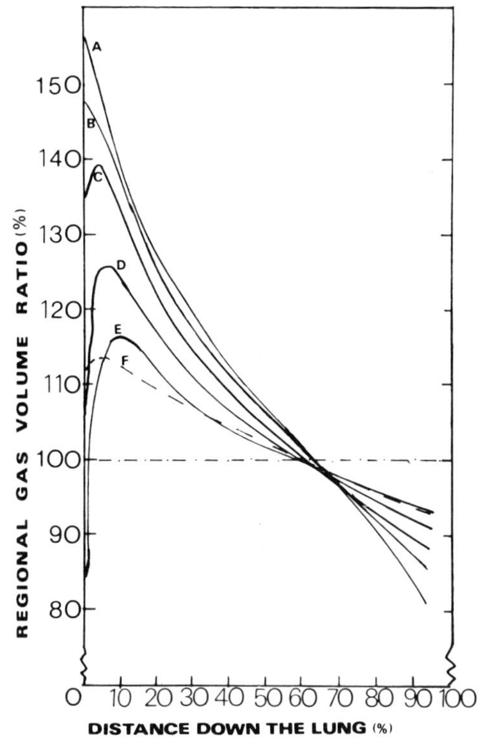

Figure 6.8 Effect of surface tension. $C/D_A = 2.45$ kPa in all curves. Values of surface tension for curves are 0, 10, 20, 40, 60 mN/m (0, 10, 20, 40, 60 dynes/cm). Curve F is repeated from Figure 6.2 and has normal surface tension

either C/D_A or γ scaled as in Figure 6.9 is plotted. Although both curves are monotonic there is little other similiarity. Figure 6.11 shows that the two mechanisms affect the distribution of pleural pressure differently. The apex-to-base pressure difference as a percentage of the apical pressure is plotted against C/D_A and γ. With increasing stiffness (increasing C/D_A) the apex–base pressure difference declines rapidly but for increasing surface tension there is a minimum difference when γ is about 0.025 N/m. For higher values the pressure difference again increases.

It is clear that surface tension affects both the regional expansion and the pressure distribution and that its effects cannot be simulated by merely altering the material constants. It remains to be shown whether the distributions are sensitive to the choice of surface tension model.

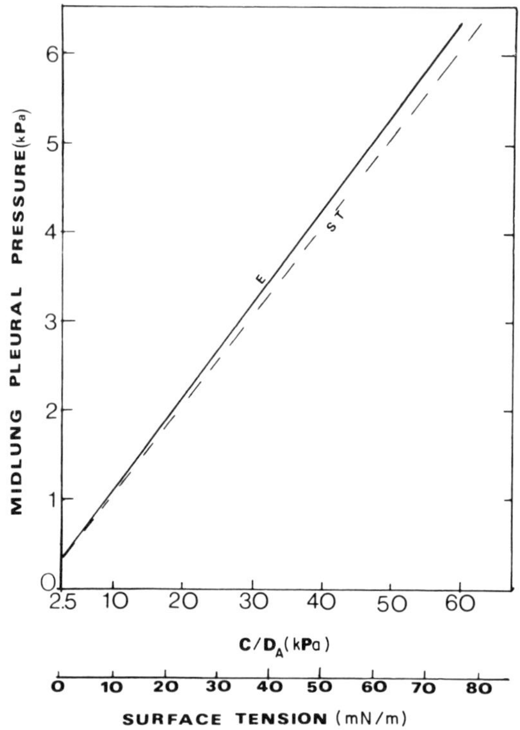

Figure 6.9 Effect of material constants on midlung pressure. Curve E is for increasing stiffness (C/D_A). Curve ST is for increasing surface tension

6.8.3 The sensitivity of the results to the choice of surface tension model

Five surface tension models have been analysed. There are two effects to be described here:

1. The effect of the surface tension model on the regional gas volume in the lung. In the previous section alveolar expansion was shown to be non-uniform and expansion at the apex to be significantly different than that at the midlung and base. We would like to determine if this finding is sensitive to the model choice.
2. The effect of the model choice on the pleural pressure distribution. At any lung volume, a negative (i.e. expanding) pressure must be applied on the pleural surface to hold the lung in equilibrium (see Figure 6.6). There is an apex-to-base gradient in pressure and these pressures will be shown to be sensitive to the model choice.

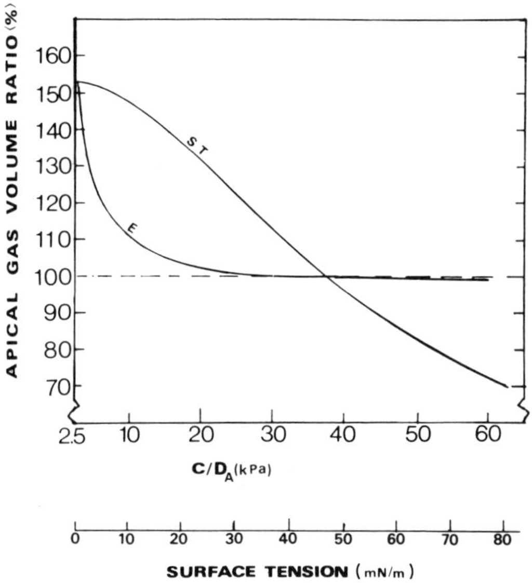

Figure 6.10 Effect of material constants on apical gas volume using the scaling of Figure 6.6. Notice that although midlung pressures are nearly identical, apical volumes are much different. Curve E is for increasing stiffness. Curve ST is for increasing surface tension

6.8.3.1 Regional gas volume

Figures 6.12, 6.13, and 6.14 show the regional gas volume ratio distribution for three overall lung volumes. In the figures overall lung volumes are 1.73, 2.74, and 4.10 times the stress-free lung volume, respectively. This correponds to air volumes 2.15, 3.74, and 5.88 times the stress-free air volume. The results presented above were for an overall volume ratio of 2.74. As lung volume increases, the lung becomes stiffer and surface tension increases. These two factors reduce the influence of gravitational loading at the higher lung volumes. For a purely elastic lung, not influenced by gravity, the gas volume distribution would be uniform. Therefore, variations in distribution at high lung volumes (Figure 6.14) are indicative of inter-facial effects.

There is a similarity among the lung wash curves (a, b, c, s) at all lung volumes despite wide variations in the surface tension (Figure 6.3). This is true whether the curves are modelled by a Fourier series (a, b, c) or by an exponential model (s). At low lung volumes (Figure 6.12) the subpleural model (sp) predicts a distribution markedly different than the other models. At higher lung volumes (Figures 6.13 and 6.14), however, the subpleural model results are quite similar to those of the other models.

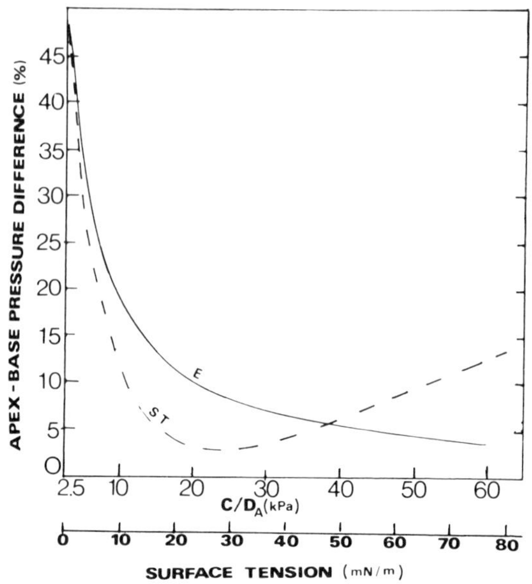

Figure 6.11 Effect of material constants on the apex to base pressure difference. As stiffness increases (curve E) the apex-to-base difference approaches zero but is nonmonotonic for increasing surface tension

Figure 6.3 indicates that for low lung volumes (corresponding to low surface areas) the surface tension in the subpleural model is quite low compared to the other models. In fact, the RGVR distribution for the subpleural model in Figure 6.12 (sp) does not differ significantly from the distribution predicted when the surface tension is absent (Vawter, 1980a). As lung volume increases the surface tension in the subpleural model rapidly increases and becomes comparable to that of the other models.

If very high volume expansions were used, the results suggest that the subpleural model would again diverge because the surface tension is unbounded. However, these high volumes are beyond the physiologic range and beyond the range of application of the subpleural model.

It appears then that, except for low lung volumes, all surface tension models predict nearly the same distribution of alveolar expansion. It should be noted, however, that for a given overall lung volume, the model with the *higher* surface tension predicts the *lower apical* expansion.

6.8.3.2 Pleural pressure

One would expect that the pressure required to keep the lung in equilibrium would increase as the lung is expanded because both the elastic stiffness of the tissue and the surface tension increase. This is verified in Figure 6.15, which plots

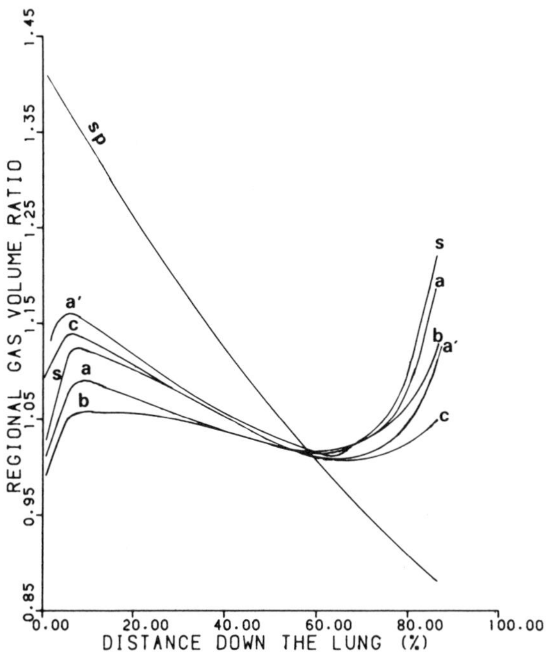

Figure 6.12 Regional gas volume distribution at low overall lung volume. Volume expansion is 1.73 which is somewhat below FRC. Symbols as in Figure 6.3

the pressure distribution for an expansion of 2.74. For a given model the pleural pressure, at any distance down the lung, increases with increasing lung volume.

The most significant finding, however, is that the pleural pressure is highly sensitive to model choice. For instance, the apical pressure is four times higher for one of the lung wash models (b) than for the subpleural mode (sp). Even at high lung volumes (not shown) the apical pressures differ by a factor of three. This is consistent with the difference in surface tension between the models. The surface area consistent with the volume expansion in Figure 6.15 (assuming no regional differences in strain) is 2.56. For this area the lung wash model (b) has a surface tension four times that of the subpleural model (sp). This further illustrates the dominance of interfacial effects over elastic effects. If the elastic forces were dominant the differences in pleural pressures between the models should be much less. The pressure distributions at other lung volumes show the same trends, (Shields and Vawter, 1980).

6.8.4 The role of the diaphragm

In the above presentation, the diaphragmatic surface of the lung was assumed flat in both the stress-free and expanded shape. This, of course, does not

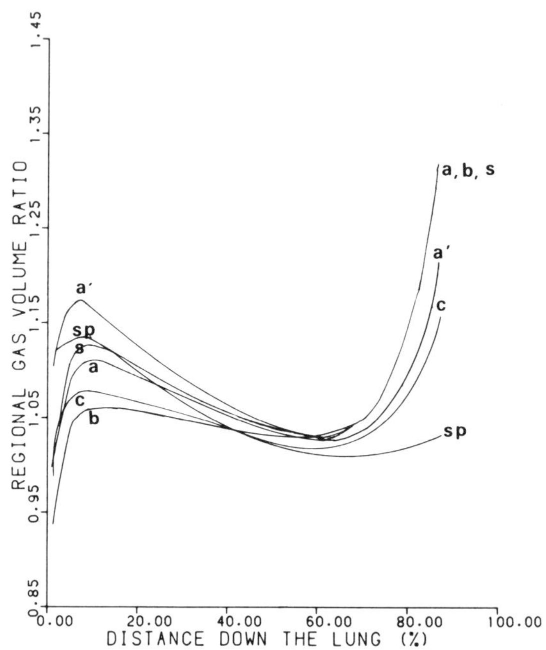

Figure 6.13 Regional expansion at moderate overall lung volume. Volume expansion is 2.74 which is close to FRC. Symbols as in Figure 6.3

accurately model the anatomy of either the human or dog lung. The regional gas volumes ratios (RGVR) were analysed when the diaphragm was shaped as shown in Figure 6.16. The diaphragm was considered rigid for a given lung expansion.

When the diaphragm is flat, RGVR is relatively insensitive to lateral position as seen in figure 6.17. Lines of constant expansion are presented for an overall gas volume ratio of 3.74. Figures 6.18 and 6.19 show that the deformation pattern is significantly altered by a curved diaphragm. Even for a shallow diaphragam (Figure 6.18), the lung slides laterally outward along the diaphragm lowering the expansion at the base of the lung near the pleural surface. For a more severe diaphragm (Figure 6.19), the compression of the basal-pleural region is more pronounced, causing an overexpansion of the region near the centre line. It is clear from these results that one cannot get an accurate picture of the deformation pattern by plotting regional gas volume ratio against 'distance down the lung' (e.g. Figures 6.5, 6.7) when a curved diaphragm is present.

The above results appear to be caused by the presence of interfacial forces. If interfacial forces are eliminated (Figure 6.20) the resulting deformation pattern again becomes relatively independent of lateral position even when a curved

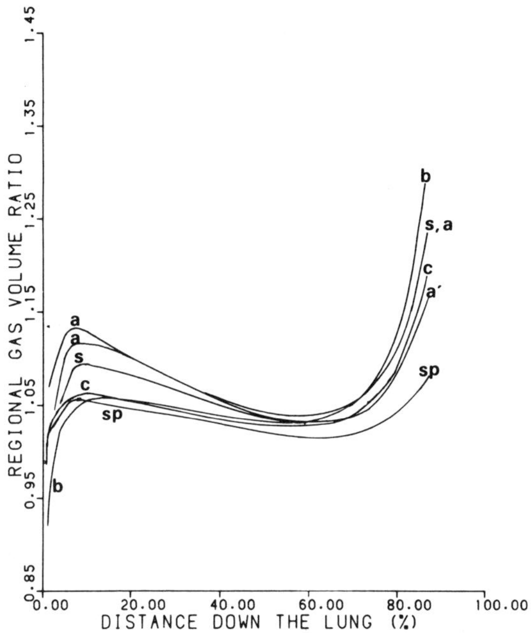

Figure 6.14 Regional expansion at high overall lung volume. Volume expansion is 4.10 which is above FRC. Symbols as in Figure 6.3

diaphragm is present. The apex-to-base gradient in expansion is much higher than when surface tension is present.

The effect of surface tension on regional expansion is further illustrated in Figure 6.21, in which interfacial forces are present but gravitational forces are not. The lung is expanded rather uniformly in the upper region with most RGVR's differing by only a few percent. Non-uniformities do appear near the base.

The non-uniform expansion persists even if the diaphragm is lowered from the curved position to flat (Figure 6.22). Lowering the diaphragm raises the overall gas volume ratio (GVR) from 3.74 to 4.1. One would expect a non-uniform expansion even if gravitational and interfacial forces were not present.

6.8.5 The role of hilar support

The only support included in the above analysis was at the pleural surface and the diaphragm. In the real lung some support is likely furnished by the trachea. A true model of the interaction between the trachea and the lung parenchyma is

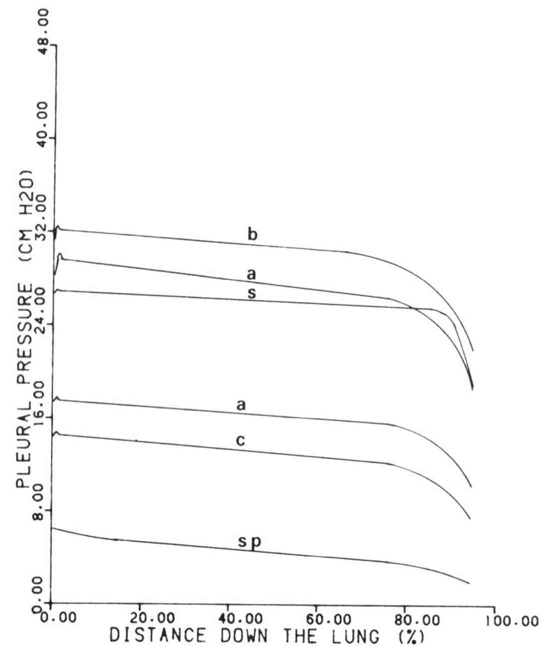

Figure 6.15 Pressure at the pleural surface for the volume distribution of Figure 6.13

beyond the scope of this work. A crude approximation was made for this support by insisting that the displacement at the hilum be zero. This, of course, models the trachea as a rigid point support. For the model studied the hilum is located 13.62 cm below the apex. The resultant displacement pattern is shown in Figure 6.23. The region immediately above the support is underexpanded as expected. When comparing the results with those for no hilar support (Figure 6.19), the regional gas volume ratios at the pleural surface are nearly the same as is the distribution of pleural pressure (not shown). This implies that hilar support effects are largely confined to regions near the hilum. An understanding of the true nature of these localized effects must await a more accurate modelling of the hilar region.

6.9 DISCUSSION

It is clear that surface tension alters regional gas distribution, especially near the apex. One of the findings of West and Matthews (1972) was that the apical gas volume ratio was quite high ($\sim 200\%$ at FRC) and was always greater than 100% regardless of the stiffness. Liu and Lee (1978) calculated lower values ($\sim 120\%$ at FRC) somewhat closer to the present values ($\sim 155\%$ at FRC). A strict

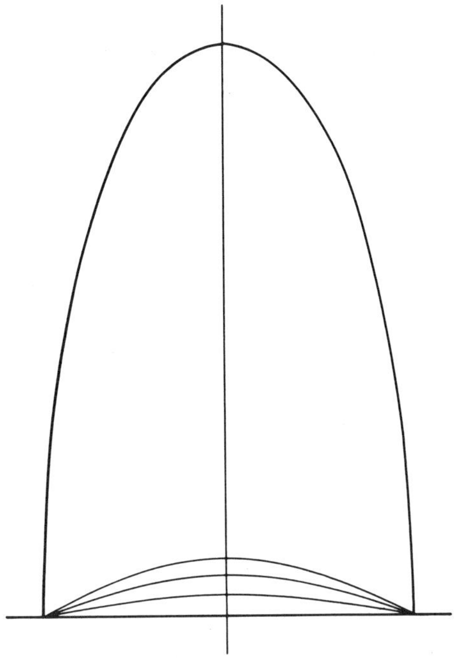

Figure 6.16 External shape of the lung for the diaphragm shapes analysed

comparison with Liu and Lee's results is not possible due to differences in model geometry and boundary conditions. Also, Liu and Lee used a different density at the stress-free state. They assumed that the gas volume at zero stress was also zero. We have previously shown that the lung is not totally collapsed when in a stress-free state (Vawter, 1977).

For the 'narrow' shape West and Matthews used 27 nodes or 54 degrees of freedom. Convergence cannot be predicted for their model based on our results, but it appears from Table 6.2 that significant errors would occur for *our* model using such a coarse grid. Liu and Lee used a 40-node model with 120 degrees of freedom. Since their results are for a three-dimensional model a comparison with Table 6.2 is not possible. However, three-dimensional problems in general require significantly more degrees of freedom for convergence.

If behaviour at low lung volumes is to be modelled accurately, a more precise description of the *in vivo* variation in surface tension will be necessary, for it is in this region that the results are sensitive to the choice of surface tension model. Behaviour at low lung volumes is important in that this is the region where West and Matthews (1972) found that apical volumes *decreased* when overall lung volume *increased*. This paradoxical breathing was not found in the present study.

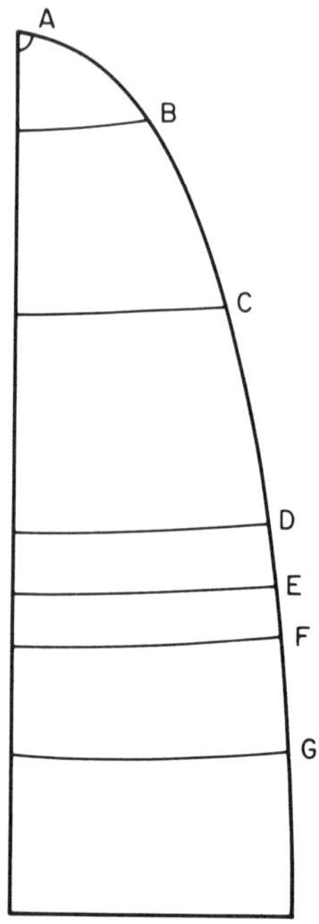

Figure 6.17 Regional gas volume contours (%) for a flat diaphragm. $A = 114$, $B = 112$, $C = 107$, $D = 102$, $E = 100$, $F = 99$, $G = 96$. Notice that contours run horizontally

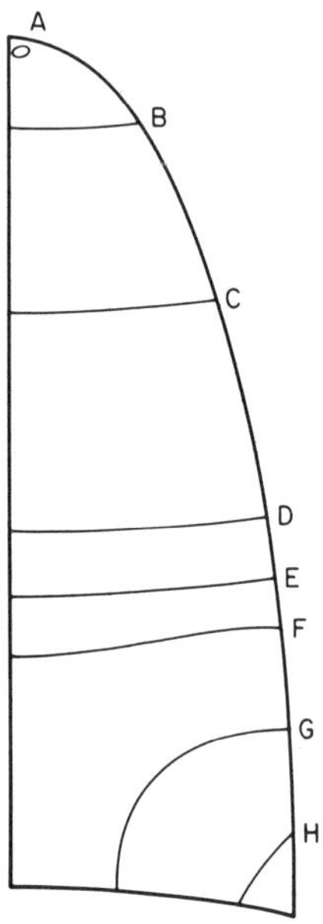

Figure 6.18 Regional gas volume contours (%) for a shallow diaphragm. $A = 114$, $B = 112$, $C = 107$, $D = 102$, $E = 100$, $F = 99$, $G = 96$, $H = 91$

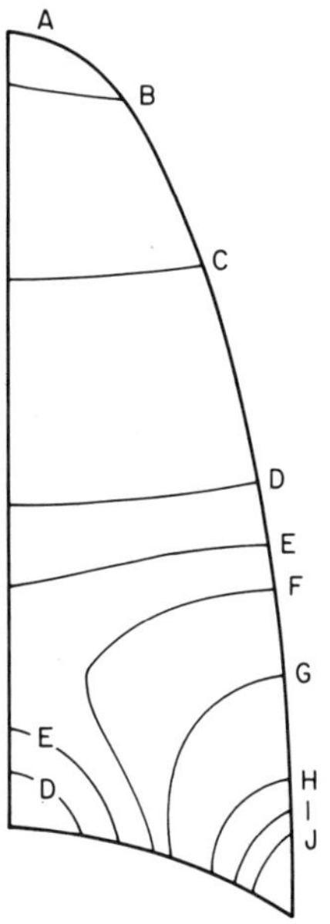

Figure 6.19 Regional gas volume contours (%) for a high diaphragm. $A = 115$, $B = 112$, $C = 107$, $D = 101$, $E = 100$, $F = 99$, $G = 96$, $H = 91$, $I = 85$, $J = 80$

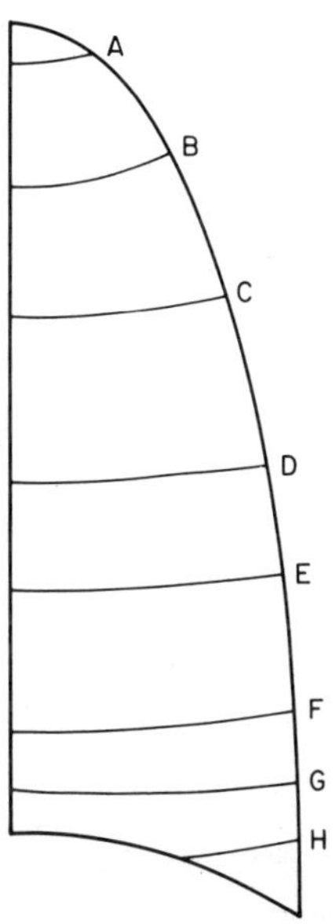

Figure 6.20 Regional gas volume contours (%) for high diaphragm but no surface tension. $A = 114$, $B = 128$, $C = 117$, $D = 107$, $E = 99$, $F = 91$, $G = 85$, $H = 80$

It is apparent that a curved diaphragm and hilar support significantly alter the regional expansion of the lung and more accurate modelling would be desirable. It is interesting that the effects of the diaphragm are accentuated by interfacial forces. The importance of a curved diaphragm suggests that perhaps finite

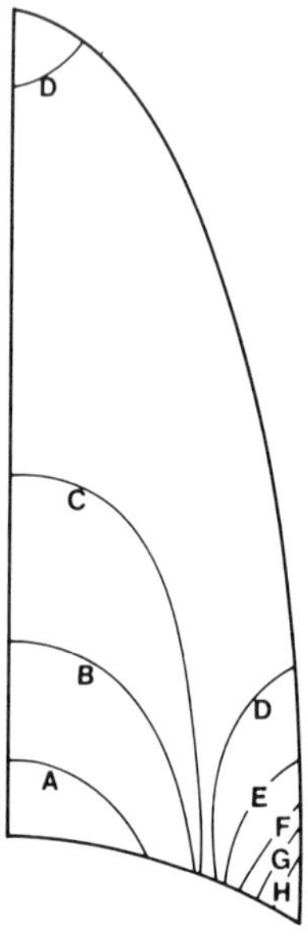

Figure 6.21 Regional gas volume contours (%) for a high diaphragm without body forces. $A = 107$, $B = 101$, $C = 100$, $D = 99$, $E = 96$, $F = 91$, $G = 85$, $H = 80$

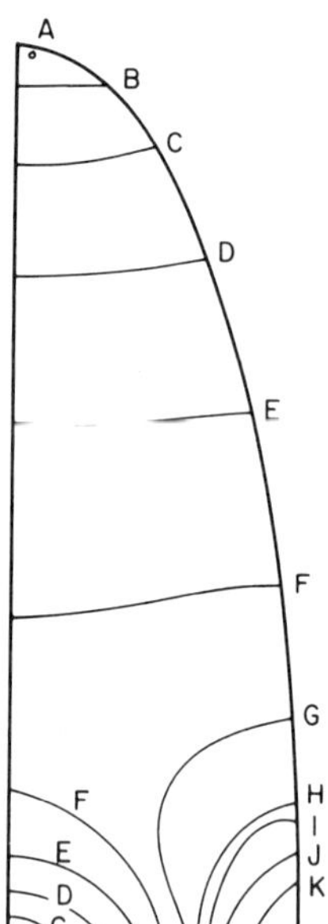

Figure 6.22 Regional gas volume contours (%) for a high diaphragm lowered to horizontal. No body forces present. Percentage based on overall gas volume before lowering diaphragm. $A = 118$, $B = 117$, $C = 116$, $D = 115$, $E = 114$, $F = 112$, $G = 107$, $H = 102$, $I = 100$, $J = 96$, $K = 86$

elements with curved boundaries should be used in the future. This may also be important at the apex a region of high curvature.

From the results shown it can be concluded that the inclusion of interfacial forces explicitly is necessary in order to model the lung numerically. It appears

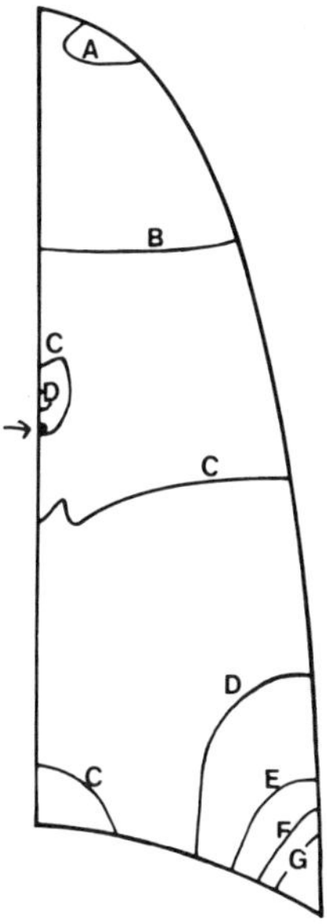

Figure 6.23 Regional expansion contours with hilar support at arrow. $A = 112, B = 107, C = 102, D = 96, E = 91, F = 86, G = 80$

that the choice of surface tension model is especially important at low lung volumes and to calculate the pleural pressure distribution.

ACKNOWLEDGEMENTS

The project was supported in part by a Biomedical Research Support Grant from NIH. The authors wish to thank Glen Dorband, William Fulcher, and Jay Humphrey for their assistance in the numerical calculations.

REFERENCES

Brown, E. S., Johnson, R. P., and Clements, J. A. (1959) Pulmonary surface tension, *Journal of Applied Physiology*, **14,** 717–720.

Clements, J. A. (1962) Rat, cat, dog lung surfactant, *Physiologist*, **5,** 11.

Flicker, E., and Lee J. (1974) Equilibrium of force of subpleural alveoli: Implications to lung mechanics, *Journal of Applied Physiology*, **36,** 366–374.

Fung, Y. C., Tong, P., and Pattitucci, P. (1978) Stress and strain in the lung, *Journal of Engineering Mechanics Division*, ASCE, **104,** (EMI), 201–223.

Liu, J. T., and Lee, G. C., (1978) Static finite deformation analysis of the lung, *Journal of the Engineering Mechanics Division,* ASCE, **104,** (EMI), 225–238.

Pian, T. H. H., and Tong, P. (1971) Variational formulation of finite displacement analysis, *High Speed Computing of Elastic Structures,* (ed. Fraeijs de Veubeke), Univ. of Liege, Belgium.

Scarpelli, E. M. (1965) *The Surfactant System of the Lung*, Lea and Febiger, Philadelphia, Pa.

Scarpelli, E. M., and Gabbay, K. H. (1965) Rat lung surfactant, *Science*, **148,** 1067.
Shields, W. H. (1979) Feasibility of using intact lung tests to determine the mechanical properties of the lung, M. S. Thesis, VPI&SU.
Shields, W. H., and D. L. Vawter (1980), The effect of surface tension on regional expansion in the lung, in: *Finite Elements in Biomechanics* Vol. 1 (ed. B. Simon), pp. 369–388.
Slama, H., Schodel, W., and Hanson, E. (1973) Goat lung surfactant, *Respiration Physiology*, **19,** 233.
Vawter, D. L. (1977) The stress free equilibrium volume of the lung, *Journal of Applied Physiology, J. of Appl. Physiol, Respirat, Environ. Exercise Physiol.*, **43** (1), 3–7.
Vawter, D. L., (1980a) A finite element model for macroscopic deformation of the lung, *Journal of Biomechanical Engineering.* **102** (1), 1–7.
Vawter, D. L. (1980b) A finite element model for macroscopic deformation of the lung, in: *Finite Elements in Biomechanics*, Vol. 1 (ed. B. Simon), pp. 389–408.
Vawter, D. L., Fung, Y. C., and West, J. B. (1978) Elasticity of excised dog's lung tissue, *Journal of Applied Physiology, Respirat, Environ. Exercise Physiology*, **45** (2), 261–269.
Vawter, D. L., and J. S. Lee (1976) A mechanical analysis of the pressure–volume behaviour of dog lungs, *Proc. 4th New England Bioengineering Conference* (ed. S. Saha) Pergamon Press, pp. 139–142.
Vawter, D. L., Matthews, F. L., and West, J. B. (1975) Effect of shape and size of lung and chest wall on stresses in the lung, *Journal of Applied Physiology*, **39,** 9–17.
West, J. B., and Matthews, F. L. (1972) Stresses, strains, and surface pressures in the lung caused by its weight, *Journal of Applied Physiology*, **32,** 332–345.

Finite Elements in Biomechanics
Edited by R. H. Gallagher, B. R. Simon, F. C. Johnson, and J. F. Gross

Chapter 7

Finite Element Analysis of Lung Parenchyma

A. D. Karakaplan, M. P. Bieniek and R. Skalak

7.1 INTRODUCTION

The theory of the mechanical behaviour of the lung is an important tool in proper understanding of the functioning of the lung. The overall lung compliance, deformations at the alveolar level, the response to local distortions, atelectasis, and many other phenomena are, to a large extent, purely mechanical processes in a soft-tissue structure interacting with a liquid surfactant and subjected to a system of forces (primarily inflation pressure, gravity, and constraints of the thoracic cavity and adjoining organs).

A number of previous analyses of the lung elasticity have been based on simplified models of the structure of the parenchyma. A survey of the pertinent literature has been given in a recent paper by Karakaplan, Bieniek, and Skalak (1980). The following brief outline is limited to the works which have been used in the formulation of the objectives and methods of the present analysis. One of the first models of the lung elasticity has been proposed by Mead, Takashima, and Leith (1970). The parenchyma in their model is represented by a two-dimensional system of springs forming a hexagonal mesh; the basic conditions of equilibrium for a pressure loading and a simulation of elementary structural defects have been formulated and discussed. The volume–pressure response of the lung has been analysed by Powell (1971) with the aid of the 'average alveolus' model in the form of a spherical bubble at the end of a rigid tube. Fung (1974) has developed a model of parenchyma consisting of a system of membranes forming cubical or spherical air sacks; the properties of the membranes are specified in terms of a non-linear stress–strain relation (derived from a strain energy function), while the effect of the surfactant is taken into account by area-dependent surface tension forces. Frankus and Lee (1974) have proposed a similar model, using a dodecahedron as the idealized geometrical shape of an alveolus.

Due to the use of the finite element method, the present analysis is based on a relatively realistic model of the lung parenchyma. The geometry of the alveolar sacks, their dimensions, and their arrangement around an air duct are based on the morphological study by Weibel (1963). In view of the statistical nature of the alveolar dimensions, in addition to an 'average' regular model, randomly distorted configurations have been constructed and analysed.

The elastic properties of the alveolar walls have been described in terms of a strain energy function of exponential type proposed by Fung (1973) for soft biological tissues. The constants of this function have been determined by matching the theoretical predictions of this analysis with a variety of experimental data obtained from tension tests on specimens of alveolar walls (Fukaya *et al.*, 1968), three-dimensional expansion of specimens of lung parenchyma (Hoppin *et al.*, 1975) and one- and two-dimensional tests on specimens of the parenchyma (Vawter *et al.*, 1978). It should be noted that the experimental curves of loading and unloading never exactly coincide; there is hysteresis. However, with loading and unloading dealt with separately, the strain-energy method has proven to be most effective in establishing the constitutive equations of the solid component of the lung parenchyma.

The last component of the present model of the lung parenchyma is the surfactant. The finite element method allows for a fairly general assumption concerning the properties of the surface tension coefficients. Its dependence on the area changes can be of an arbitrary nature, with different values for the inflation and the deflation processes.

7.2 STRUCTURAL MODEL OF LUNG PARENCHYMA

An extensive discussion of the modelling problems of lung parenchyma is given in a recent paper by Karakaplan *et al.*, (1980), which also contains a survey of the literature on the subject. The structural model of the lung parenchyma which has been adopted for the present analysis is as follows.

The basic unit of the parenchyma consists of a cluster of alveoli surrounding an air duct. Figure 7.1 shows the stacking of the alveoli as seen from the inside of the duct, where the honeycomb pattern of the hexagonal openings of the alveolar sacks is visible. Figure 7.2 shows the cross-section of the duct and alveoli in the

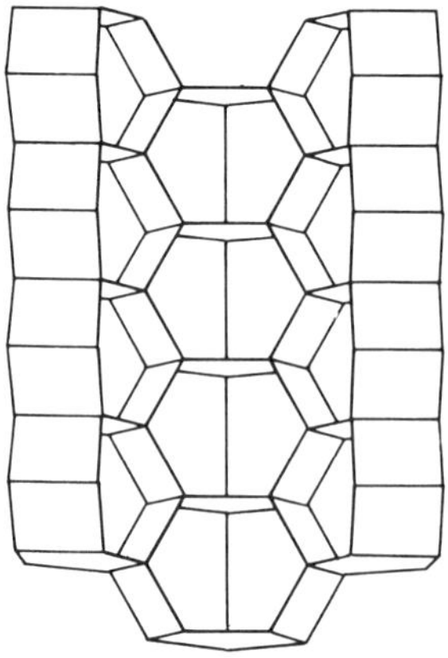

Figure 7.1 The proposed geometric model of a cluster of alveoli around an air duct; view from inside of the duct

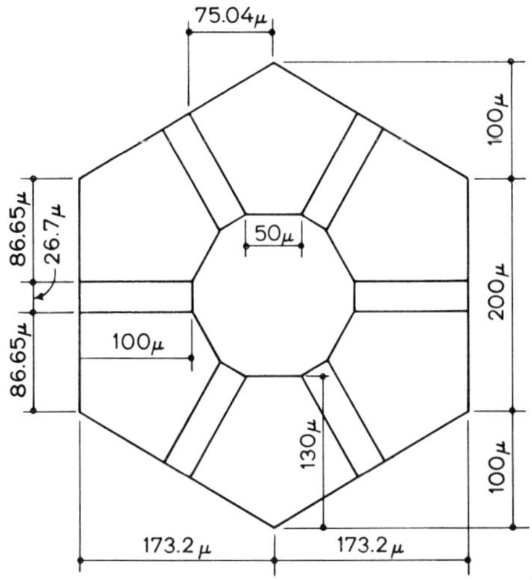

Figure 7.2 The proposed geometric model of a cluster of alveoli around an air duct; cross-section normal to the axis of the duct. The dimensions refer to the configuration at zero pressure (i.e., no inflation)

plane perpendicular to the axis of the duct. The dimensions shown in Figure 7.2 represent the average, regular system. The average wall thickness has been assumed at 10 μm. Again, deviations from this average configuration, which can be observed in a normal lung, as well as various pathological changes in the lung morphology (e.g. tissue loss in an emphysemic lung) can be imposed on the pattern shown in Figures 7.1 and 7.2.

The elastic properties of the alveolar walls are defined in terms of the strain energy W given as a function of the material strain tensor, e_{ij}. With the assumption of plane stress in each membrane, the pertinent strain components are e_{11}, e_{22}, and e_{12}. With the notation

$$I_1 \equiv e_{11} + e_{22} \tag{7.1}$$

$$I_2 \equiv e_{11}^2 + 2e_{12}^2 \tag{7.2}$$

the expression for W used in this analysis is

$$W = \frac{1}{2}(\lambda\delta_{ij}\delta_{kl} + \mu\delta_{ki}\delta_{lj})e_{ij}e_{kl}\exp(aI_1^2 + bI_2) \tag{7.3}$$

where λ, μ, a, and b are material constants, δ_{ij} is the Kronecker delta; the indices assume values 1 and 2, with summation convention with respect to repeated indices. The components of the material (Piola–Kirchhoff) stress tensor are obtained as

$$s_{ij} = \partial W/\partial e_{ij} \tag{7.4}$$

The tangent elastic moduli, D_{ijkl}, such that

$$ds_{ij} = D_{ijkl} de_{kl} \tag{7.5}$$

are obtained as

$$D_{ijkl} = \partial^2 W / \partial e_{ij} \partial e_{kl} \tag{7.6}$$

For the purposes of the finite element method, Equation (7.5) can be easily written as

$$\mathbf{ds} = \mathbf{Dde} \tag{7.7}$$

where $\mathbf{s}^T \equiv (s_{11}, s_{22}, s_{12})$, $\mathbf{e}^T \equiv (e_{11}, e_{22}, 2e_{12})$, and $\mathbf{D}$ is a 3×3 matrix formed from the components D_{ijkl}. The following values of the constants λ, μ, a, and b appear to approximate the mammalian lung:

$\lambda = 7355\,\text{dynes/cm}^2$, $\mu = 12\,258\,\text{dynes/cm}^2$, $a = 0.50$, $b = 0.50$.

Figure 7.3 illustrates the behaviour of a membrane with the strain energy function (Equation (7.3)), the above values of the material constants and thickness of $10\,\mu$m. (The membrane force T_1 is the force per unit deformed length of the edge perpendicular to the x_1 axis, while the ratio $\lambda_1 \equiv L_1/L_0$ determines the extension along the x_1 axis.)

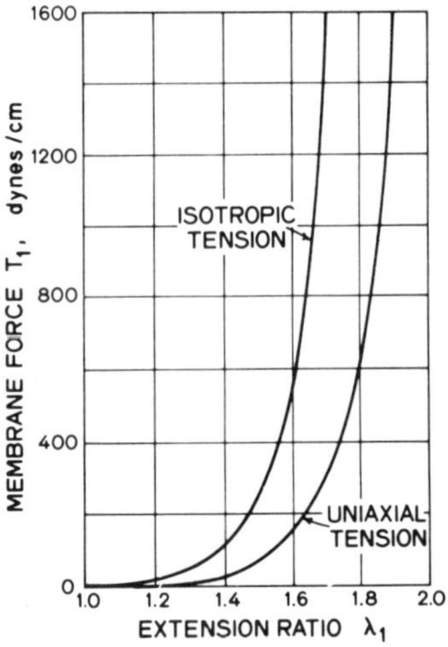

Figure 7.3 The behaviour of an alveolar wall under isotropic tension and uniaxial tension

The surface tension effect, due to the presence of the thin liquid layer of surfactant covering the alveolar walls, is determined in terms of the work of the surface tension forces. If the area changes from A_0 to A, the work U and its variation δU are

$$U = \int_{A_0}^{A} \gamma(A)\,\mathrm{d}A \tag{7.8}$$

and

$$\delta U = \gamma(A)\,\delta A \tag{7.9}$$

where $\gamma(A)$ is the surface tension coefficient of the lung surfactant. The dependence of γ on the area changes, as deduced in the course of this analysis, is shown in Figure 7.4. The two curves in Figure 7.4 represent γ as a function of the average area change for the whole block of alveoli and as a function of the 'local' area change over a single triangular element.

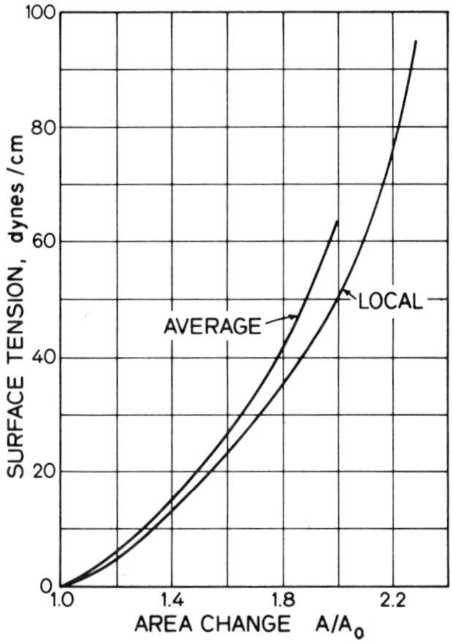

Figure 7.4 The variation of the surface tension as a function of surface-area changes

7.3 FINITE ELEMENT FORMULATION

A constant-strain triangular element has been chosen for this analysis. The finite element mesh is shown in Figure 7.5. In order to test the adequacy of this relatively coarse mesh, a single wall has been investigated with the aid of a finer mesh. The results corresponding to the finer mesh, in terms of the nodal forces and the strains and stresses in the wall, are very close (within a few percent) to the results from the mesh as in Figure 7.5. It seems that the fact that the strains and stresses are fairly uniformly distributed within a wall (except at the corners adjacent to the free edges) justifies the use of a coarse mesh and a simple element.

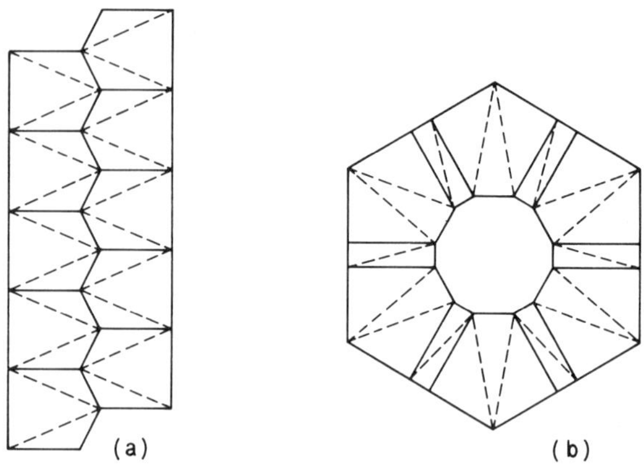

Figure 7.5 The finite element mesh used in this analysis; (a) side view, (b) plan

Although the walls are assumed to be subjected to plane stress, due to the geometric non-linearities all three components of displacement must be considered within each element and in the nodal degrees of freedom. The equations of the elastic membrane element follow from the principles of the finite element method (Zienkiewicz, 1971). For the nth element as shown in Figure 7.6, we have

$$\mathbf{u}_n = \mathbf{N}_n \mathbf{q}_n \tag{7.10}$$

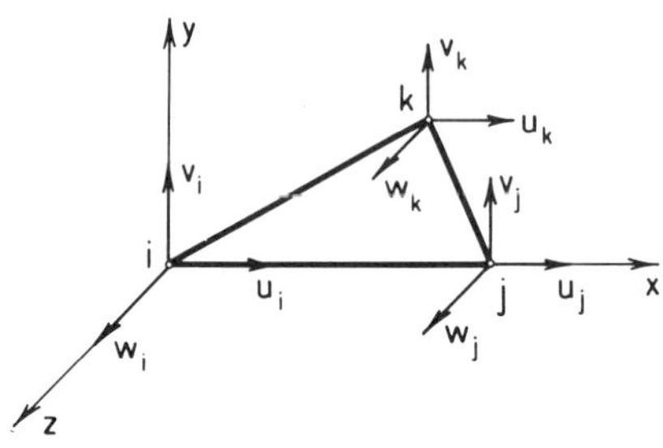

Figure 7.6 A typical triangular element and the local coordinate system

where

$$\mathbf{u}_n^{\mathrm{T}} \equiv (u, v, w)_n$$
$$\mathbf{q}_n^{\mathrm{T}} \equiv (u_i, v_i, w_i; u_j, v_j, w_j; u_k, v_k, w_k)$$

and N_n are linear approximating functions of x and y. Utilizing the strain–displacement relations, we find

$$\mathbf{e}_n = (\mathbf{B}_n' + \tfrac{1}{2}\mathbf{B}_n'')\mathbf{q}_n \tag{7.11}$$

and

$$d\mathbf{e}_n = (\mathbf{B}'_n + \mathbf{B}''_n)\, d\mathbf{q}_n = \overline{\mathbf{B}}_n\, d\mathbf{q}_n \tag{7.12}$$

where $\mathbf{B}'_n$ and $\mathbf{B}''_n$ correspond to the linear and the non-linear terms in the strain displacement relations. The nodal force vector of the membrane element is then computed as

$$\mathbf{Q}^s_n = \int_{V_{0n}} \overline{\mathbf{B}}^{\mathrm{T}}_n\, \mathbf{S}_n\, d V_0 \tag{7.13}$$

with the integration performed over the initial volume.

In the Newton–Raphson method of solution of the equations for $\mathbf{q}_n$, the first derivatives of the generalized forces with respect to generalized displacements will be needed. These derivatives form square matrices

$$\mathbf{K}^s_n = \frac{d\mathbf{Q}^s_n}{d\mathbf{q}_n} \tag{7.14}$$

known as the *tangent stiffness matrices.* Using Equation (7.13), we obtain

$$\mathbf{K}^s_n\, d\mathbf{q}_n = \int_{V_{0n}} (d\overline{\mathbf{B}}^{\mathrm{T}}_n\, \mathbf{s}_n + \overline{\mathbf{B}}^{\mathrm{T}}_n\, d\mathbf{s}_n)\, d V_0 \tag{7.15}$$

The first term of Equation (7.15) can be written as

$$\int_{V_{0n}} d\overline{\mathbf{B}}^{\mathrm{T}}_n\, \mathbf{s}_n\, d V_0 = \mathbf{K}'_n\, d\mathbf{q}_n \tag{7.16}$$

where $\mathbf{K}'_n$ is usually called the *initial stress matrix* or the *geometric matrix* and it is dependent on the Piola–Kirchhoff stresses. Relatively simple steps lead to

$$\mathbf{K}'_n = (\mathbf{T}^{\mathrm{T}}_n \mathbf{S}_n \mathbf{T}_n)\, V_{0n} \tag{7.17}$$

where

$$\mathbf{S}_n = \begin{bmatrix} s_{xx} & 0 & 0 & s_{xy} & 0 & 0 \\ 0 & s_{xx} & 0 & 0 & s_{xy} & 0 \\ 0 & 0 & s_{xx} & 0 & 0 & s_{xy} \\ s_{xy} & 0 & 0 & s_{yy} & 0 & 0 \\ 0 & s_{xy} & 0 & 0 & s_{yy} & 0 \\ 0 & 0 & s_{xy} & 0 & 0 & s_{yy} \end{bmatrix} \tag{7.18}$$

$$\mathbf{T}_n = \frac{1}{2A_{0n}} \begin{bmatrix} b_i & 0 & 0 & b_j & 0 & 0 & b_k & 0 & 0 \\ 0 & b_i & 0 & 0 & b_j & 0 & 0 & b_k & 0 \\ 0 & 0 & b_i & 0 & 0 & b_j & 0 & 0 & b_k \\ c_i & 0 & 0 & c_j & 0 & 0 & c_k & 0 & 0 \\ 0 & c_i & 0 & 0 & c_j & 0 & 0 & c_k & 0 \\ 0 & 0 & c_i & 0 & 0 & c_j & 0 & 0 & c_k \end{bmatrix} \tag{7.19}$$

and

$$\begin{aligned} a_i &= x_j y_k - x_k y_j \\ b_i &= y_j - y_k \\ c_i &= x_k - x_j \end{aligned} \tag{7.20}$$

From the second term of Equation (7.15), it is obvious that the differential relationship between stress increment and strain increment has to be established. We can write it as

$$\mathbf{ds}_n = \mathbf{D}_n \mathbf{de}_n \tag{7.21}$$

where $\mathbf{D}_n$ is the incremental or tangent elasticity matrix which follows from Equation (7.7). Using Equations (7.21) and (7.12), we obtain

$$\int_{V_{0n}} \overline{\mathbf{B}}_n^{\mathrm{T}} \mathbf{ds}_n \, \mathrm{d}V_0 = \int_{V_{0n}} \overline{\mathbf{B}}_n^{\mathrm{T}} \mathbf{D}_n \mathbf{B}_n \mathbf{dq}_n \, \mathrm{d}V_0 = \mathbf{K}_n'' \mathbf{dq}_n \tag{7.22}$$

or

$$\mathbf{K}_n'' = (\mathbf{B}_n'^{\mathrm{T}} \mathbf{D}_n \mathbf{B}_n' + \mathbf{B}_n''^{\mathrm{T}} \mathbf{D}_n \mathbf{B}_n''^{\mathrm{T}} + \mathbf{B}_n''^{\mathrm{T}} \mathbf{D}_n \mathbf{B}_n') V_{0n} \tag{7.23}$$

Finally, the tangent stiffness matrix due to the elastic membrane is

$$\mathbf{K}_n^s = \mathbf{K}_n' + \mathbf{K}_n'' \tag{7.24}$$

We will now include the effect of the surface tension phenomena into the finite element method. The surfactant is assumed to be uniformly distributed over the area of the membrane element. According to Equation (7.9), the surface tension coefficient γ, which has the unit of force per unit length, is the only physical property that represents the state of the surfactant, and the variation of the work done by surfactant is directly proportional to the variation of the surface area. If δU is the virtual work of the surfactant and δA is the change in the deformed area of the surface of an element n, then

$$\delta U_n = \gamma_n \delta A_n \tag{7.25}$$

and by definition of the generalized forces, we can write

$$\mathbf{Q}_n^l = \gamma_n \frac{\delta A_n}{\delta \mathbf{q}_n} \tag{7.26}$$

where $\mathbf{Q}_n^l$ represent generalized force vectors due to the liquid surfactant.

We can simply write the expression for the deformed area of element n at any state of deformation as

$$A_n = \tfrac{1}{2} (E_1^2 + E_2^2 + E_3^2)^{1/2} \tag{7.27}$$

with

$$E_k = (\mathbf{v}_{ij} \times \mathbf{v}_{ik}) \cdot \mathbf{i}_k$$

where $\mathbf{v}_{ij}$, $\mathbf{v}_{ik}$ are the vectors from nodes i to j, and i to k, respectively in the deformed state. In the above expressions the coordinates and the components of displacement are all referred to the coordinate system of the undeformed membrane.

The surface tension coefficient will be assumed to be a function of the deformed area and based on this assumption, the tangent stiffness matrix which is the variation of generalized displacements has the form of

$$(K_n^l)_{ij} = \frac{\partial \gamma_n}{\partial A} \frac{\partial A_n}{\partial q_j} \frac{\partial A_n}{\partial q_i} + \gamma_n \frac{\partial^2 A_n}{\partial q_i \partial q_j} \tag{7.28}$$

where

$$\frac{\partial A_n}{\partial q_i} = \frac{1}{4A_n}\left(E_k \frac{\partial E_k}{\partial q_i}\right) \tag{7.29}$$

and

$$\frac{\partial^2 A_n}{\partial q_i \partial q_j} = \frac{1}{4A_n}\left\{\frac{\partial E_k}{\partial q_i}\frac{\partial E_k}{\partial q_j} + E_k \frac{\partial^2 E_k}{\partial q_i \partial q_j} - \frac{1}{16A_n^3}\left(E_k \frac{\partial E_k}{\partial q_i}\right)\left(E_k \frac{\partial E_k}{\partial q_j}\right)\right\} \tag{7.30}$$

After substituting Equations (7.30) and (7.31) in (7.29), we obtain the elements of the tangent stiffness matrix

$$(K_n^l)_{ij} = \frac{\partial \gamma}{\partial A}\frac{1}{16A_n^2}\left(E_k \frac{\partial E_k}{\partial q_j}\right)\left(E_k \frac{\partial E_k}{\partial q_j}\right) + \frac{\gamma}{4A_n}\left\{\frac{\partial E_k}{\partial q_j}\frac{\partial E_k}{\partial q_j} + E_k \frac{\partial E_k}{\partial q_i \partial q_j} - \frac{1}{16A_n^3}\left(E_k \frac{\partial E_k}{\partial q_i}\right)\left(E_k \frac{\partial E_k}{\partial q_j}\right)\right\} \tag{7.31}$$

In matrix notation we can write

$$K_n^l = \frac{\partial \gamma}{\partial A}\frac{1}{16A_n}\bar{\mathbf{L}}\bar{\mathbf{H}}\bar{\mathbf{H}}^{\mathrm{T}}\bar{\mathbf{L}}^{\mathrm{T}} + \frac{\gamma}{4A_n}\left(\bar{\mathbf{L}}\mathbf{L}^{\mathrm{T}} + G - \frac{1}{16A_n^3}\bar{\mathbf{L}}\bar{\mathbf{H}}\bar{\mathbf{H}}^{\mathrm{T}}\bar{\mathbf{L}}^{\mathrm{T}}\right) \tag{7.32}$$

where

$$\bar{\mathbf{H}} = \begin{bmatrix} E_1 \\ E_2 \\ E_3 \end{bmatrix}$$

and

$$\overline{\mathbf{L}} = \begin{bmatrix} 0 & -w_j + w_k & y_j + v_j - y_k - v_k \\ w_j - w_k & 0 & -x_j - u_j + x_k + u_k \\ y_k + v_k - y_j - v_j & x_j + u_j - x_k - u_k & 0 \\ 0 & -w_k + w_i & y_k + v_k - v_i \\ w_k - w_i & 0 & -x_k - u_k + u_i \\ -y_k - v_k + v_i & x_k + u_k - u_i & 0 \\ 0 & -w_i + w_j & v_i - y_j - v_j \\ w_i - w_j & 0 & -u_i + x_j + u_j \\ -v_i + y_j + v_j & u_i - x_j - u_j & 0 \end{bmatrix} \tag{7.33}$$

$$\overline{\mathbf{G}} = \begin{bmatrix} 0 & 0 & 0 & 0 & E_3 & -E_2 & 0 & -E_3 & E_2 \\ 0 & 0 & 0 & -E_3 & 0 & E_1 & E_3 & 0 & -E_1 \\ 0 & 0 & 0 & E_2 & -E_1 & 0 & -E_2 & E_3 & 0 \\ 0 & -E_3 & E_2 & 0 & 0 & 0 & 0 & E_3 & -E_2 \\ E_3 & 0 & -E_1 & 0 & 0 & 0 & E_3 & 0 & E_1 \\ -E_2 & E_1 & 0 & 0 & 0 & 0 & E_2 & -E_1 & 0 \\ 0 & E_3 & -E_2 & 0 & -E_3 & E_2 & 0 & 0 & 0 \\ -E_3 & 0 & E_1 & E_3 & 0 & -E_1 & 0 & 0 & 0 \\ E_2 & -E_1 & 0 & -E_2 & E_1 & 0 & 0 & 0 & 0 \end{bmatrix} \tag{7.34}$$

The total tangent stiffness matrix due to elastic membrane and liquid surfactant for a single element is

$$\mathbf{K}_n = \mathbf{K}_n^s + \mathbf{K}_n^l \tag{7.35}$$

The nodal forces due to external loads are determined consistently with the assumed interpolation functions $\mathbf{N}_n$ for the displacement field within the elements. In the case of loading by internal pressure, specified per unit deformed area, the nodal forces are

$$\mathbf{R}_n = \int_{A_n} \mathbf{N}_n \mathbf{p} \, \mathrm{d}A \tag{7.36}$$

7.4 SOLUTION OF EQUILIBRIUM EQUATIONS

The equation of equilibrium in the global form are

$$\mathbf{Q}(\mathbf{q}) = \mathbf{R}(\mathbf{q}) \tag{7.37}$$

where $\mathbf{Q}(\mathbf{q})$ represents the nodal forces due to the elastic tissue and the surfactant while $\mathbf{R}(\mathbf{q})$ represents the nodal forces due to external loads. In view of the fact that $\mathbf{Q}(\mathbf{q})$ and, for some loadings, $\mathbf{R}(\mathbf{q})$ are non-linear functions of the nodal displacements, $\mathbf{q}$, Equations (7.37) are non-linear algebraic equations. Their solution has been obtained by the Newton–Raphson method, combined with incremental application of external loads. Thus at some external load level, $\mathbf{p}_m$, the interaction step i requires solution of the system of equations

$$\mathbf{K}(\mathbf{q}_m^{i-1})\Delta\mathbf{q}^i = \Delta\mathbf{R}_m^i \tag{7.38}$$

where

$$\Delta\mathbf{R}_m^i = \mathbf{R}_m(\mathbf{q}_m^{i-1}, \mathbf{p}_m) - \mathbf{Q}(\mathbf{q}_m^{i-1}) \tag{7.39}$$

The ith approximation to the displacements corresponding to the load level $\mathbf{p}_m$ is

$$\mathbf{q}_m^i = \mathbf{q}_m^{i-1} + \Delta\mathbf{q}^i \tag{7.40}$$

The convergence of the above iteration procedure depends on a number of factors. In problems of the type analysed in this paper, lack of convergence may develop if:

(a) The system of equations and the tangent stiffness matrices do not satisfy the known convergence requirements of the Newton–Raphson method, even if in the vicinity of the load level $\mathbf{p}_m$ a regular and unique equilibrium path exists (Ortega and Rheinbolt, 1970).
(b) The equilibrium path contains bifurcation and/or limit points in the vicinity of the load level $\mathbf{p}_m$. The existence of singular points on the equilibrium path may signify actual non-uniqueness or instability of the deformation process of the parenchyma structure. Singular points may also be caused by the fact that the mathematical formulation of the problem admits non-unique solutions although the actual deformation of the parenchyma follows a unique path.

In the examples investigated in this work, no actual bifurcations of instabilities seem to exist. Spurious bifurcations, however, have been observed in problems involving large distortions of the alveolar shapes. Figure 7.7a shows a deformed configuration of alveoli in which two membranes penetrate each other in a physically inadmissible manner. Once the system is constrained from inadmissible displacements, a physically acceptable and apparently unique deformation process is obtained. Figure 7.7b shows a continuation of the same deformation as in Figure 7.7a, except that membrane penetration has been prevented.

7.5 APPLICATIONS

The purpose of presenting a number of specific examples (analysed by the method described in this chapter) is a better and more detailed understanding of the mechanical processes in the parenchyma. So far, the following problems have been investigated.

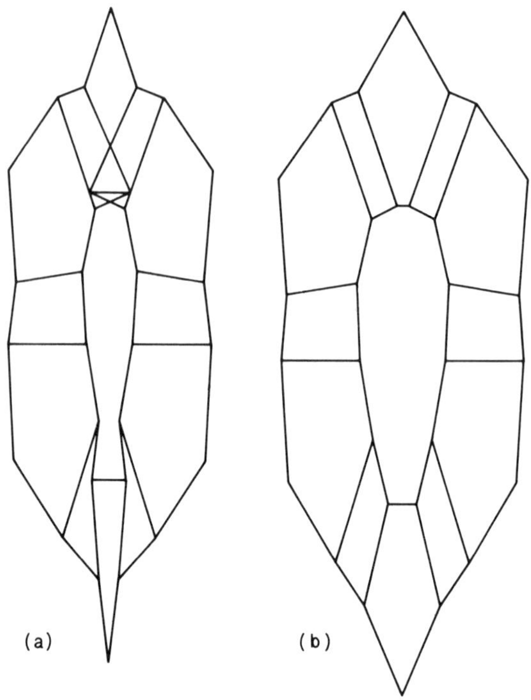

Figure 7.7 Deformation of alveolar sacks in uniaxial extension of a specimen of parenchyma;
(a) spurious deformation pattern at $s_{11} = 6122$ dynes/cm^2
(b) compatible deformation pattern at $s_{11} = 9310$ dynes/cm^2

7.5.1 Volume–pressure response

The volume–pressure relation of the parenchyma has been obtained by finding the displacements of the system and the corresponding volumes at various stages of alveolar pressure. The results of this computation are shown in Figure 7.8 for the deflation branch of the 'breathing' cycle. The theoretical predictions of the volume–pressure behaviour have been compared with the experimental data from tests on excised lung lobes or three-dimensional specimens of parenchyma. Such comparisons have been used to verify the serviceability of the model and to deduce the values of the material constants of the solid tissue. By analysing the differences between the compliances of the saline-filled and the air-filled lungs, the properties of the surface tension of the surfactant have been inferred.

One of the most important questions of lung mechanics is the problem of stability of the inflation and deflation processes in the lung. Some of the earlier investigations (based on the average-alveolus geometrical model, a simplified

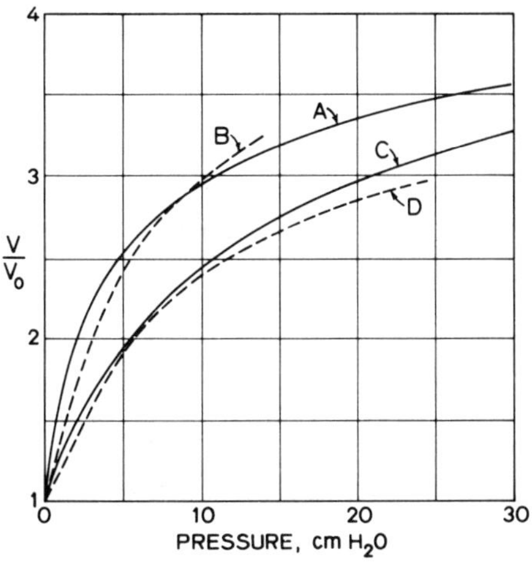

Figure 7.8 The pressure–volume behaviour of specimens of parenchyma. Saline-filled specimens: A – theory, B – experiment. Air-filled specimens: C – theory, D – experiment

elastic behaviour of the tissue, and a constant value of the surface tension) concluded that the inflation–deflation cycle contains unstable points. The presence of such points was held responsible for the hysteresis loop in the volume–pressure diagram of the lung in cyclic expansion. The present analysis indicates that due to the hardening elastic characteristics of the tissue, the area-dependence of the surface tension, and the connectivity of the alveolar walls, the inflation and deflection processes in the lung are stable. The hysteresis in the volume–pressure diagram appears to be due to the hysteretic behaviour of the surfactant and the presence of visco-elasticity in the solid tissue.

7.5.2 Geometrical changes in the parenchyma

The following aspects of the changes of the geometry of alveolar sacks and air ducts have been studied: (a) Deformed dimensions of alveoli and air ducts under internal pressure. Figure 7.9 shows the super-imposed images of the duct and alveoli at zero pressure and at internal pressure of 10 cm H_2O. (b) The ratio of the depth D, and the opening width, MD, as a function of pressure. Figure 7.10 shows the results of the present analysis as well as the data obtained by Klingele and Staub (1970), and Weibel (1963). (c) The variation of the area of the alveolar walls, i.e., the area of the blood–air gas exchange, as a function of the lung expansion (Figure 7.11). (d) Distortion of the parenchyma geometry in uniaxial texts (Figure 7.7) discussed earlier.

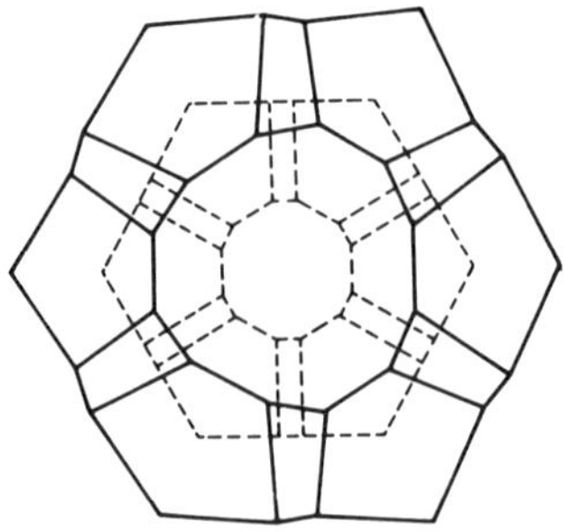

Figure 7.9 Alveolar cluster at internal pressure of 10 cm H_2O, air-filled state. Broken lines show the same cluster at zero pressure

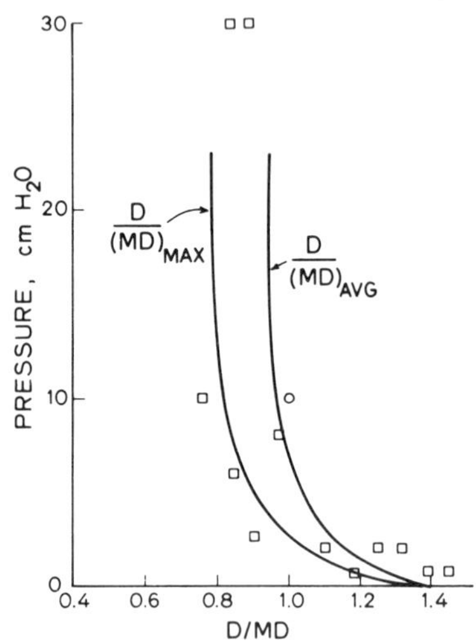

Figure 7.10 The ratio of the depth, D, to the opening diameter, MD, of alveolar sacks as a function of pressure. Continuous lines: present model; squares and circle: experimental data

7.5.3 Stress and strain conditions in the solid tissue

Figure 7.12 shows the values of maximum extensions, λ_1, and membrane forces, T_1, in a typical alveolar wall. Similar curves have been obtained for several cases of irregularities in the alveolar dimensions and emphysemic deterioration of the lung tissue.

7.5.4 Abnormal changes in the tissue

Among various pathological changes in an emphysemic lung, loss of the solid tissue (i.e. alveolar walls) is readily observed in morphological studies. The

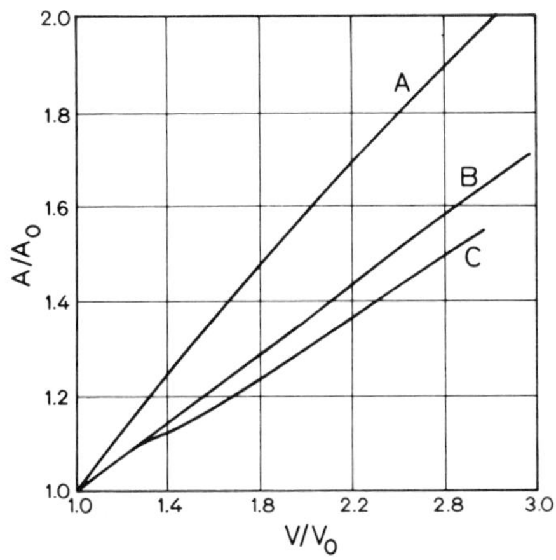

Figure 7.11 Area change vs. volume change curves for: (A) a sphere or cube; (B) the proposed model; (C) the proposed model with several alveolar walls missing

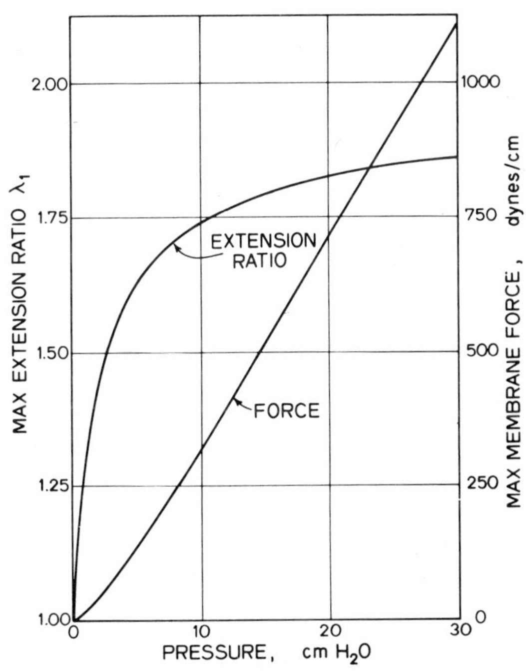

Figure 7.12 The larger principal extension ratio and the corresponding membrane force in a typical alveolar wall as a function of internal pressure

mechanical effects of missing alveolar walls manifest themselves in the change of the volume–pressure characteristics of the parenchyma and in the local strain and stress conditions. The stress concentrations in the vicinity of a structural defect may lead, conceivably, to progressive tearing of the solid tissue. In the present study, an analysis of this effect has been performed by assuming that two external walls in the alveolar cluster in Figure 7.1 are missing. This increase in the membrane forces at locations adjacent to one of the missing walls is of the order of 250 to 300%.

REFERENCES

Frankus, A., and Lee, G. C. (1974) A theory for distortion studies of lung parenchyma based on alveolar membrane properties, *J. Biomechanics*, **7,** 101–107.
Fukaya, H., Martin, C. J., Young, A. C., and Katsura, S. (1968) Mechanical properties of alveolar walls, *J. Appl. Physiol*, **25** (6), 689–695.
Fung, Y. C. (1973) Biorheology of soft tissues, *J. Biorheology*, **10,** 139–155.
Fung, Y. C. (1974) A theory of elasticity of the lung, *J. Applied Mechanics*, **41,** 8–14.
Hoppin, F. G., Lee, G. C., and Dawson, S. V. (1975), Properties of lung parenchyma in distortion, *J. Applied Physiology* **39,** 742–751.
Karakaplan, A. D., Bieniek, M. P., and Skalak, R. (1980) A mathematical model of lung parenchyma, *J. Biomechanical Engrg.*, **102,** 124–136.
Klingele, T. C. and Staub, N. C. (1970) Alveolar shape changes with volume in isolated, air-filled lobes of cat lung, *J. Applied Physiology*, **28,** 411–414.
Mead, J., Takishima, T., and Leith, D. (1970) Stress distribution in lungs: a model of pulmonary elasticity, *J. Applied Physiology*, **28,** 596–608.
Ortega, J. M., and Rheinbolt, W. C. (1970) *Iterative Solution of Nonlinear Equations in Several Variables*, Academic Press.
Powell, W. R. (1971) Experimental and theoretical studies of the lung as an elastic structure, Ph.D Dissertation, Stanford University.
Vawter, D. L. Fung, Y. C., and West, J. B. (1978) Elasticity of excised dog lung parenchyma, *J. Applied Physiology*, **45,** 261–269.
Weibel, E. R. (1963) *Morphometry of the Human Lung*, Academic Press, New York.
Zienkiewicz, O. C. (1971). *The Finite Element Method in Engineering Science*, McGraw-Hill, Maidenhead, Berkshire, England.

Finite Elements in Biomechanics
Edited by R. H. Gallagher, B. R. Simon, P. C. Johnson, and J. F. Gross

Chapter 8

Finite Elements in Stress Analysis and Estimation of Mechanical Properties of the Working Heart

Y. C. Pao

8.1 INTRODUCTION

Why apply the finite element method for studies on Cardiomechanics? Cardiomechanics deals with the motion and functioning of the heart. How the heart performs as a pump depends largely on the contractile activity of its wall muscle, the *myocardium*. The myocardial fibre lengths are stretched and shortened during diastolic and systolic phases, respectively, of a cardiac cycle. The oxygen supplied by the intramyocardial blood flow provides the energy required for the myocardial fibre bundles to contract. Since the flow in a blood vessel is affected by the external pressure acting normally on the vessel and the large intramural arteries and capillaries are known to run radially and circumferentially in the myocardial wall, noninvasive quantitation of the stresses throughout the heart wall is required so that the contractile activity of the myocardium can be adequately assessed.

The finite element method is ideally suited for cardiac analyses. It provides, by partitioning of the heart into elements of simple geometric shapes, several versatile capabilities: (1) spatially, enabling different anisotropic, non-elastic passive properties and contractile forces to be assigned to different myocardial regions for examining the asymmetric nonhomogeneity connected with the regional myocardial perfusion, ischema, and infarction; and (2) temporally, making instant-to-instant changes in the regional passive properties and contractile forces, and use of the updated geometric data enables the time-dependency of the muscle properties, asynchronous fibre contraction, and large deformation of the heart during successive cardiac cycles to be investigated by a quasi-static, incremental-loading, iterative approach.

The objectives of finite element cardiac analyses are to quantitate regional myocardial stresses, to examine regional myocardial mechanical properties for both normal and diseased states, and ultimately to provide useful information in diagnosis and procedure for surgical and medical intervention of patients with heart diseases.

Emphasis of this review is placed on the application of finite element analyses of dynamic changes in stresses, strains, and mechanical properties of the left ventricle of the working heart throughout a complete cardiac cycle. Attention is focused on the left ventricle because it is the main pumping chamber of the heart.

Only the solid mechanics aspects of the ventricular myocardial analyses are selectively delineated. Although the possible relationship between regional myocardial perfusion and stress distribution is discussed herein, no attempt is made to include the fluid mechanics aspects of cardiovascular analyses in this review. Methods for analytical solution of the ventricular myocardial stress problems based on simplified geometry models which reduce the problems to one- or two-dimensional and thereby permit tractable solution in simple mathematical expressions are not covered in this review.

Currently available computer programs specifically designed for finite element analyses of the myocardial stresses and mechanical properties are described. Explanations are given about the underlying theory and geometric simplification used, material properties assumed, and computer software and hardware required in these computer programs. In anticipation of further development of refined computer programs for finite element cardiac analyses, requirements for acquisition of more accurate ventricular and valvular shapes, and their mechanical properties, formulation of new finite element methods, and adaptation of new computer software and hardware are also discussed.

8.2 ANALYSES OF LEFT VENTRICULAR STRESSES AND STRAINS

Woods (1892) applied Laplace's law to compute the mean wall stresses in the left ventricle by approximating it as a thin-walled pressure vessel. Since then many thick-walled models of axisymmetric myocardial shapes have been proposed, as a hollow sphere, circular cylinder, ellipsoid, or spheroid, and all with uniform wall thickness. These geometrically simplified models allow for analytic estimation of the stress variation across the myocardial wall and for consideration of non-homogeneity, anisotropy, myocardial fibre angle, and large deformation of the left ventricle. For these analytical solutions, based on idealized axisymmetric geometry of the left ventricle, readers are referred to the publications by Mirsky (1973), Streeter and Hanna (1973), Mirsky *et al.*, (1974), and Glantz and Kernoff (1975).

As far as the application of finite element method for the left ventricular myocardial stress analysis is concerned, Gould *et al.*, (1972) appear to be the first group of investigators to publish work in this area. They treated the left ventricle as a stack of shell ring elements with varying radius but all with the same uniform wall thickness. The irregular shape of one angiographic left ventricular silhouette was used for estimating the stresses in the left and right walls (Figure 8.1). High-order displacement polynomials were employed in the stress analysis which

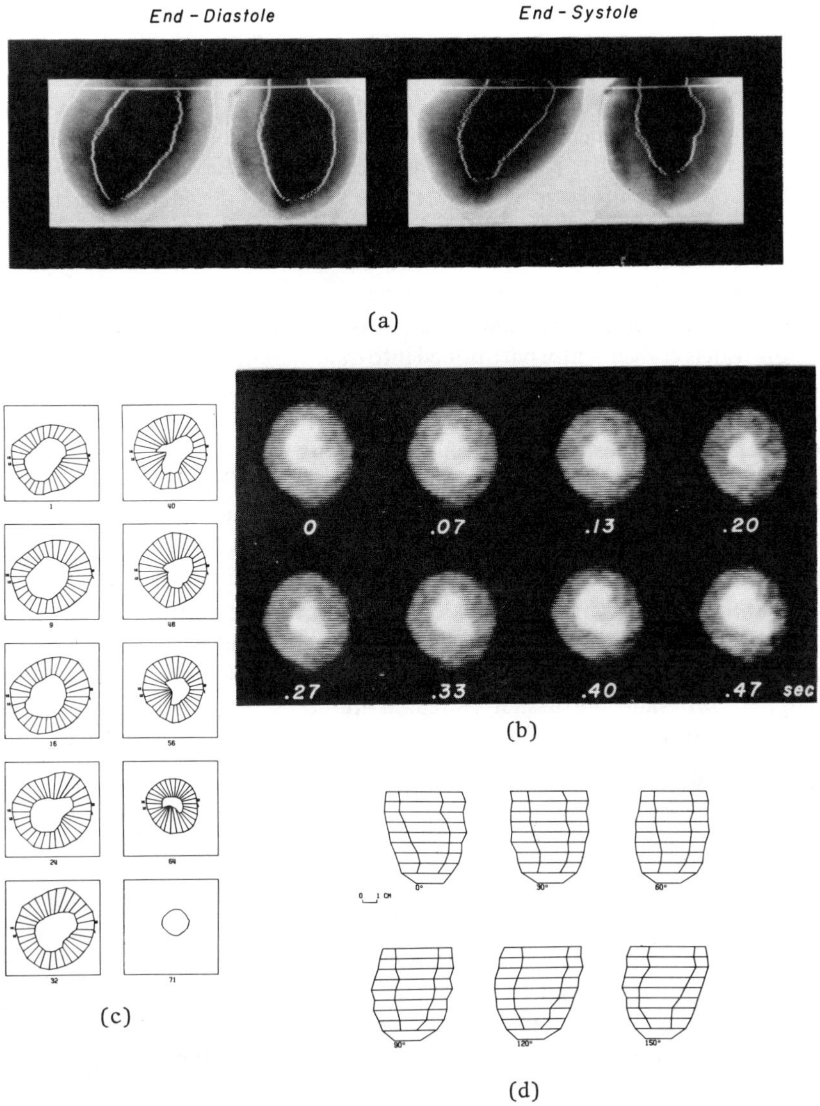

Figure 8.1 (a) Typical biplane silhouettes at two different instants of a cardiac cycle, (b) typical temporal shape changes of a cross-section midway between base and apex of an isolated, working canine left ventricle, (c) typical shape variation from base to apex, and (d) 'reconstructed' silhouettes formed by stacking the cross-sections shown in (c) viewed from different angles. The cross-sections shown in (b) and (c) were obtained by application of the computer-tomographic reconstruction technique

included the effects of bending and transverse shear deformation. Gould *et al.*, claim that their proposed method of analysis provides a more accurate left ventricular stress distribution and a more accurate magnitude of the stress concentration at the apex than those obtained by use of idealized geometry models. In the comparative study, they have assumed the myocardium as a homogeneous, isotropic, elastic, and incompressible medium.

In an internal report, Ross (1972) has described the development of a finite element computer program which idealizes the left ventricular geometry as a truncated shell of revolution generated by confocal ellipses representing the inner and outer outlines of the left ventricular mono-plane silhouette and using them as the generator to rotate about the left ventricular basal-to-apical axis. Wall thickness along the longitudinal direction is appropriately thinner at the apex and the wall is conveniently partitioned into quadrilateral elements assigned with material properties of a two-phase (fibre and binder) composite. Ross used the incremental pressure loadings and included the effects of material and geometric non-linearities in determining the time course of the wall stresses during diastole.

Making direct use of the wall thickness variation along the longitudinal direction of the left ventricular silhouette, Pao, Ritman, and Wood (1974) introduced a computer program FEAXIS (Figure 8.2), which follows the approach of Gould *et al.*, using axisymmetric shells but calculates the wall stresses in the left and right halves of the silhouette by partitioning them into coaxial rings of triangular elements and by adopting the well-established, conventional, axisymmetric finite element analysis (Zienkiewicz, 1977). This computer program can assign increased numbers of nodes across the wall thickness in the regions of the left ventricular wall where particular attention is needed, to permit estimation of the stresses at any point within the wall. It has also enabled the non-linear stress distribution through the wall thickness of the left ventricle to be estimated, a feature not hitherto made available by the other models.

Hamid and Ghista (1974) have attempted the use of 20-node isoparametric elements to calculate the stresses in the left ventricular wall. They approximated the left ventricle as a stack of basal-to-apical cross-sections which were of elliptical shapes with major-to-minor axis length ratios ranging from 0.9 to 1.0 based on the monoplane angiocardiogram data reported by Sandler and Dodge (1968). They obtained a peak wall stress which was three times that predicted by previous idealized geometrical analyses not specified by them.

Janz *et al.*, (1974) have developed a finite element model similar to that of Ross as far as the geometric shape for the left ventricle is concerned. However, the non-homogeneous composition of the ventricular wall is represented by considering the truncated ellipsoid of revolution as consisting of two different homogeneous, isotropic layers. They have applied finite deformation elasticity theory and studied the effects of non-linear geometry on the fibre extension. A difference of as much as 100 % in prediction of the fibre extensions in certain portions of the

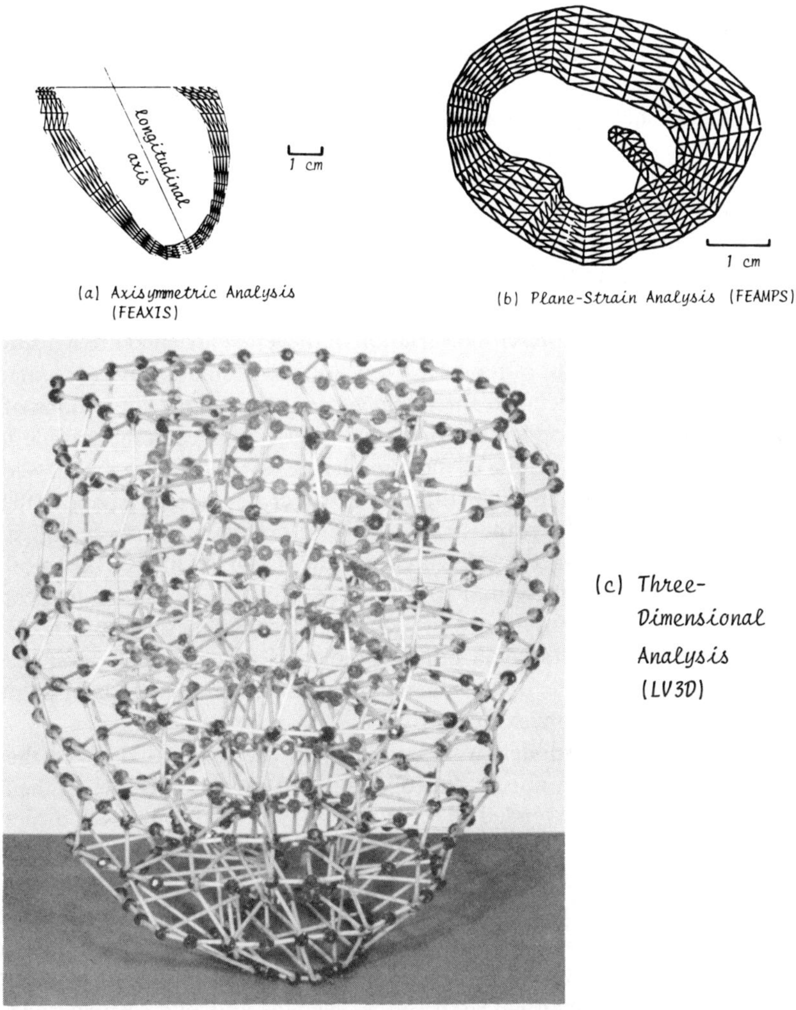

Figure 8.2 Three finite element models, FEAXIS, FEAMPS, and LV3D, developed by Pao, Ritman, and co-workers

ventricular wall was found in application of both the finite deformation elasticity theory and the classical theory of elasticity.

All of the above-mentioned models assume uniform wall thickness in the circumferential direction of the left ventricle, even though the wall thickness may be changed in the meridional direction according to the mono-plane cineangiogram. Neckyfarow and Perlman (1976) have reported the investigation of trigonometrically varying the left ventricular wall thickness in the circumferential direction. They found a better conformity with the observed asymmetric deformation of the ventricle using the variable-thickness model than using the uniform-thickness model. Curved isoparametric elements were used to generate a thick-walled shell but the three-dimensional geometry was based on the data obtained from mono-plane cineangiocardiograms and approximated by suitable selection of assumed circular cross-sections. The elements were considered as linear, homogeneous, incompressible and equivalently orthotropic with the principal material directions conforming to the circumferential and meridional directions of the ventricle. Bergel and Hunter (1979) have also reported the use of 18 tricubic Hermite elements to model the geometry of the left ventricle.

Concerning the circumferential variation of the left ventricular cross-sectional wall thickness, the advent of modern X-ray and ultrasonic scanners has made it possible to obtain the detailed three-dimensional shape and dimensions of the heart (Figure 8.1) by application of the computer-assisted tomographic technique (Wood, 1978). The finite element model FEAMPS (Figure 8.2) developed by Pao, Robb, and Ritman (1976) utilizes the true cross-sectional shape of the left ventricle and applies the plane-strain theory to calculate the stress and strain distributions throughout the cross-section. The major finding of this model is that compressive circumferential stresses develop at the innermost myocardial layers where the wall curvatures are small and sometimes even turn convex inward. Such a stress state can not be predicted by idealized left ventricular models having a cross-sectional inner border either of circular or of elliptical shape that is entirely convex outward.

Up to this time, all models were designed for the stress analysis of the left ventricular myocardium during diastolic phase of cardiac cycle when the fibre bundles are passively extended. Pao, Wang *et al.*, (1976) proposed a method analogous to thermal shrinkage analysis for the contractile study of the myocardium during systolic phase of cardiac cycle when its fibre bundles become active and contract. In the course of assessing the average cross-sectional contractile strength by application of the developed computer program FEAMPS, they have utilized the computer-tomographically reconstructed true left ventricular cross-sectional shapes at systolic instants of a cardiac cycle and incorporated with fibrous analysis. A sample stress distribution throughout the myocardial wall of a left ventricular cross-section obtained by FEAMPS is illustrated in Figure 8.3.

Chen *et al.*, (1978) have also reported finite element contractile myocardial analysis by introducing an activation function and adopting a truncated ellipsoid geometry for the left ventricle. This model is represented by 72 isoparametric elements with two orthotropic layers having distinct myocardial fibre angles of $\pm 45^\circ$. They find that their model has enabled them to perform the simulated study of the side-twisting motion of the left ventricle during systole and obtained results indicating that substantial deformation of the left ventricle during a cardiac cycle is largely due to the active contraction.

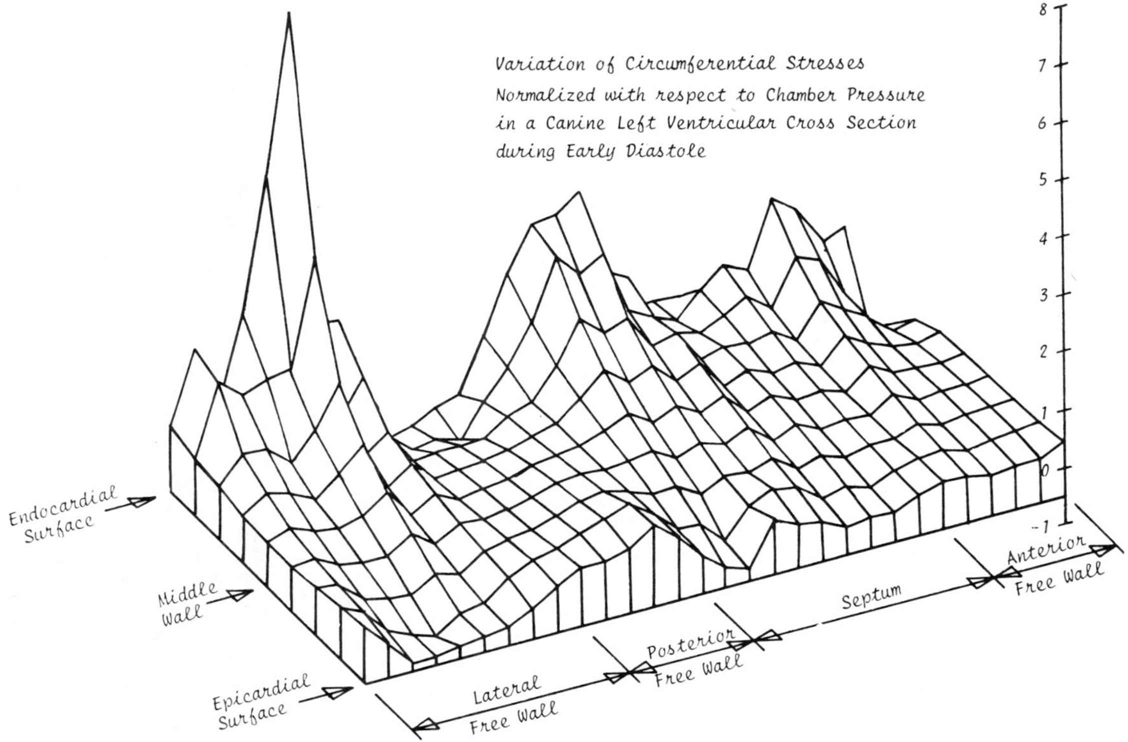

Figure 8.3 Typical stress distribution throughout a left ventricular cross-section calculated by the finite element model FEAMPS

By stacking the computer-tomographically reconstructed basal-to-apical cross-sections of the left ventricle, it is possible to obtain the true shape and dimensions of the left ventricle (Figures 8.1 and 8.2) and similarly of the entire heart. Heethaar *et al.*, (1977) have described the problems encountered in the three-dimensional finite element stress analysis of the true three-dimensional geometry of the left ventricle and of the connected right and left ventricles. More details on the three-dimensional finite element computer program LV3D using tetrahedral elements are given in an internal report by Pao (1978a). At the present time, a major problem to be overcome is the economical feasibility of computer reconstruction of the ventricular shapes and computer solution of the finite element results. This is further addressed in the section on software and hardware (Section 8.7). Figure 8.4 illustrates some typical results of the principal stress contours on the epicardial and endocardial surfaces of an isolated canine left ventricle obtained from a preliminary application of the computer program LV3D.

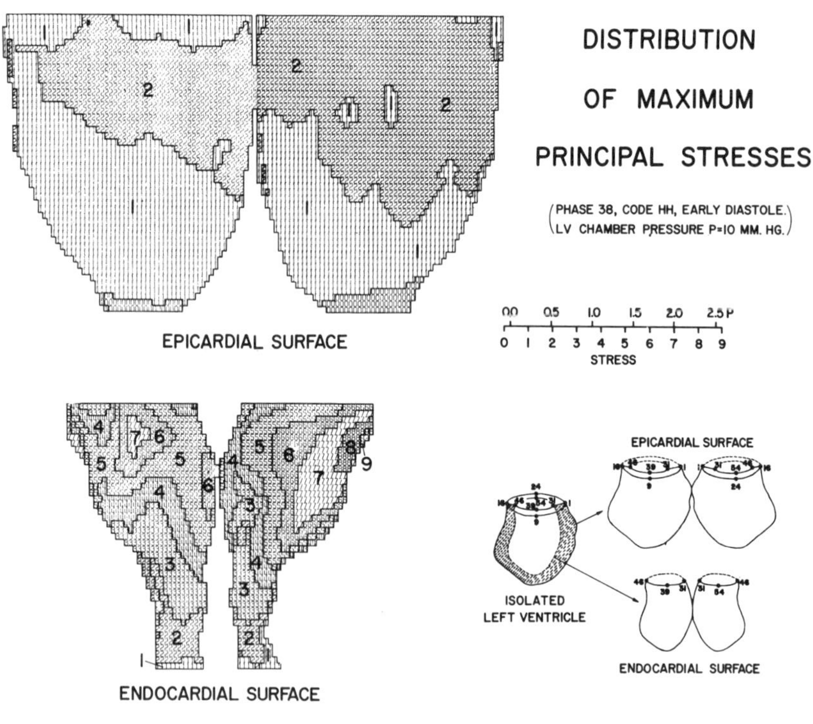

Figure 8.4 Typical distribution of maximum principal stresses on the epicardial and endocardial surfaces of an isolated canine left ventricle calculated by the finite element model LV3D

Only the procedure for viscoelastic finite element analysis of the left ventricular myocardium has been addressed by Pao and Ritman (1977a); actual implementation of quasi-static approach for the dynamic study cannot be carried out until: (1) the temporal geometric change data of the ventricular shape during cardiac cycle become more accessible; (2) the damping coefficient of the myocardium can be reasonably estimated; and (3) the incremental-loading solution by a new-generation computer is economically feasible.

8.3 LOADING AND BOUNDARY CONDITIONS, AND MYOCARDIAL MECHANICAL PROPERTIES

All of the aforementioned finite element models consider the left ventricle as subjected to uniform internal pressure only. To account for large deformation of the left ventricle, use of incremental loading, deformed geometry, and iterative scheme is a common approach adopted by most models. Unless a complete, rather than truncated, shell of rotational symmetry is used as a model, such as the one proposed by Gould *et al.*, (1972), either the equator where the shell has the largest diameter, or the base of the left ventricle where the aortic and mitral valves are assumed to be located, is selected as the reference plane. In addition to restraining the axial displacements, other boundary conditions are often imposed on the in-plane displacements at a selected number of points on this reference plane for the purpose of suppressing rigid-body motion of the left ventricle. Since the deformations of the aortic and mitral valves are neglected in left ventricular analysis, their effects on the myocardial stress results are yet to be investigated.

The earlier finite element models for the left ventricular analysis assume the myocardium to be a homogeneous, isotropic, elastic, and incompressible material, such as in Gould *et al.*, (1972), Pao, Ritman and Wood (1974), Hamid and Ghista (1974), and Pao, Robb, and Ritman (1976). Poisson's ratios of 0.49 and 0.45 were taken to approximate the incompressible material in Pao, Ritman, and Wood (1974) and Hamid and Ghista (1974), respectively. These values were used for avoiding the singularity problem which would arise in defining the elasticity matrix if a value of exactly 0.5 were assigned to be Poisson's ratio of the isotropic material. Regarding the Young's moduli of the assumed isotropic myocardium, rather than assigning a specific value to be used in the finite element model analysis, it is the intention in developing these models to calculate them inversely and to use them as one of the indices for comparing normal and diseased cardiac muscles.

Again, there are many idealized geometry models for estimating the passive stiffness, E, of the left ventricular myocardium (see, e. g., Mirsky *et al.*, 1974). It is generally accepted that the instantaneous stiffness E (tangent modulus) is stress-dependent. The idealized models, such as approximating the left ventricle as a sphere or an ellipsoid, enable the stiffness E to be expressed as a linear function of the mean wall stress. For normal human hearts and those with congestive cardiomyopathy, $E = 246 \times 10^3$ and 1348×10^3 dyne/cm^2, respectively, were calculated. Pao and Ritman (1976) have employed the axisymmetric finite element computer program FEAXIS to iteratively evaluate the global passive stiffness of an isolated canine left ventricle during the diastolic phase of a cardiac cycle. Based on the monoplane roentgenangiographic silhouettes and the associated chamber pressures of the left ventricle recorded at consecutive instants of the entire diastolic phase, they have obtained a pressure-dependent stiffness equation of the form $E = 0.166 - 0.013\mathrm{P} + 0.032P^2$, where E is in 10^4 dyne/cm^2 and P is the left ventricular chamber pressure recorded in mm. Hg. Ray *et al.*, (1976) have reported a similar approach of matching the displacement vectors of the finite element modeal and the cineangiograms of a normal and an infarcted left ventricle. They reported an approximate seven-fold increase in E from 5×10^3 to 35×10^3 dyne/cm^2 in the infarcted region. Most recently, Ghista (1979), in an invited lecture, has presented the upper and lower bounds of the constitutive relative for a normal myocardium as $E = 0.785 + 7.996\sigma$ and $E = 0.109 + 2.116\sigma$, respectively, where σ is the myocardial maximum principal stress and both E and σ are in 10^5 dyne/cm^2.

Utilization of computer-tomographically reconstructed left ventricular cross-sectional shape changes during diastolic phase of cardiac cycle has also enabled the iterative estimation of the average left ventricular cross-sectional passive stiffness (Pao, Ritman *et al.*, 1975). By derivation of an E–P equation through a multi-step diastolic iterative estimation scheme and using this equation for extrapolation of E values at systolic pressures, they performed subsequent iteration for the average cross-sectional contractile strength of the myocardium with the plane-strain finite element program FEAMPS. The simulated study by Pao, Wang *et al.*, (1976) showed that a sustained contraction of shortening the instantaneous myocardial fibre length by approximately 13% is necessary for attaining an increase in the left ventricular chamber pressure from 12 to 120 mm Hg during systolic phase of a cardiac cycle. More details on iteration of myocardial mechanical properties are given in the ensuing sections on computer programs GLOBLE and CROSEC (Sections 8.4 and 8.5).

Incorporation of the variation of myocardial fibre orientation across the left ventricular wall thickness makes possible to consider the left ventricle as layered shell (Ross, 1972; Janz *et al.*, 1974; Pao, Wang *et al.*, 1976; Neckyfarow and Perlman, 1976; Panda and Natarajan, 1977; Pao and Ritman, 1977a and 1978). However, only limited information is available on both the myocardial fibre orientation change and the anisotropic properties of the fibre bundles. Streeter and Hanna (1973) have published data on the myocardial fibre angle changes not only across the left ventricular wall from the endocardial to epicardial surfaces and from base to apex, but also during the diastolic and systolic phases of the cardiac cycle. Numerous stretching tests of cardiac papillary muscle and wall muscle strips (Brady, 1968; Glantz and Kernoff, 1975; Pao *et al.*, 1980) involve uniaxial tests for determination of the fibre's longitudinal stiffness E_1. Yamada

(1970) seems to be the only source that gives an E_1/E_2 ratio of 2, where E_2 is the stiffness in the plane transverse to the fibre direction. Neckyfarow and Perlman (1976) have used $E_2/E_1 = 0.5, 0.8$, and 1.0 to investigate the orthotropic property effect of myocardial fibres on the deformation of the human left ventricle. They observed better agreement in deformation with the cineangio shape when E_2/E_1 was assumed as less than 0.8. Concerning the question of how many layers would be an adequate representation of the true physical variation of the myocardial fibre orientation across the left ventricular wall, Pao, Nagendra *et al.*, (1979a), Pao *et al.*, (1980) have shown that use of three layers of equal thickness and with fibre angles $60^\circ, 0^\circ, -60^\circ$ measured circumferentially is a reasonable good approximation for practical analyses of the left ventricular myocardium. Based on the stretching test of myocardial wall strips excised from the canine left ventricle and on the application of the theory of laminated composites, they have also found that E_1/E_2 ratio does not significantly affect the derivation of the fibre's stiffness equation, in terms of the longitudinal stress, from the strip's stiffness equation.

It is apparent that there is a considerable lack of information on mechanical properties of the intact myocardium. Development of methodologies based on finite element modelling so far has barely covered the static analysis, extension to dynamic analysis is presently deterred by the high costs of: (1) reconstructing the dynamic cardiac geometry at successive cardiac cycles, and (2) computer solution of the three-dimensional cardiac structures. Also, whether the constituent relationships of the myocardium determined from testings of the small isolated heart muscle preparations are readily applicable for analysis of the intact working heart remains as an unresolved controversial issue (Elzinga and Westerhof, 1979). While the scanners and the computer-tomographic technique will continue to be improved for acquiring speedily the accurate cardiac geometry changes during cardiac cycles, and application of this combined method to trace the dynamics of the implanted markers (Ingles *et al.*, 1975; Pao, Nagendra *et al.*, 1979) will enable the measurements of regional strains, how to measure accurately the stresses inside the myocardial wall is still a challenging problem to be resolved.

8.4 COMPUTER PROGRAM 'GLOBLE'

'Globle' is a computer program developed specifically for estimating the global elastic stiffness E of left ventricular wall muscle during the diastolic phase of a cardiac cycle when the muscle is passively stretched. This program requires left ventricular geometric shape and chamber pressure, and an existing finite element computer program for analysis of axisymmetric thick shells, FEAXIS.

Figure 8.5 illustrates the antero-posterior silhouettes of an isolated canine left ventricle recorded at early and end of diastole of a cardiac cycle. Pao, Ritman, and Wood (1974) have reported the development of an axisymmetric finite element computer program FEAXIS for approximate analysis of the dynamic geometric shapes of the left ventricle. The experimental preparation of the

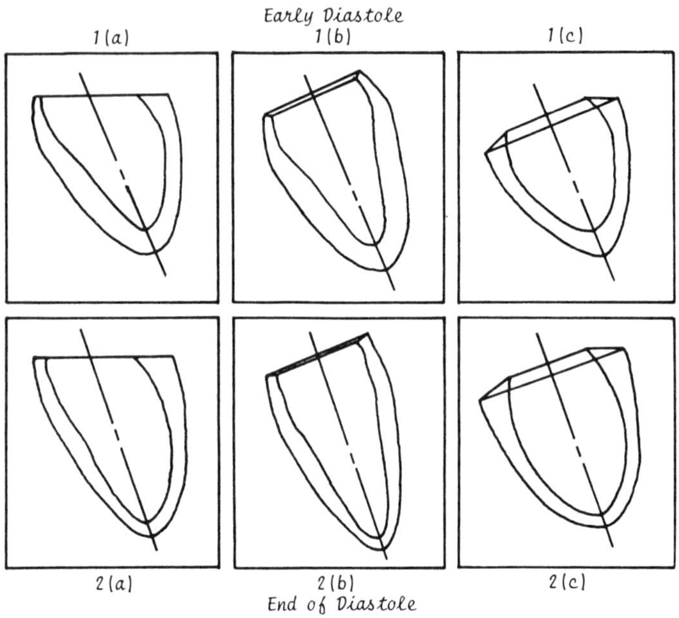

Figure 8.5 Generation of two axisymmetric shells (b) and (c) by rotating the left and right halves, respectively, of the mono-plane angiographic silhouettes of the inner and outer surfaces of the heart muscle about its basal-to-apical axis. The outlines of mono-plane angiographic silhouettes of an isolated canine left ventricle recorded at early diastole and at end of diastole of a cardiac cycle are shown in 1(a) and 2(a), respectively

working isolated canine left ventricle, the biplane video fluoroscopic image intensifier system for videoangiography of the left ventricular geometry during cardiac cycle, and the finite element formulation of the approximated axi-symmetric analysis presented therein, will not be repeated here. Emphasis of this presentation is placed on the programming procedure, structure, and usage.

Basically, FEAXIS is a special-purpose computer program developed for the finite element analysis of axisymmetric thick shells. Triangular elements and linear displacement functions are adopted. The mathematical formulation and notation used follow those in Zienkiewicz (1977). The reasons for not adopting the commercially available general-purpose computer programs are two-fold: (1) for reduction of computer cost, program statements can be economized to those needed for the specific purpose of analysing left ventricular silhouettes, and (2) the in-house developed subroutines are much easier for continuous refinement of incorporating with the newly available analysis techniques and physiological data.

The main program of FEAXIS carries out the following sequential steps:

1. Read in the number of stations, N, along the generator of the shell and the radial, $R(I, J)$, and axial, $Z(I, J)$, coordinates of the nodes on the inner ($J = 1$) and outer ($J = 2$) boundaries, for $I = 1, 2, \ldots, N$.
2. Read in the numbers of nodes across the wall thickness at each station, $NH(I)$ for $I = 1, 2, \ldots, N$.
3. Read in the stiffness E, chamber pressure P, and boundary conditions $NB(I)$ for $I = 1, 2, \ldots, NBT$, where NBT is the total number of restrained displacements.
4. Sequentially call the subroutines ELEMEN, ELASTY, LOAD, ASK, BANDEQ, and NSTRES. Detailed explanation of these subroutines is presented in Tables 8.1 and 8.2.
5. Optional printout and plot arrangements of the elemental nodes, their coordinates, loads, displacements, and stresses, and the structural stiffness matrix are provided for in the main program.

Only details of the programming aspect of FEAXIS are described here. Interested readers are referred to our earlier work (Pao, Ritman, and Wood, 1974) for more information on mathematical formulation, sample applications, and numerical results.

The iterative procedure for calculation of the average stiffness E is schematically described in Figure 8.6 using a typical diastolic phase of a cardiac cycle when the left ventricle is passively expanded by an increase in the chamber pressure from P_1 to P_2. As input, the geometric shape of the left ventricular silhouette at a selected instant in diastole is defined by specifying N_L and N_R stations on the left and right halves of the silhouette, respectively. At each station, the radial and axial coordinates, R and Z, of two points, one on the inner (endocardial) surface and the other on the outer (epicardial) surface, are measured from silhouettes. As shown in Figure 8.5, each half of the silhouette is considered as the generator of a shell of rotational symmetry with respect to the basal-to-apical axis of the left ventricle. FEAXIS proceeds to partition the generators into a desired number of triangular elements, and performs the deformation, strain, and stress analyses.

Also by application of FEAXIS, the beginning and ending diastolic geometric shapes G_1 and G_2 and their respective chamber pressures P_1 and P_2 enable the strain energy per unit E value to be calculated assuming the heart muscle as a homogeneous, isotropic, elastic, and incompressible material. Meanwhile, the external work done on the left ventricle during the diastolic phase can easily be computed using the pressure data and the consequent change in the chamber volume from V_1 to V_2. By equating the strain energy and the external work, an average stiffness value E can be estimated and used as initial guess in the incremental-loading iteration diagrammed in Figure 8.6. For calculation of the

Table 8.1 Subroutines of 'FEAXIS'*

Subroutine name	Description	Input arguments	Output arguments	Call subroutine(s)
ASK	Assembles the structural stiffness matrix SK	X, Y, NN, NODE, NEQ, NBAND, D	SK	EKORDB
BANDEQ	Solves the matrix equation AX = C by Gaussian elimination where A is a 2K − 1 diagonal-band, NXN matrix but stored as a NXK rectangular matrix	A, N, K, C	C (X exits as C.)	none
EKORDB	For KODE = 2, calculates the elemental stiffness EK and the product DB of matrices D and B; for KODE = 1, calculates DB only	XT, YT, D, KODE	DB, EK	MTXMTP
ELASTY	Calculates the elasticity matrix D	E, PR	D	none
ELEMEN	Partitions the generator of axisymmetric shell into triangular elements	R, Z, NH, N	X, Y, NN, NODE, NE	none
ELMNDS	Calculates and accumulates the stresses at the three nodes of IEth element in SNODAL to be averaged in Subroutine NSTRESS	XT, YT, IE, D, NODE, DS, SNODAL	SNODAL	EKORDB, MTXMTP
LOAD	Calculates the nodal forces R1 equivalent to the internal pressure loading	X, Y, NEQ, N, NH, P	R1	none
MTXMTP	Multiplies matrices F(I, J) and G(J, K) to obtain H(I, K)	F, G, I, J, K	H	none
NSTRES	Calculates the average nodal stresses	X, Y, NN, NODE, D, DS, SNODAL	none	none

* See Table 8.2 for the FORTRAN variables which are not explained herein.

Table 8.2 Glossary of FORTRAN variables used in FEAXIS* and FEAMPS

FORTRAN variables	Description	Dimensioned
B	strain–displacement matrix	4×6*, 3×6#
D	elasticity matrix	4×4*, 3×3#
DS	displacement vector	NEQ
E	Young's modulus	non-subscripted
N	number of stations along the generator of shell	non-subscripted
NBAND	semi-band width of SK, equal to 2x NHMAX which is the maximum among all NH's	non-subscripted
NE	total number of triangular elements	non-subscripted
NEQ	$= 2 \times$ NN, total number of equations	non-subscripted
NH(I)	number of nodes across the thickness of shell at station I	N
NN	total number of nodes	non-subscripted
NODE	node number of triangular elements	$3 \times$ NE
P	internal pressure	non-subscripted
PR	Poisson's ratio	non-subscripted
R1	nodal load vector	NEQ
R, Z	radial and axial coordinates of stations on inner and outer boundaries of the shell generator	$N \times 2$
SK	structural stiffness matrix	NEQ $\times$ NBAND
SNODAL	nodal stress vector	$4 \times$ NN*, $3 \times$ NN#
X, Y	coordinates of nodes	NEQ
XT, YT	temporary storages for the coordinates of the three nodes of an element under consideration	3

adjustment of E values to be used for the continuing iterations, the well-known Newton–Raphson method (see for example, James *et al.*, 1977) is employed.

8.5 COMPUTER PROGRAM 'CROSEC'

The advent of modern computer-tomographic (CT) techniques has enabled the reconstruction of the true shape and dimensions of the left ventricular cross-section of working isolated canine hearts by application of an X-ray scanner under a precisely controlled gated laboratory environment. For reconstruction of the cross-sections of a working intact heart, a multi–source and detector scanner is needed for capturing the fast dynamic changes of the cardiac shape. The dynamic spatial reconstructor (DSR) system, reported by Ritman *et al.*, (1978), eliminates the need for 'gated' CT imaging. We have already taken the true shape and dimensions of reconstructed canine left ventricular cross-sections and conducted finite element analyses (Pao, Ritman *et al.*, 1975; Pao, Robb and Ritman, 1976) and simulation studies (Pao, Wang, *et al.*, 1976). Described below is the computer program developed for finite element analysis of myocardium by the plane-strain theory, called FEAMPS.

Except for the two major changes of: (1) utilizing the computer-tomographically reconstructed true cross-sectional shape of the left ventricle, and (2)

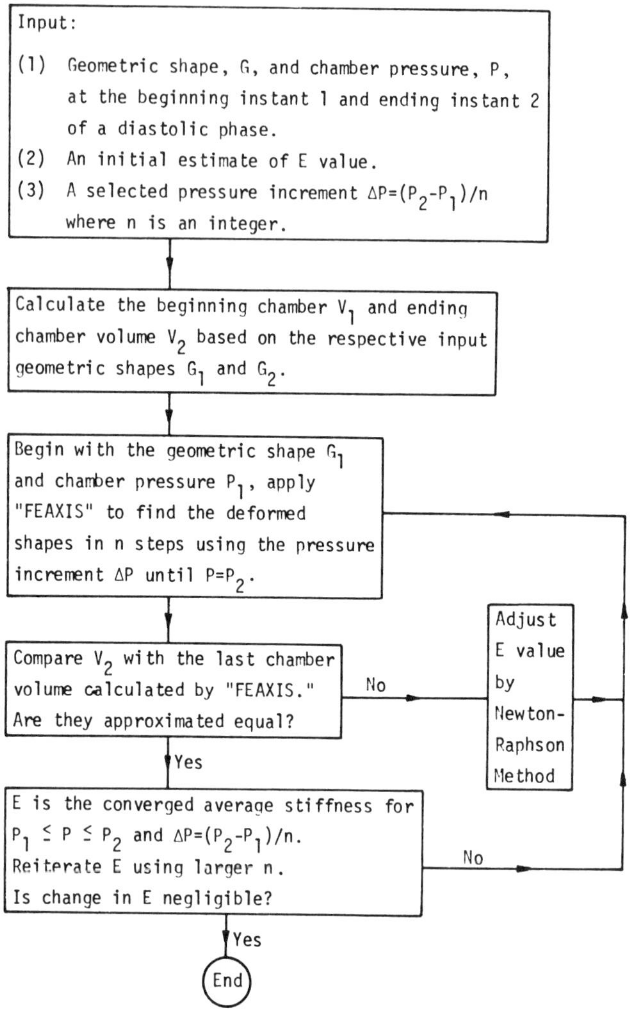

Figure 8.6 Flow chart of 'GLOBLE'

employing plane-strain theory, the procedure of analysis for FEAMPS is analogous to that for FEAXIS delineated in the preceding section. The consequent alterations needed to be made in the development of FEAMPS from FEAXIS are as follows:

1. Use of true cross-sectional shape requires
   (a) N to be redefined as the number of stations around the circumference of the cross-section;

(b) equal number of nodes, NH, to be assigned across the wall thickness at each station so as to facilitate the calculation of the semi-bandwidth NBAND being simply equal to $2 \times (N + 1)$.

2. Application of the plane-strain theory requires
   (a) the strain-displacement matrix B to be redefined in the subroutine EKORDB and dimensioned 3×6;
   (b) the elasticity matrix D to be redefined in the subroutine ELASTY and dimensioned 3×3;
   (c) the nodal stress matrix SNODAL to be dimensioned $3 \times (NN)$;
   (d) the addition of a new subroutine PSTRES designed for computation of principal stresses, circumferential and radial stresses, and the axial stress.

An iterative procedure, analogous to that of GLOBLE, for estimation of the average cross-sectional stiffness E during a diastolic phase of cardiac cycle has been worked out and incorporated into a computer program called CROSEC. This computer program not only estimates the cross-sectional passive elastic stiffness E during diastolic phases but also calculates the active contractile strength C of the heart wall muscle during systolic phases of cardiac cycle.

As illustrated in Figure 8.7, we propose that an E–P equation be obtained by the least-square curve fitting of the E and P data determined from a multi-phase diastolic analysis method and this E–P equation be utilized for extrapolation of E values at higher left ventricular chamber pressure during systolic phases. A 6-instant, 5-phase diastolic analysis is demonstrated as an example in Figure 8.7. Consider a typical systolic phase when the chamber pressure is increased from P_6 to P_7 due to the contraction of the heart wall muscle. Let C be the contractile strength of the muscle fibre, in percent of the muscle length, with a minus sign indicating shortening. The lower panel of Figure 8.7 depicts that a typical systolic phase 6–7e is considered as composed of two steps: (1) passive expansion of the chamber cavity area from A_6 to $A_{7'}$ responding to the increase of the left ventricular chamber pressure from P_6 to $P_{7'}$; and (2) active contraction of the heart muscle causing the cavity area to reduce from $A_{7'}$ to A_{7e}. The contractile strength C is iteratively determined by CROSEC with a procedure similar to that described in Figure 8.6 but with the following changes:

1. Utilize the computer-tomographically reconstructed cross-sectional shape at the beginning instant of a systolic phase, and analyse the cross-sectional deformation by n-step application of FEAMPS with a chamber pressure increase of P from the recorded pressure value at the beginning instant, e.g. P_6 in Figure 8.7.
2. Continually adjust the value of the muscle stiffness E according to the E–P equation derived from diastolic analysis until the chamber pressure is equal to that recorded at the ending instant of the systolic phase under consideration, e.g. P_7 in Figure 8.7.

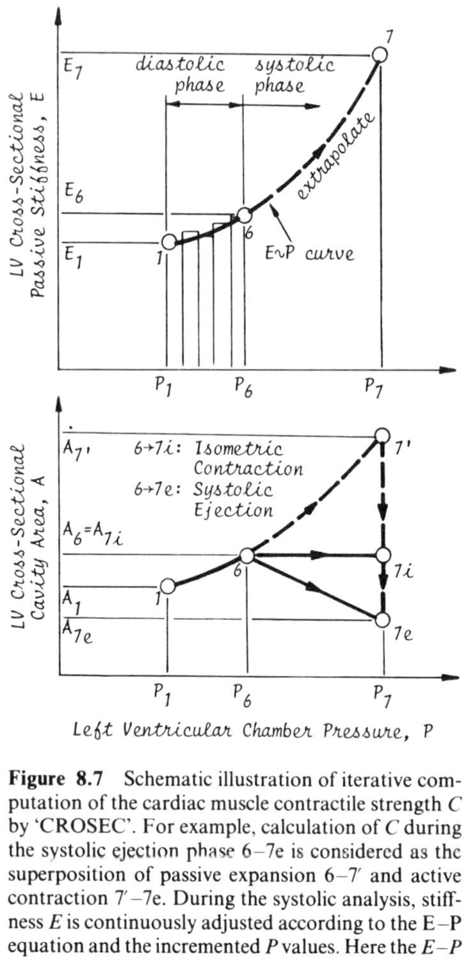

Figure 8.7 Schematic illustration of iterative computation of the cardiac muscle contractile strength C by 'CROSEC'. For example, calculation of C during the systolic ejection phase 6–7e is considered as the superposition of passive expansion 6–7′ and active contraction 7′–7e. During the systolic analysis, stiffness E is continuously adjusted according to the E–P equation and the incremented P values. Here the E–P equation is, for illustration, obtained by a five-phase diastolic analysis with P_1 and P_6 being the early and end diastolic pressure, respectively

3. Modify the elemental load vector by inclusion of the effect of contractile forces determined with the assumed muscle contractile strength C and with a procedure similar to that for computation of thermal shrinkage (see Pao *et al.*, 1976c for detailed mathematical formulation). For estimation of an initial guess of the C value needed to commence the iterative process, a formula of the simple form $C = [(U_p - W)/u_C]^{1/2}$ has been derived from a consideration of the external work, W, and strain energies due to passive expansion, U_p, and due to active contractions per unit C value, u_C. The work and energies are computed by FEAMPS using the geometric and haemodynamic data of the beginning and ending instants of the systolic phase being investigated.

4. Terminate the C iteration when the cross-sectional cavity area calculated by FEAMPS is approximately equal to the recorded counterpart at the ending instant of the systolic phase.

8.6 SAMPLE RESULTS

Use of a computer controlled system for pacing and rotation of a metabolically maintained working isolated canine left ventricle has enabled the video-roentgenographically recording of its dynamic shape changes during cardiac cycles (Pao *et al.*, 1974). An E-P equation of the parabolic form $E = 1660 - 139P + 320P^2$, where E is in dyne/cm^2 and P is in mm Hg, has been obtained by the least-square curve-fit of five pairs of E and P data iteratively determined by GLOBLE. For obtaining the E and P data, the single-plane silhouettes recorded at six consecutive diastolic instants were analysed (Pao and Ritman, 1976a).

Since GLOBLE uses approximated axisymmetric geometry of the left ventricle, we have also investigated the effect of not adopting the true three-dimensional geometry of the left ventricle by employing a urethane plastic cast of the left ventricle. The E values estimated by GLOBLE were compared with the E values of the strips of the same urethane material measured directly from stretching tests. The $\pm 15\%$ variability of GLOBLE-calculated values was found attributable primarily to variation in silhouette shape taken at different angle of views (Pao and Ritman, 1977b).

The true shape and dimensions of the cross-section midway between base and apex of an isolated canine left ventricle at eight instants of a cardiac cycle have also been made available for application of CROSEC by the computer-tomographic reconstruction technique. Details of the geometric shapes and associated haemodynamic data are shown in Figure 8.8. The three diastolic shapes were merely sufficient for the calculation of the stiffness E values for two steps of the diastolic pressure increase. Consequently, an exact fit only by a linear equation of the form $E = 24.9P - 330\,000$, where E and P are both in dyne/cm^2, was possible. This equation was later used for extrapolation of E values during the iterative calculations of the contractile strength C for the five systolic phases of the cardiac cycle. The calculated C values for the five systolic phases are shown in Figure 8.8. In order to assess the applicabilities of CROSEC, it needs to be substantively tested. However, the time-consuming computer-tomographic reconstruction of the left ventricular cross-sections currently impedes the availability of more geometric shapes as inputs for analysis by CROSEC.

8.7 SOFTWARE AND HARDWARE

Application of the finite element analysis technique for cardiac studies leads to the mathematical problem of solving a large system of linear algebraic equations. For example, when 10 basal-to-apical cross-sections are stacked to form the

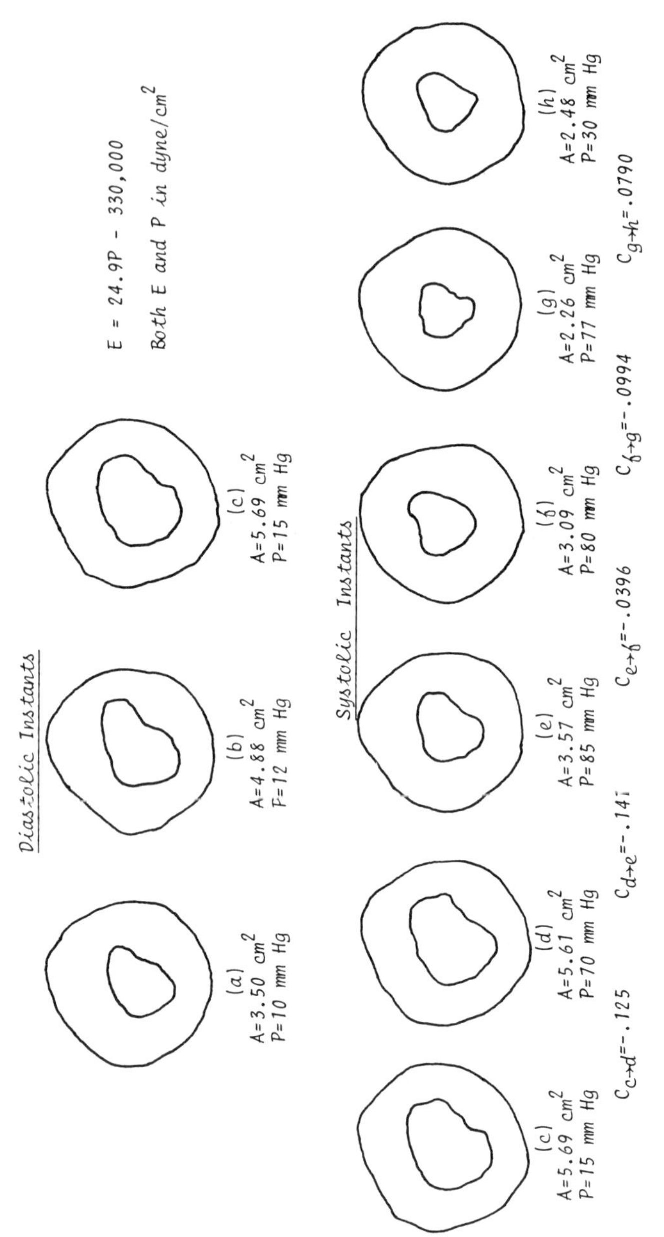

Figure 8.8 Sample results of 'CROSEC' for a 7-phase analysis of the shape changes of the cross-section midway between base and apex of an isolated canine left ventricle

three-dimensional left ventricle and 30 nodes each are selected to approximate the geometry of the endocardial and epicardial borders of the cross-sections plus 2 additional nodes at the apex, the resulting matrix equation contains 1806 unknowns that are the displacements of the 602 nodes being used for partitioning of the left ventricle into 1710 tetrahedral elements (Pao, 1978a). Because the coefficient matrix is symmetric and has non-zero elements located only in a finite number of diagonal bands, to expedite the computer solution and to reduce the computer storage sizes, both require some special techniques and algorithm to be employed.

Schemes for condensing the square coefficient matrix into a rectangular matrix containing only the elements in the right upper-diagonal bands, and formulas for the external work and other computations based on the condensed storage scheme have been reported by Zienkiewicz (1977) and Pao (1975). Further condensation by storing the non-zero elements of the coefficient matrix can be accomplished by deleting the zero elements inside the diagonal bands and rearranging all the non-zero elements to be stored as a vector. As a result of storage saving, Gauss–Seidel iterative solution of the matrix equation needs to be accordingly modified (Pao and Ritman, 1975). The use of a pointer matrix is required in the retrieval, from the compacted vector, of non-zero coefficient elements throughout the iterative solution process.

A scheme of resumable, segmented Gaussian solution of large matrix equations has been developed by Pao (1979) that allows for a large finite element problem to be solved in several segments using a direct-access storage disk. Considerable savings in computer time can be achieved by application of direct-access programming in the backward-substitution step of the Gaussian elimination. In the paper by Pao (1978b), examples have been presented to show how a system with a symmetric coefficient matrix of $2N_b - 1$ diagonal bands can be solved by use of N_b^2 *or* $2N_b$ core storages along with the direct-access programming. Of course, it certainly is most economical and efficient to solve the system using $N_b \times N$, where N is the number of equations, core spaces in a computer. Otherwise, it has been demonstrated that use of N_b^2 core spaces should be attempted to minimize the number of input/output transfers between the computer main storage and disk files. This also manifests the importance of minimizing the semi-band width N_b. For that, Gibbs *et al.*, (1976), and Akhras and Dhatt (1976) have developed algorithms and schemes.

Another popular method of solving symmetric stiffness matrix equations arising from finite element analyses is Cholesky's square-root method (Salvadori and Baron, 1961). In application of this method to finite element cardiac analyses, it has been found that complex FORTRAN programming is sometimes necessary for decomposition of matrices which do not possess the positive definite property (Pao, 1980). This happens when the stiffness matrices are derived from using displacement field functions which are simpler than those required to meet the interelement compatibility conditions.

Comparative studies on the computer cost have also been made by use of the specially developed cardiac computer programs FEAMPS and LV3D, and the commercially available finite element computer programs NASTRAN, NONSAP, and STRESS (see, e.g., Pilkey *et al.*, 1974 for explanation of these programs). Two test cases of: (1) a plane-strain problem, and (2) a three-dimensional problem, were adopted and the results (Pao, Ritman, and Wang, 1975; Pao, Chargin, *et al.*, 1975) show that FEAMPS and LV3D used approximately only 1/10 to 1/3 of the computer time required by the other general-purpose computer programs while achieving the same accuracy in the computed stresses and strains.

The computer program LV3D has also been tested on the IBM 360/370, CDC 35000, and Cray 1 (Russel, 1978) computers for calculation of the stresses and strains in a 602-node, three-dimensional left ventricle. The required computer times of 40–50, 20–25, and 1–1.5 minutes, respectively, for the three types of computers mentioned, indicate that availability of multibatch (CDC 35000) and vector processors (Cray 1) makes it possible to expedite the complex cardiac analyses.

8.8 CLOSING REMARKS

It is evident that future progress will be based on the availability of reliable geometric and physiologic data of the heart and on further development of finite element models for stress analysis and inverse estimation of the mechanical properties of the heart. Simultaneous efforts should be made to: (1) continually refine a three-dimensional heart model; (2) introduce new models of simplified geometry; and (3) develop new models for assessing regional cardiac performance. While the simplified-geometry models may provide easy-to-apply methods for the clinicians to assess *average* cardiac properties, the elaborated three-dimensional model, with capabilities of considering true three-dimensional shape and dimensions of the heart, anisotropy, non-homogeneity, and other complex aspects, will be useful for evaluating the applicabilities and limitations of the results produced by the simplified-geometry models. Use of implanted markers on the surfaces and inside the myocardial wall has enabled the regional dynamics to be studied (Ingles *et al.*, 1975; Pao *et al.*, 1979b) by tracing the changes in the three-dimensional coordinates of these markers during cardiac cycles. Finite element analyses of the cardiac regions bounded by myocardial markers have been conducted and should be further explored. Rather than getting involved with the entire heart, it is economically more feasible to investigate, for example for ischemic and infarction studies, the effects of myocardial anisotropy and non-homogeneity on the regional myocardial mechanical properties at a local level.

In the three-dimensional finite element model analysis of the heart, the loading and boundary conditions need to be more accurately defined. Presently, the chamber pressure is the only load considered during diastole, along with the muscle contraction during systole, and the basal cross-section is assumed to undergo no axial displacement. Efforts should therefore be made to incorporate such variables as valve dynamics and shear loading due to the blood flow in the ventricular chambers.

To extend the inverse method of estimating the myocardial mechanical properties, calculation of the beat-to-beat energy dissipation could be a feasible way for estimating the global and regional damping characteristic of the heart. The measured regional myocardial perfusion and regional stresses and strains computed by a three-dimensional finite element computer program are potential determinants of spatial variation of myocardial damping properties required in the ultimate viscoelastic analysis of myocardium.

Continual incorporation of more advanced features for cardiac analyses into the existing finite element computer programs will inevitably require more and more computer time and capabilities for implementing the solution of a problem. As has been mentioned earlier, use of vector processing devices such as the Cray 1 computer will help alleviate the problem of increasing computer time and cost. But Knight *et al.*, (1976) have pointed out that use of vector processors capable of sustaining computing rates of 10^8 arithmetic results will require improved peripheral hardware and improved operating system to keep such processors supplied with data. They have presented an algorithm specifically designed for solving a banded linear system on vector computers. This and other algorithms, and new commercial finite element computer programs that will become available in the coming years, should all be tested for potential application to cardiovascular studies.

ACKNOWLEDGEMENT

This work was supported in part by research grants HL4664, RR7, and RR7055 from the National Institutes of Health, and by the Engineering Research Center of the University of Nebraska-Lincoln. The author wishes to acknowledge the cooperation and assistance of the Biodynamics Research Unit, Department of Physiology & Biophysics, Mayo Clinic and Mayo Foundation, Rochester, Minnesota for the collaborative works reported herein, particularly to Drs. E. H. Wood and E. L. Ritman.

REFERENCES

Akhras, G., and Dhatt, G. (1976) An automatic node relabelling scheme for minimizing a matrix or network bandwidth, *Int. J. Num. Meth. Engrg.*, **10,** 787–797.

Bergel, D. A., and Hunter, P. J. (1979) The mechanics of the heart, *Quantitative Cardiovascular Studies, Clinical and Research Applications of Engineering Principles* (eds. N. H. C. Hwang, D. R. Gross, and D. J. Patel), University Park Press, Baltimore, Maryland, pp. 151–213.

Brady, A. J. (1968) Active state in cardiac muscle, *Physiol. Rev.*, **48,** 570–600.

Chen, C. J., Falsetti, H. L., Chen, C. H., and Rim, K. (1978) Simulation of an active LV by a two-layer orthotropic model, *Proc. 31st ACEMB.*, **20,** 315.

Elzinga, G., and Westerhof, N. (1979) How to quantify pump function of the heart, *Circ. Res.*, **44,** 303–308.

Ghista, D. N. (1979) Biomechanics of cardiac mechanical performance assessment, *Developments in Mechanics* (eds. S. C. Sinha, P. G. Kirmser, and C. L. D. Huang), Kansas State University Press, Manhattan, Kansas, pp. 299–321.

Gibbs, N. E., Poole, W. G., Jr., and Stockmeyer, P. K. (1976) An algorithm for reducing the bandwidth and profile of a sparse matrix, *SIAM J. Numer. Analysis*, **13,** 235–250.

Glantz, S. A., and Kernoff, R. (1975) Muscle stiffness determined from canine left ventricular pressure curves, *Circ. Res*, **37,** 787–794.

Gould, P. L., Ghista, D. N., Brombolich, L., and Mirsky, I. (1972) *In vivo* stresses in the human left ventricular wall: analysis accounting for the irregular 3-dimensional geometry and comparison with idealized geometry analyses, *J. Biomechanics*, **5,** 521–539.

Hamid, M. S., and Ghista, D. N. (1974) Finite element analysis of human cardiac structures, *Finite Element Methods in Engineering*, The University of New South Wales, pp. 337–348.

Heethaar, R. M., Pao, Y. C., and Ritman, E. L., (1977) Computer aspects of three-dimensional finite-element analysis of stresses and strains in the intact heart, *Computers & Biomed. Res.*, **10,** 271–285.

Ingles, N. B., Daughters, G. T., Stinson, E. B., and Alderman, E. L. (1975) Measurement of midwall myocardial dynamics in intact man by radiography of surgically implanted markers, *Circulation*, **52,** 859–867.

James, M. L., Smith, G. M., and Wolford, J. C. (1977) *Applied Numerical Methods for Digital Computation with FORTRAN and CSMP*, IEP-A Dun-Donnelley Publisher, New York.

Janz, R. F., Kubert, B. R., Moriarty, T. F., and Grimm, A. F. (1974) Deformation of the diastolic left ventricle. II. nonlinear geometric effects, *J. Biomechanics*, **7,** 509–516.

Knight, J. C., Poole, W. G., Jr., and Voigt, R. G. (1976) System balance analysis for vector computers, *Proc. ACM Minneapolis Conf.*, 163–168.

Mirsky, I. (1973) Ventricular and arterial wall stress based on large deformation analysis, *Biophys. J.*, **13,** 1141–1159.

Mirsky, I., Ghista, D. N., and Sandler, H. (eds.) (1974) *Cardiac Mechanics: Physiological, Clinical, and Mathematical Considerations*, Wiley-Interscience, New York.

Neckyfarow, C. W., and Perlman, A. B. (1976) Deformation of the human left ventricle: material and geometric effects, *Proc. 4th New England Bioengrg. Conf.*, 169–172.

Panda, S. C., and Natarajan, R. (1977) Finite element method of stress analysis in the human left ventricular layered wall structure, *Med. & Biol. Engrg. & Computing*, **15,** 67–71.

Pao, Y. C. (1975) On computations involving stiffness matrices stored in rectangular form, *Int. J. Num. Meth. Engrg.*, **9,** 250–252.

Pao, Y. C. (1978a) *Users' Manual of LV3D Computer Program (Three-Dimensional Finite-Element Analysis of the Left Ventricle)*, Tech. Rept. UNL/MAYO 7801, University of Nebraska-Lincoln.

Pao, Y. C. (1978b) Algorithms for direct-access Gaussian solution of structural stiffness matrix equation, *Int. J. Num. Meth. Engrg.*, **12,** 751–764.

Pao, Y. C. (1979) *An Algorithm for Resumablė, Segmented Gaussian Solution of Large Structural Stiffness Matrix Equations*, Tech. Rept. UNL/MAYO-7901.

Pao, Y. C. (1980) On triangular decomposition of nonpositive definite symmetric matrices using complex FORTRAN programming, *Int. J. Num. Meth. Engrg.* (in press).

Pao, Y. C., Chargin, M. K., Ghista, D. N., and Sandler, H. (1975) *Application of MSC/CDC NASTRAN for left ventricular analyses*, Tech. Rept. UNL/NASA-ARC 7501, University of Nebraska-Lincoln.

Pao, Y. C., Nagendra, G. K., Padiyar, R., and Ritman, E. L., (1979a) Analysis of heart wall muscle as a three-layer laminated composite. *Proc. 1979 CANCAM*, **2,** 817–818.

Pao, Y. C., Nagendra, G. K., Padiyar, R., and Ritman, E. L. (1980) Derivation of myocardial fiber stiffness equation based on theory of laminated composite. Submitted to *J. Biomech. Engrg.*

Pao, Y. C., Padiyar, R., and Ritman, E. L. (1979) Regional stresses and strains in a urethane plastic cast of left ventricular myocardium calculated from computer-tomographically recorded displacements of affixed surface markers. *Developments in Mechanics* (ed. S. C. Sinha, P. G. Kirmser, and C. L. D. Huang), Kansas State University Press, Manhattan, Kansas, pp. 219–223.

Pao, Y. C., and Ritman, E. L. (1975) On storage, retrieval, and Gauss–Seidel iteration involving nonzero elements of a sparse stiffness matrix compacted as a vector, *Computers & Biomed. Res.*, **8,** 232–243.

Pao, Y. C., and Ritman, E. L. (1976) An energy method for evaluating the effective global stiffness of the left ventricle in the passive state, *Proc. 4th New England Bioengg. Conf*, 209–212.

Pao, Y. C., and Ritman, E. L. (1977a) Viscoelastic, fibrous, finite-element, dynamic analysis of beating heart, *Proc. Symp. on Application of Computer Meth. in Engrg*, The University of Southern California Press, Los Angeles, pp. 477–486.

Pao, Y. C., and Ritman, E. L. (1977b) Global Young's modulus of urethane plastic cast of left ventricular myocardium calculated from roentgen angiographic silhouette, *J. Biomechs.*, **10,** 133–140.

Pao, Y. C., and Ritman, E. L. (1978) Phase-by-phase determination of the stress–strain relationship of working heart muscle throughout a cardiac cycle, *Advance in Bioengineering* (eds. R. C. Eberhart and A. H. Burstein), ASME Publication H00114, New York, pp. 11–14.

Pao, Y. C., Ritman, E. L. Robb, R. A., and Wood, E. H. (1975) A finite-element method for evaluating cross-sectional Young's modulus of diastolic left ventricle, *Proc. 28th ACEMB Annual Conf.*, **17,** 385.

Pao, Y. C., Ritman, E. L., and Wang, H. C. (1975) Application of NASTRAN for stress analysis of left ventricle of the heart, *Compendium of the 4th NASTRAN Users' Colloquium*, NASA Tech. Memo. TMX–3278, 201–219.

Pao, Y. C., Ritman, E. L., and Wood, E. H. (1974) Finite-element analysis of left ventricular stresses, *J. Biomechs.*, **7,** 305–318.

Pao, Y. C., Robb, R. A., and Ritman, E. L. (1976) Plane-strain finite-element analysis of reconstructed diastolic left ventricular cross section, *Annals Biomed. Engg.*, **4,** 232–249.

Pao, Y. C., Wang, H. C., Ritman, E. L., Robb, R. A., and Wood, E. H. (1976) Computer simulation of the cross-sectional shape change of a beating heart, *Simulation of Systems* (ed. E. Dekker), North-Holland Company, Amsterdam, pp. 609–615.

Pilkey, W., Saczalski, K., and Schaeffer, H. (1974) *Structural Mechanics Computer Programs, Surveys, Assessment, and Availability*, University Press of Virginia, Charlottesville.

Ray, G., Chandran, K. B., Nikravesh, E. P., Ghista, D. N., and Sandler, H. (1976) Estimation of the local modulus of the normal and infarcted left ventricle from angiographic data, *Proc. 4th New England Bioengrg. Conf.*, 173–176.

Ritman, E. L., Robb, R. A., Johnson, S. A., Chevalier, P. A., Gilbert, B. K., Greenleaf, J. F., Sturm, R. E., and Wood, E. H. (1978) Quantitive imaging of the structure and function of the heart, lungs, and circulation, *Mayo Clinic Proc.*, **53,** 3–11.

Ross, A. L. (1972) *A Finite Element Computer Program for the Nonlinear Structural Analysis of the Heart*, Tech. Rept. 72SD213, General Electric Company, Philadelphia, Penn., 29 pages.

Russel, R. M. (1978) The Cray-1 computer system, *Comm. ACM*, **21,** 63–72.

Salvadori, M. C., and Baron, M. L. (1961) *Numerical Methods in Engineering*, Prentice-Hall, Englewood Cliffs, New Jersey.

Sandler, H., and Dodge, H. T. (1968) The use of single plane angiocardiogram for the calculation of left ventricular volume in man, *Am. Heart J.*, **75,** 325–334.

Streeter, D. D., Jr., and Hanna, W. T. (1973) Engineering mechanics for successive state in canine left ventricular myocardium. I. Cavity and wall geometry. II. Fiber angle and sarcomere length, *Circ. Res.*, **33,** 639–664.

Wood, E. H. (1978) Evolution of instrumentation and techniques for the study of cardiovascular dynamics from the thirties to 1980 (Alza lecture, April 10, 1978), *Annals Biomed. Eng.*, **6,** 250–309.

Woods, R. N. (1892) A few applications of a physical theorem to membranes in the human body in a state of tension, *J. Anat. Physiol.*, **26,** 362–370.

Yamada, H. (1970) *Strength of Biological Materials*, Williams & Wilkins, Baltimore.

Zienkiewicz, O. C. (1977) *The Finite Element Method in Engineering Science*, McGraw-Hill, New York.

Finite Elements in Biomechanics
Edited by R. H. Gallagher, B. R. Simon, P. C. Johnson, and J. F. Gross

Chapter 9

A Non-linear Finite Element Stress Analysis of Bioprosthetic Heart Valves

G. W. Christie and I. C. Medland

9.1 INTRODUCTION

When one of the natural valves of the heart begins to display a significant degree of dysfunction, the cardiac surgeon will generally replace it with one of a large range of commercially available prosthetic devices. Here we present an analysis of the static mechanics of one class of such devices, that is *bioprosthetic valves* modelled on the basic anatomy of the natural trileaflet *aortic valve*. The term 'bioprosthetic', while not universally accepted, is widely used in the medical literature to distinguish between the purely prosthetic valves such as 'ball-in-cage' or 'tilting disc' devices, and the wide assortment of trileaflet designs in which the leaflet material is of biological origin.

Fundamentally, the bioprosthetic valve consists of three approximately equal tissue leaflets sutured to a tricornute frame called a *stent*. The stent itself may be rigid or slightly flexible (Reis *et al.*, 1971; Thomson and Barratt-Boyes, 1977) and always has a tight cloth covering, usually of Dacron, to provide a firm substrate for suturation as well as protecting the stent material from direct contact with the blood with its attendent risks of thromboembolism. A description of the assembly of a Hancock porcine xenograft bioprosthesis may be found in Zuhdi *et al.* (1974) and further details are given by Cohn and Collins (1979). The leaflet material most commonly encountered consists of an entire porcine aortic valve preserved in a solution of gluteraldehyde although the use of antibiotic-treated human aortic valves (allografts) and various non-differentiated connective tissues such as bovine pericardium is also common.

The tissue of natural aortic valves displays a significant degree of elastic anisotropy with the fibres of the collagen network preferentially aligned in the circumferential rather than radial directions within each leaflet (Clark and Finke, 1974; Broom, 1978). Whereas human tissues only require to be treated with antibiotic solutions having a negligible effect on its elastic properties, all the non-human tissues are treated with a gluteraldehyde solution which acts as preservative, an antibiotic, and greatly reduces the antigenicity of the preserved tissue. However, treatment with gluteraldehyde has a profound effect on the compliance of the tissue, the degree of which depends on the state of stress the tissue was in at the time fixation took place (Broom and Thomson, 1979).

The fluid mechanics of heart valves has been studied by, amongst others, Bellhouse and Talbot (1969), Lee and Talbot (1979), Peskin (1972), van Streenhoven and van Dongen (1979), and Swanson and Clark (1973). However, for the purposes of this study, only the mechanics of a *statically* loaded valve will be considered, and hence the position of the valve within the heart (aortic, mitral, pulmonary, or tricuspid) is of no consequence except in so far as the magnitude of the total pressure will be different in each case. For this reason the total pressure load applied during all calculations reported in this work was taken to be 16 kPa, which is approximately the normal pressure for the mitral position and the maximum pressure the bioprosthesis is likely to sustain in a normal individual. For hypertensive recipients, however, the applied pressure could exceed this value by more than 50 %.

With natural valves, as wear and damage take place in the tissue the damaged components are continually replaced to maintain the full structural integrity of the material. With the non-viable tissue of the bioprosthetic valves, this process cannot take place and a process of steady degradation of the tissue structure starting from the time of implantation has been observed to occur (Ferrans, Spray, Billingham, and Roberts, 1978). Although no experimental studies have been undertaken as yet, it seems to be most likely that the rate of structural degradation or *fatigue* of the tissue components is closely related to the stress the tissue is required to sustain. Biochemical processes are also an important factor but it must be recognized that any biochemical process which acts to weaken the tissue structure will tend to make it even more susceptible to damage by mechanical forces. It must therefore be considered advantageous for the recipient to receive a bioprosthesis which is so designed that the stress in the tissue components of the valve is as low as possible. Before this state can be achieved a detailed knowledge of the mechanics, both static and dynamic, of these structures is essential. Certainly, the nature of the failure of the bioprosthesis is such that the progress of its deterioration can be traced clinically and a suitable moment chosen for the implantation of a new device. In this respect the bioprosthesis offers considerable advantages over the catastrophic and generally fatal mode of failure of the prosthetic valves. Nevertheless, the risk of an early, post-operative death is significantly increased on reoperation (Rossiter *et al.*, 1979) so the effective, long-term survival rate of the recipient should be significantly enhanced by reducing the stress levels in the first implant.

Binet and associates (1965) were the first to implant a bioprosthetic valve in a human and since that time considerable advances and diversification in design has taken place. The first porcine xenografts in the late 1960s were fixed in formaldehyde with highly unfavourable results, most failing with leaflet perforation in less than one year. Gluteraldehyde was found to give much more satisfactory long-term results. Numerous long-term follow-up studies of porcine xenografts (Stinson, Griepp, Oyer, and Shumway, 1977; Carpentier *et al.*, 1976), heterologous tissue valves (Ionescu *et al.*, 1974), and allografts (Roche, Barratt-Boyes, Heng and Rutherford, 1977) have shown that the expected lifetimes for these valves appears to be in the range of five to ten years. Results from accelerated cycling experiments have been reported by Clark *et al.*, (1978) for whole valves and by Broom (1977, 1980) for uniaxial stretching of fresh and treated heart valve tissue. Such tests provide information on the mechanism of the fatigue process but unreliable estimates of valvular longevity *in vivo*.

To the knowledge of the authors, no theoretical work has previously been reported on bioprosthetic valves. Analyses of stress in unstented, natural human (Cataloglu, Clark, and Gould, 1977) and natural porcine (Chong and Missirilis, 1978) aortic valves have been presented, but both differ substantially from the method adopted for this analysis. The approach used for this work has been to develop a 'model' for the mechanical behaviour of the heart valve, based on anatomical observations of the structure of the leaflet material. The first model treated the leaflets as a thin isotropic membrane such as would be the case for many heterologous tissue valves. The second model refined this by allowing the tissue to have anisotropic elastic properties as observed experimentally in natural, differentiated aortic valve tissue. This was achieved by embedding hyperelastic fibres into the isotropic membrane and aligning them circumferentially in accordance with the data given by Clark and Finke (1974). Since some of the parameters for the models were either not available or not reliably known, numerical 'experiments' were conducted using the finite element program holding all parameters constant except one. By varying the remaining free parameter over assumed physiological limits, its effect on the mechanics of the whole valve could be estimated.

In what follows, the foundation of the theory of thin membrane structures is outlined, together with a brief discussion of the choice of strain-energy function. The formulation of the finite element equilibrium equations is given and the boundary conditions used are justified. Results are presented for both the isotropic and anisotropic models which show the anisotropic properties of natural aortic valve tissue substantially improve the effectiveness of the seal formed between two adjacent leaflets.

9.2 DEFINITIONS OF STRESS AND STRAIN TENSORS

The leaflets of aortic heart valves exhibit both large displacements and finite strains when subjected to hydrostatic pressure loading. As observed by Swanson and Clark (1973), bending effects are probably negligible compared with the in-plane forces so the assumption of purely membrane behaviour provides a good mathematical description. In addition, we assume the existence of a strain-energy density function, W, which characterizes the elastic properties of the tissue and gives the strain energy per unit of the undeformed volume. Such materials have been termed *hyperelastic*. A detailed account of a general theory for the finite deformation of hyperelastic membranes has been given by Green and Adkins (1960), and it is on their work that the current formulation is based.

In what follows the usual tensor notation has been used with summation assumed over repeated indices unless otherwise stated; Greek indices take the values 1, 2 while Latin indices take the values 1, 2, 3 with upper- and lower-case symbols representing quantities in the deformed and undeformed states respectively. The reference state is taken to be the undeformed state, so the Green-St. Venant tensor, γ_{ij}, provides a consistent definition of strain and the corresponding symmetric stress tensor, τ^{ij}, is called the second Piola–Kirchhoff tensor.

A system of convected coordinates, θ, may be defined such that θ_α lie in the middle surface of the membrane and θ_3 is directed normal to the surface. The covariant metric tensors associated with the deformed and undeformed states are G_{ij} and g_{ij}, while those associated with the respective middle surfaces are $A_{\alpha\beta}$ and $a_{\alpha\beta}$. Green and Adkins (1960) have shown that the Green-St. Venant strain tensor may be defined as

$$\gamma_{\alpha\beta} = \frac{1}{2}(A_{\alpha\beta} - a_{\alpha\beta}), \qquad \gamma_{33} = \frac{1}{2}(\lambda^2 - 1) \tag{9.1}$$

where λ is the extension ratio normal to the middle surface. The remaining terms, $\gamma_{\alpha 3}$, are zero because θ_3 is normal to the middle surface and hence $G_{\alpha 3} = g_{\alpha 3} = 0$. The strain tensor, γ_{ij}, is symmetric so there exist three strain invariants which, on the middle surface, take the form

$$I_1 = \lambda^2 + (A/a)\, a_{\alpha\beta} A^{\alpha\beta} \tag{9.2}$$

$$I_2 = \lambda^2 (I_1 - \lambda^2) + A/a \tag{9.3}$$

$$I_3 = \lambda^2 A/a \tag{9.4}$$

where

$$A^{\alpha\mu} A_{\mu\beta} = \delta^\alpha_\beta, \quad \alpha^{\alpha\mu}\alpha_{\mu\beta} = \delta^\alpha_\beta \tag{9.5}$$

and

$$A = \det(A_{\alpha\beta}), \quad a = \det(a_{\alpha\beta}) \tag{9.6}$$

The second Piola–Kirchhoff stress tensor, τ^{ij}, is related to the tensor γ_{ij} for incompressible materials (i.e. where $I_3 = 1$) by means of the expression

$$\tau^{ij} = \frac{1}{2}\left(\frac{\partial W}{\partial \gamma_{ji}} + \frac{\partial W}{\partial \gamma_{ij}}\right) + pG^{ij} \tag{9.7}$$

where p is a scalar function required to ensure satisfaction of the incompressibility condition. If the material can be assumed to be isotropic then the strain energy function takes the form

$$W = W(I_1, I_2) \tag{9.8}$$

which on substitution into Equation (9.7) leads to

$$\tau^{\alpha\beta} = (\phi + \lambda^2 \psi) a^{\alpha\beta} + (p + \psi/\lambda^2) A^{\alpha\beta} \tag{9.9}$$

$$\tau^{33} = \lambda^2 \phi + \lambda^2 (I_1 - \lambda^2) \psi + p \tag{9.10}$$

$$\tau^{\alpha 3} = 0 \tag{9.11}$$

where

$$\phi = 2\frac{\partial W}{\partial I_1} \quad \text{and} \quad \psi = 2\frac{\partial W}{\partial I_2} \tag{9.12}$$

Green and Adkins have shown that $\lim_{h \to 0} \tau^{\alpha 3} = 0$ and $\lim_{h \to 0} \tau^{33} = O(P)$ where P is the applied pressure load. As $P \ll \tau^{\alpha\beta}$ for thin membranes, it is customary to set $\tau^{33} = 0$ and to use this condition with Equation (9.10) to eliminate the unknown scalar function, p, from Equation (9.9). This gives

$$\tau^{\alpha\beta} = (\phi + \lambda^2 \psi) a^{\alpha\beta} - [\phi + (I_1 - \lambda^2 - 1/\lambda^2)\psi] A^{\alpha\beta} \tag{9.13}$$

with $\tau^{\alpha 3} = \tau^{33} = 0$. The extension ratio, λ, is determined from Equation (9.4) since, by virtue of the assumption of incompressibility, $I_3 = 1$.

The definition of the strain tensor given in Equation (9.1) could alternatively have been written explicitly in terms of the displacement vector, **u**. For the special case of cartesian coordinates, the positions of the same material point in the deformed and undeformed states are related by the expression

$$X_i = x_i + u_i \tag{9.14}$$

which in turn leads to

$$\gamma_{\alpha\beta} = \frac{1}{2}(u_{\alpha,\beta} + u_{\beta,\alpha} + u_{k,\alpha} u_{k,\beta}), \tag{9.15}$$

for the components of the strain tensor on the middle surface of the membrane. The comma is used to denote partial differentiation with respect to the indicated cartesian coordinate. The more general strain–displacement relations for convected coordinates may be found in Green and Zerna (1954).

9.3 THE STRAIN-ENERGY FUNCTION

Equation (9.13) gives an expression for the second Piola–Kirchhoff stress tensor components on the membrane middle surface applicable to incompressible, isotropic, hyperelastic materials. To complete the characterization of the constitutive law for the leaflet material, an expression for the strain-energy density of a soft, biological tissue in terms of strain invariants is required. Over the past ten years numerous forms have been suggested which are consistent with the results of uniaxial testing experiments, all being based on the work of Fung (1967) (e.g. Blatz, Chu, and Wayland, 1969; Veronda and Westmann, 1970; Gou, 1970; Snyder, 1972; Vito, 1973; Demiary and Vito, 1976). However, no general biaxial experimental data appears to be currently available with which to compare the consistency of these proposed strain-energy density functions under general two-dimensional deformation. A practical choice would therefore be the simplest form of W which is not inconsistent with available experimental data. Thus the form suggested by Blatz, Chu, and Wayland (1969), a special case of the forms used by Veronda and Westmann (1970) and Vito (1973), has been adopted. The same form has been used by Hunter (1975) for the passive behaviour of myocardium and may be expressed as

$$W(I_1) = \frac{c_2}{2c_1}\left\{\exp[c_1(I_1-3)]-1\right\} \tag{9.16}$$

where c_1 and c_2 are elastic parameters, c_1 being dimensionless and c_2 having dimensions of stress. The substitution of Equation (9.16) into Equations (9.12) and (9.13) gives

$$\tau^{\alpha\beta} = \{c_2 \exp[c_1(I_1-3)]\}(a^{\alpha\beta}-A^{\alpha\beta}), \qquad \tau^{\alpha 3} = \tau^{33} = 0 \tag{9.17}$$

with representative values for the elastic parameters being obtained from available experimental data.

Comparison with the data of Broom (1978) yields the approximate values of $c_1 = 57$ and $c_2 = 0.775\,\text{kPa}$ for fresh, untreated, porcine pulmonary valve tissue. The principal advantage of this method of specifying the constitutive law of soft bio-materials is that the number of elastic parameters is kept to a minimum while at the same time allowing a ready comparison with experimental results.

Even though the strain-energy density function used has only two free parameters, very good fits are obtained to one dimensional experimental results. Figure 9.1 shows the solution for the stretching of a narrow strip of tissue in simple elongation for a range of values for c_1 and c_2. It may be observed that c_1 primarily determines the post-transition modulus while c_2 primarily determines the pre-transition modulus, so by varying these two parameters a wide range of isotropic tissue types can be represented. Anisotropic properties are treated by fibre-reinforcing the isotropic membrane in the desired direction in preference to using an anisotropic form of the strain-energy function because such a composite structure is anatomically more realistic.

9.4 FINITE ELEMENT FORMULATION

A discussion of the formulation and solution of highly non-linear membrane problems using simplex finite elements may be found in Oden (1972). The

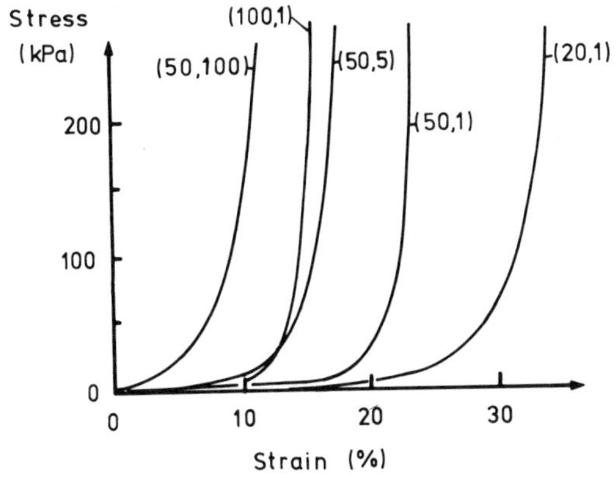

Figure 9.1 Stress–strain curves as a function of the parameters (c_1, c_2) where c_1 is dimensionless and c_2 is in kPa

elements used in this study were a three-noded constant-strain triangle and a two-noded constant-strain fibre element, both conforming and having C^0 continuity. Because the undeformed state of the membrane element is that of a flat triangle and the formulation is Lagrangian, the metric tensor $a_{\alpha\beta}$ becomes $\delta_{\alpha\beta}$. Upper-case subscripts take the range from 1 to the number of nodes per element.

The displacement field, $u_i(\mathbf{x})$ may be discretized in terms of the modal displacement values, u_k^N and *linear* basis functions

$$\psi_N(\mathbf{x}) = a_N + b_{Nj} x_j$$

to give

$$u_i(\mathbf{x}) = u_i^N \psi_N(\mathbf{x}) \tag{9.18}$$

Substitution of Equation (9.18) into (9.15) yields the discrete form of the strain tensor on the middle surface of the element,

$$\gamma_{\alpha\beta} = \frac{1}{2}(\psi_{N,\alpha} u_\beta^N + \psi_{N,\beta} u_\alpha^N + \psi_{N,\alpha} u_k^N \psi_{M,\beta} u_k^M) \tag{9.19}$$

The highly non-linear force–displacement relations for this element may be obtained from a variational approach, the precise details being given by Oden (1972), and can be expressed as

$$f_k^N = \int_{v_0^{(e)}} \tau^{\alpha\beta} \psi_{N,\alpha} (\delta_{\beta k} + \psi_{M,\beta} u_k^M) \, dv_0^{(e)} \tag{9.20}$$

where f_i^N are the generalized nodal force components, $v_0^{(e)}$ is the undeformed volume of the element e, and δ_{ij} is the Kronecker delta.

These generalized nodal force components must be in equilibrium with the forces exerted on the element by the hydrostatic pressure of the blood which will be directed normal to the surface of the *deformed* element. If P is the applied static pressure, X_k^N is the global kth cartesian coordinate of the Nth node of the membrane element, and ε^{ijk} is the usual third-order permutation symbol, then the nodal force components due to P, transformed to the plane of the undeformed element, are

$$p_k^N = \frac{1}{6} P \varepsilon^{ijk} (X_i^2 - X_i^1)(X_j^3 - X_j^1) \tag{9.21}$$

provided X_k^1 coincides with the origin of the local elemental coordinate system. For the element to be in equilibrium we require that the internal and external force components be equal, hence $p_k^N = f_k^N$. No external loads are applied directly to the fibre elements. The force–displacement relations for the fibre element are identical to those of the membrane element except that the index, N, takes the values of 1, 2 instead of 1, 2, 3.

The non-linear Equations (9.20) and (9.21) may be assembled, element by element, into a global system of equations and solved for the unknown nodal displacements by any one of a variety of well-known methods. The modified Newton–Raphson method was used with the derivatives in the Jacobian matrix being calculated numerically. The extension ratio, λ, directed normal to the middle plane of the membrane element was obtained from the requirement that the material be incompressible.

9.5 BOUNDARY CONDITIONS

To obtain the correct solutions for the leaflet stress field it is essential to correctly specify the boundary conditions. This is especially true in the zone of inter-leaflet coaptation where a contact boundary condition must be applied, the area in contact depending on a number of the problem parameters. A schematic representation of the geometry of the bioprosthesis as viewed from directly above the outflow side of the valve and from a more oblique view is shown in Figures 9.2a and 9.2b. The leaflet is firmly attached to the stent along the line ADC so at all nodes on this boundary the three displacement components in the right-handed coordinate system, (u_r, u_ϕ, u_z), were set to zero. The assumption has been made that the cloth and the sutures are rigid and while this may not be exactly the case the deformation of these components can reasonably be expected to be negligible compared with that of the soft tissue components.

The free (or coapting) edge ABC is not quite so simple. In the fully relaxed state for the leaflet, the free edge is not necessarily in contact with its neighbouring leaflets. Under the applied pressure loading, however, the leaflets deform considerably and expand until the free edge comes into contact with the corresponding edges of the other two leaflets. It is not known in advance for a given problem how much of the leaflet will coapt; in fact this is one of the

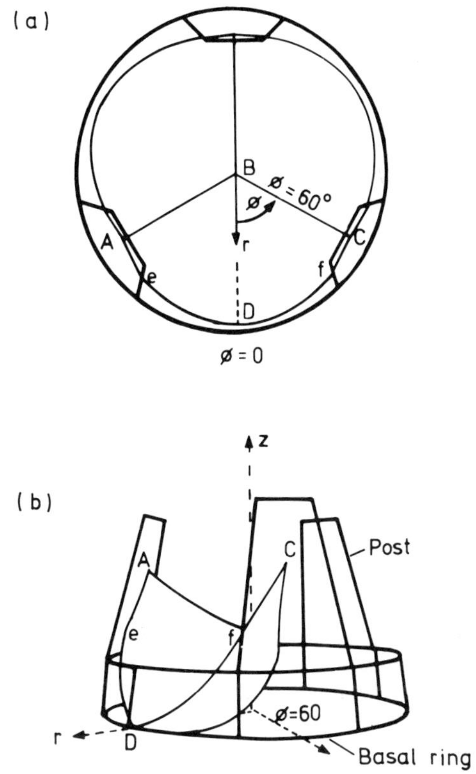

Figure 9.2 Geometry of the bioprosthesis from two perspectives showing a single leaflet mounted on the stent

quantities which the model should predict. Consequently, it has been assumed that no node may move past the (r, z) planes at which $\phi = \pm 60°$. No restriction is placed on the node to move in that (r, z) plane so that the contact is taken to be frictionless and hence the contact forces are directed normal to the plane of the element. At each incrementation of the global displacements a check is made on the satisfaction of this boundary condition and the u_ϕ component restricted if necessary. For nodes already in contact with the boundary, a check is made on the direction of the contact forces, i.e., if they are tensile the corresponding u_ϕ component is released and if they are compressive, the node remains in contact. This ensures that the contact is not 'adhesive'.

In natural aortic valves, the three leaflets are usually slightly different sizes and seldom exactly symmetrical about the line BD. Nevertheless, to a good level of approximation, it may be assumed that the lack of symmetry does not significantly affect the mechanics of statically loaded valves. Thus along the line

BD, u_ϕ is set to zero, thereby modelling only half of one of the three leaflets. The pressure loading is applied to the outflow surface of the leaflet while the inflow side is free of any applied surface tractions.

9.6 INITIAL CONDITIONS AND SOLUTION PROCEDURE

9.6.1 The undeformed leaflet shape

There is a wide variation in the anatomy of natural aortic valves, as has been noted by Angell, Angell, and Kosek (1979), so it is not possible to establish any particular shape for the unloaded leaflet as being anything more than approximately representative of actual leaflet geometry. This is especially true when considering bioprosthetic valves because of deformations inevitably introduced during the handling and mounting of the tissue.

In light of these factors, a relaxed leaflet shape was produced which was free from surface wrinkles and consistent with clinical observations of the unloaded valve. This initial finite element mesh was then smoothed out by applying a small hydrostatic pressure of less than 0.01 kPa. Failure to smooth out the surface in this way results in very erratic distributions in the stress field, and is the most likely explanation for the values obtained by Hamid and Ghista (1974). Cataloglu, Clark, and Gould (1977) used silicone rubber moulds taken at a range of pressures starting from about 2.8 kPa, together with a micro-photogrammetry technique (Karara, 1974) to determine the deformed shape of loaded natural aortic valves. Unfortunately, such moulds are not suitable for defining the unloaded leaflet shape because of the large degree of deformation induced by the moulding process itself. Whatever shape is used for the analysis, it must be considered one of the unknown parameters of the problem and varied in a systematic manner to determine its actual effect of the final solutions.

9.6.2 Leaflet thickness

The thickness of the natural heart valve leaflet, $h = h(\theta_1, \theta_2)$, varies across the surface with the thinnest parts being in the leaflet belly (about 0.175 mm) and the thickest near the centre of the free edge (about 1.4 mm). The only data which appears to be available to define the variation in thickness across the leaflet has been obtained by Clark and Finke (1974). No data appears to be available yet to define the natural range of variability for this quantity so elemental thicknesses for the finite element mesh used were chosen to be consistent with Clark and Finke's data. No attempt has been made to vary these values since it is already well known (Green and Adkins, 1960) that for thin elastic membranes, the stress is inversely proportional to the thickness.

9.6.3 Incremental loading

The equilibrium equations resulting from the finite element formulation are too non-linear to permit the application of the full physiological pressure loading in one step. The usual procedure is to apply the final load in a series of small increments with the equilibrium equations being satisfied at each load level. When dealing with soft biological tissues having strain-energy functions exhibiting the strong exponential character of Equation (9.16), the most efficient strategy for determining the pressure to be applied at the nth loading increment, P_n, will be given by

$$P_n = P_0 \exp\left[\frac{n-1}{N_L-1}\log_e\frac{P}{P_0}\right], \tag{9.22}$$

where $P_0 \neq 0$ is the starting pressure, P is the final pressure, and $n = 1, 2, \ldots, N_L$ for a total of N_L load steps. The use of Equation (9.22) ensures that between successive load steps there will be uniform increments in membrane strains.

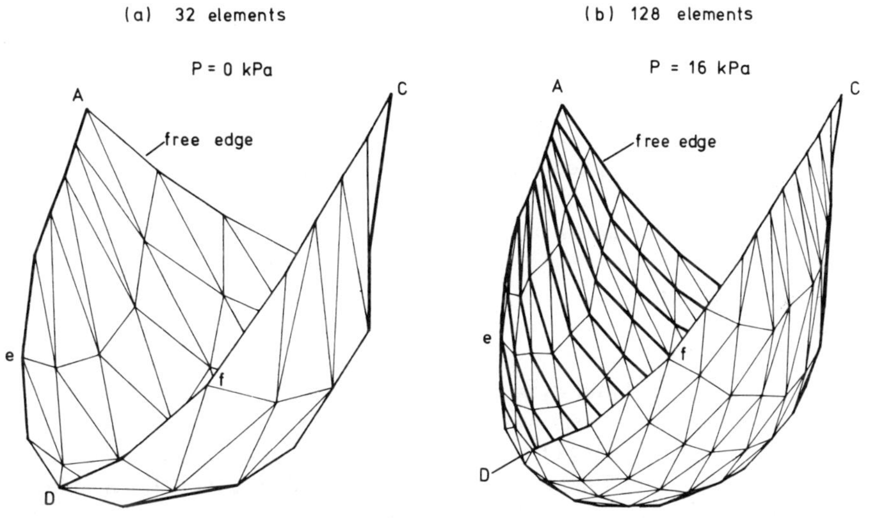

Figure 9.3 (a) The 32 element and (b) the reinforced 128 element finite element meshes

9.6.4 Membrane element subdivision

The mesh prepared was uniformly subdivided into 32 elements (Figure 9.3a) and then progressively refined to produce a uniform 128 element mesh (Figure 9.3b). The coarse mesh was used for an initial analysis by Christie and Medland (1980) while all results reported in this paper have been obtained using the 128 element mesh. The surface shapes were the same for both coarse and fine meshes. Tests have been conducted which establish that the finite element solution of this problem converges with refinement of the mesh. The main advantage in using the refined 128 element mesh is that it allows adequate resolution of the stress field and improves the ability of the mesh to reproduce the observed thickness distribution, $h(\theta_1, \theta_2)$. When conducting convergence tests, it is necessary to use the same thickness distribution for all subdivisions.

9.6.5 Fibre element subdivision

The 128 membrane element representation of the leaflet was reinforced with 66 fibre elements shown as heavy lines in Figure 9.3b. They all lie around the circumferential direction in the leaflet and use the same form for the strain-energy function as the membrane element. For each element, a value was specified for the fibre radius together with two elastic parameters which have in general, different values to those in the membrane elements. To simplify interpretation of the results all the fibre elements were assumed to have the same radius, r_f.

The relative fraction of the load carried by the membrane and fibre elements will depend on the relative cross-sectional areas of the two kinds of elements. As it was not known in advance what value this ratio should take, values of r_f ranging from 0.1 mm to 3.0 mm for anisotropic membranes were used.

The results presented below demonstrate that at a radius of 3.0 mm nearly all the load is transmitted to the stent by the fibres, the membrane elements acting solely to transmit the applied pressure load on to the fibre elements. It should be noted that the values of r_f only represent the nominal radius of the fibre elements, not the actual fibre radius in real leaflets. In other words, they should be interpreted as fibre densities which determine the degree of fibre reinforcing in a particular direction.

In the results for anisotropic models presented below, all cases have been analysed using elastic constants of $c_1 = 20$ and $c_2 = 1$ kPa for the membrane elements and $c_1 = 50$ and $c_2 = 1$ kPa for the fibre elements. Choice of these particular elastic parameters was by necessity arbitrary but is consistent with stress–strain curves taken in the circumferential and radial directions by Clark (1973) and Tan (1974). Figure 9.1 shows that at large fibre radii, the membrane will be substantially more compliant in the radial direction than the circumferential direction.

9.7 FINITE ELEMENT ANALYSIS OF AN ISOTROPIC MEMBRANE

The first model considered for the tissue leaflets of a bioprosthetic valve consisted of an isotropic, hyperelastic membrane attached to a rigid stent. Virtually all heterologous tissue valves such as the Ionescu–Shiley bovine pericardial xenograft have isotropic leaflets. However, there are also several important geometric differences between this model and the commercially available heterologous tissue valves, the significance of which remains to be properly assessed (Ionescu and Tandon, 1979).

The results presented earlier by Christie and Medland (1980) were obtained using the 32 element mesh. The 128 element mesh used in this study has allowed refined results to be obtained, particularly in the coaptive zone, and has also allowed refinements in the specification of the thickness distribution bringing the model into closer agreement with real leaflets. Nevertheless, the main conclusions obtained using the 32 element mesh remain unaltered.

9.7.1 The stress field in the leaflet

Figure 9.4a shows the magnitude of the first principal stress as contours over the leaflet surface. The view is as the leaflet would appear if it was removed from the stent and laid out flat; the free (or coapting) edge is the line ABC and the line ADC is the suture line at the stent frame.

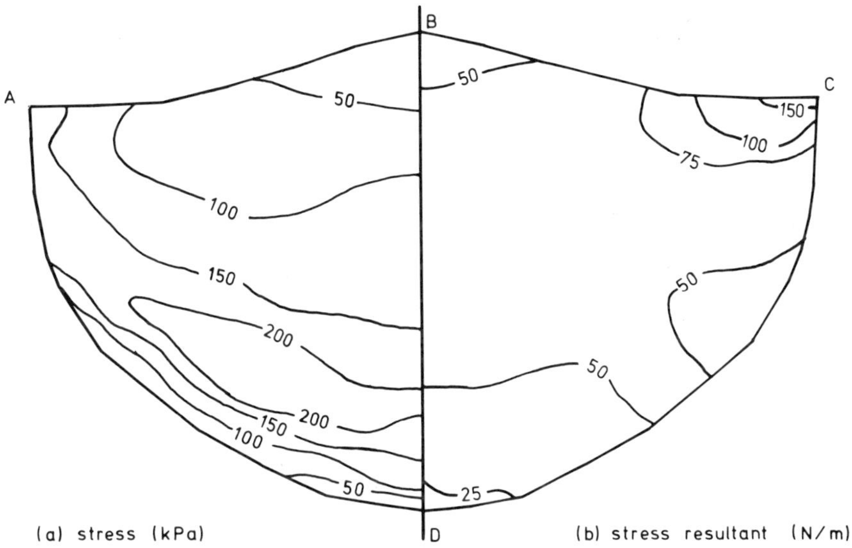

Figure 9.4 Contours of (a) first principal stress magnitude and (b) first principal stress resultant magnitude

The maximum stress is in the central belly zone of the leaflet. From the stresses in the membrane, $\tau^{\alpha\beta}$, the corresponding stress resultants, $\eta^{\alpha\beta}$, were calculated by means of

$$\eta^{\alpha\beta}_{(e)} = h_{(e)} \cdot \tau^{\alpha\beta}_{(e)} \qquad \text{(no sum on } e\text{)} \tag{9.23}$$

for each element *e*, and plotted as contours of the first principal stress resultant, given in Figure 9.4b. This demonstrates that the pattern of the stress distribution is strongly influenced by the elemental thickness values. Carpentier *et al*., (1976), after pathological examination of excised valves, have found the leaflet's belly zone to be a common site of tissue rupture and disruption.

The other site of stress concentration predicted by this model is at the commissures, the point of attachment of the leaflet corner to the top of the stent post demarked by points A and C in Figure 9.4. This concentration arises from the geometry of the valve transmitting a disproportionately large percentage of the applied forces to this point on the stent, as may be seen from the stress resultant magnitudes in Figure 9.4b. The stress field in the membrane is almost uniaxial in the close proximity of the commissures so very high values of shear stress are also found there (see Figures 9.5a and 9.5b). This high stress concentration at the commissures shows up clinically as dehiscence of the tissue from the top of the stent post (Thomson, 1976). It has been suggested by Thomson and Barratt–Boyes, (1977) that improved mounting procedures such as the avoidance of direct suturation into the tissue can assist the alleviation of this problem. The use of a flexible stent may contribute to stress reduction under dynamic, pulsatile loading conditions (Reis *et al*., 1971).

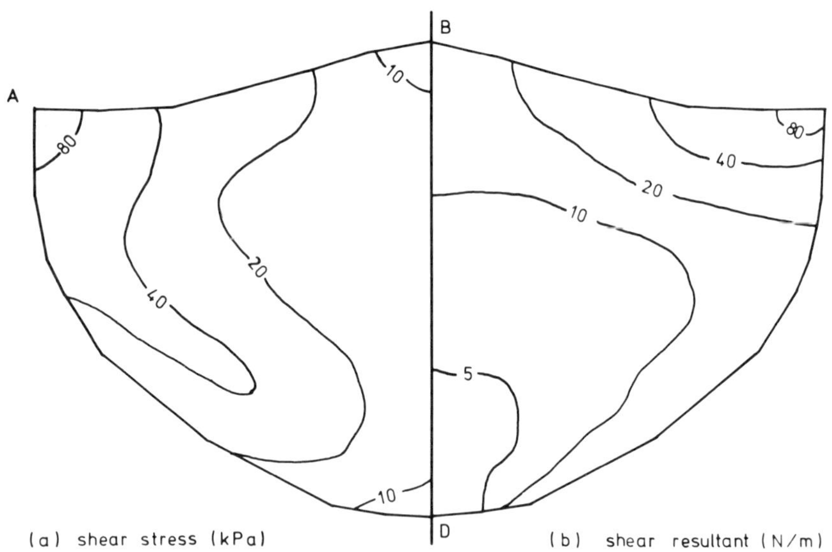

Figure 9.5 Contours of (a) first principal shear stress and (b) first principal shear stress resultants

The magnitude of the components of the stress field are almost linearly dependent on the applied pressure, the small departures from linearity being attributable to the non-linear dependence of the deformed thickness and leaflet curvature on pressure. Furthermore, it should be noted that at all points in the leaflet, the first principal components of stress are tensile. The only slightly compressive component was the second principal stress in the close proximity of the commissures arising from the uniaxial nature of the stress field.

Over most of the leaflet surface the stress field is closely aligned in the circumferential direction the only exceptions being at points close to the stent boundary. This result is in good agreement with the observations of Swanson and Clark (1974) which showed that the alignment of the fibre network in the tissue is predominantly circumferential, thus implying most of the forces transmitted by the leaflet are circumferential.

9.7.2 Influence of elastic properties on the stress field

To investigate the influence of tissue properties on the stress field, all quantities in the model were held fixed while the elastic parameters c_1 and c_2 were varied systematically over a wide range of possible tissue types. Figure 9.6 shows the percentage change in the central, belly stress value at the pressure of 16 kPa measured relative to that model with $c_1 = 20$ and $c_2 = 1$ kPa.

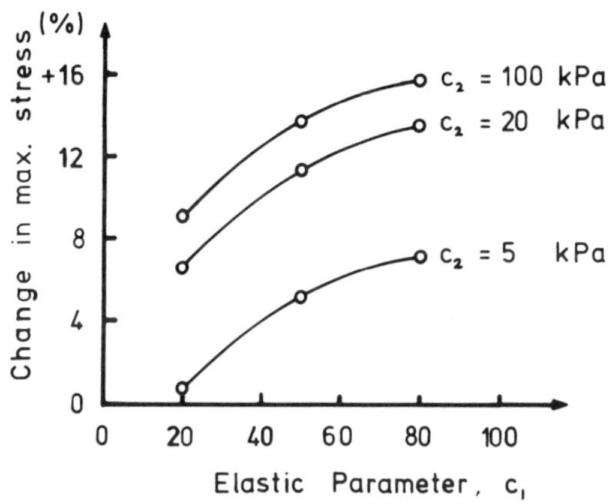

Figure 9.6 Change in maximum stress with tissue parameters

There is a clear relationship indicating that the maximum stress increases with decreasing tissue compliance and this may become an important factor with gluteraldehyde-treated tissues, particularly those treated while in a stressed state. Virtually all the porcine xenografts in use at the present time are constructed from porcine aortic valves fixed under high static heads of gluteraldehyde. Broom and Thomson (1979) have found, however, that such treatment drastically reduces the compliance of the leaflets although it does guarantee adequate inter-leaflet coaptation.

The range of elastic parameters considered here does not include the very stiff regime of such high-pressure, gluteraldehyde-treated tissue but the trends in Figure 9.6 show that for such leaflets, higher stress levels can be expected than those in fresh, untreated or low-pressure gluteraldehyde-treated valves.

9.7.3 Effect of pressure and tissue properties on leaflet shape

Heart valve models were analysed at 15 pressure steps between 0.01 kPa and 16 kPa. Note that these are *quasi-static* solutions, all dynamic terms having been neglected, so the deformation of the leaflets as a function of pressure should not be interpreted as that which could necessarily be expected under dynamic, pulsatile loading conditions.

Figure 9.7 shows the symmetry line (BD) profile of the leaflet in the (r, z) plane as a function of pressure with the elastic constants $c_1 = 20$ and $c_2 = 1$ kPa. Whereas in the centre of the leaflet there were significant radial and downward components of displacement, in the vicinity of the coapting edge the displacement was almost directly downwards and hence the coapting area remaining almost constant over a wide range of pressures in isotropic leaflets.

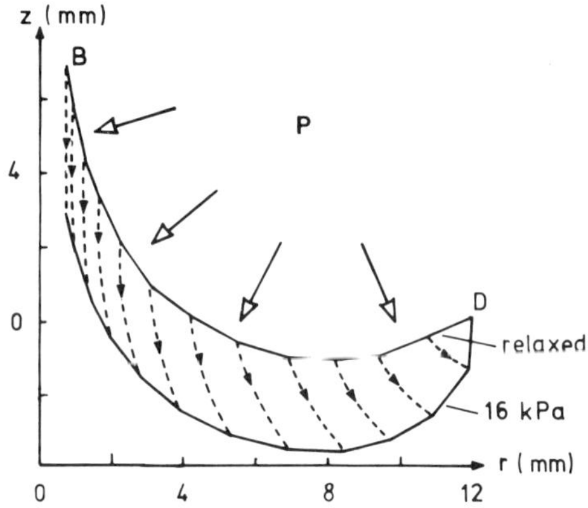

Figure 9.7 Displacement of the line BD for an isotropic leaflet with properties ($c_1 = 20$, $c_2 = 1$ kPa) as a function of pressure

Figure 9.8 shows the same leaflet profile at a pressure of 16 kPa for a series of different values for the elastic parameters. It is seen that with increasing tissue compliance, there is a greater degree of downward motion but the area of coaptation still remains effectively constant with respect to changes in c_1 and c_2.

9.7.4 Variation of thickness with pressure

As the leaflet stretches under applied loads, the thickness of the membrane must decrease in keeping with the incompressibility condition. Figure 9.9 shows the

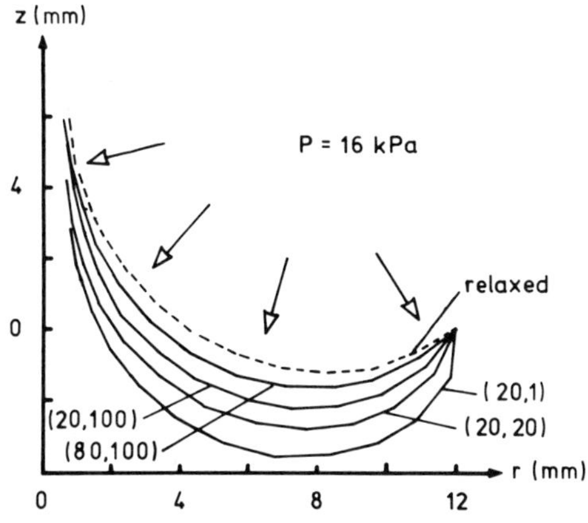

Figure 9.8 Isotropic leaflet displacement of line BD as a function of tissue properties (c_1, c_2) at a pressure of 16 kPa

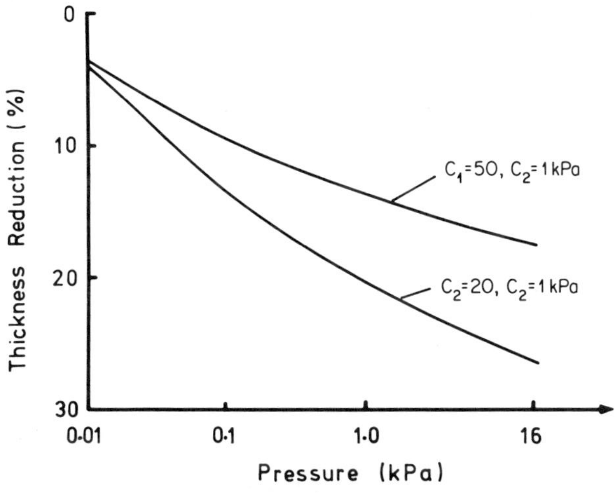

Figure 9.9 Change in leaflet thickness as a function of pressure ($c_2 = 1$ kPa)

percentage reduction in the leaflet thickness as a function of pressure, measured at a point close to the belly on the centre-line. Swanson and Clark (1974) have experimentally measured the thickness changes on the coapting edge as a function of pressure and obtained the value of 30 % for the thickness reduction at high pressures. Since the thickness reduction for incompressible membranes will be greatest at points where the in-plane strains are greatest, the point chosen in the belly will show larger changes than on the free edge. The comparatively small discrepancy between the experimental and theoretical results may be explained by noting the dependence of the thickness reduction on tissue compliance. The possibility that the tissue becomes compressible on the coaptive surface has been noted by Swanson and Clark (1974), but this seems unlikely in view of the fact that the compressive stress normal to the coaptive surface is only the same order of magnitude as the applied pressure.

9.8 FINITE ELEMENT ANALYSIS OF AN ANISOTROPIC MEMBRANE

The analyses using the isotropic membrane model for the heart valve leaflet were repeated using a leaflet with anisotropic elastic properties. Swanson and Clark (1974) have previously shown that the natural heart valve leaflet is reinforced in the circumferential direction by larger bundles of collagen fibres. To provide a model which better approximated the actual structural arrangement of natural leaflets, in the present study the isotropic mesh was reinforced with fibre elements aligned solely in the circumferential direction. The elastic constants for the membrane elements were held constant at $c_1 = 20$ and $c_2 = 1$ kPa, and those of the fibre elements held constant at $c_1 = 50$ and $c_2 = 1$ kPa, thus giving the leaflet more compliance radially than circumferentially which is at least in qualitative agreement with natural, untreated tissue.

The variable factor in this analysis then became simply the radius of the fibres, r_f, which for added simplicity was taken to be the same for each fibre element. At very small fibre radii, say $r_f < 0.01$ mm, the solutions obtained were, for practical purposes, indistinguishable from the isotropic case and the mechanical behaviour was dominated by the membrane component of the structure. At very large fibre radii, say $r_f > 3.0$ mm, the fibre elements become the dominant load-bearing structural component and produced marked changes in the mechanical behaviour of the valve.

9.8.1 Distribution of stress in anisotropic leaflets

The change in distribution of stress for $r_f = 0.1$ mm was most pronounced in the vicinity of the commissures, as may be seen in Figure 9.10, were there has been a reduction in the stress by a factor of ten. A stress reduction takes place over the entire leaflet but the change away from the commissures is not as pronounced. The reason for the very large reduction at the commissures lies in the fact that the stress field is almost uniaxial at these points whereas elsewhere there is a significant biaxial component. The fibre elements are, of course, ideal for transmitting uniaxial forces and so can be expected to produce the greatest benefits at points in the leaflet where such forces are present.

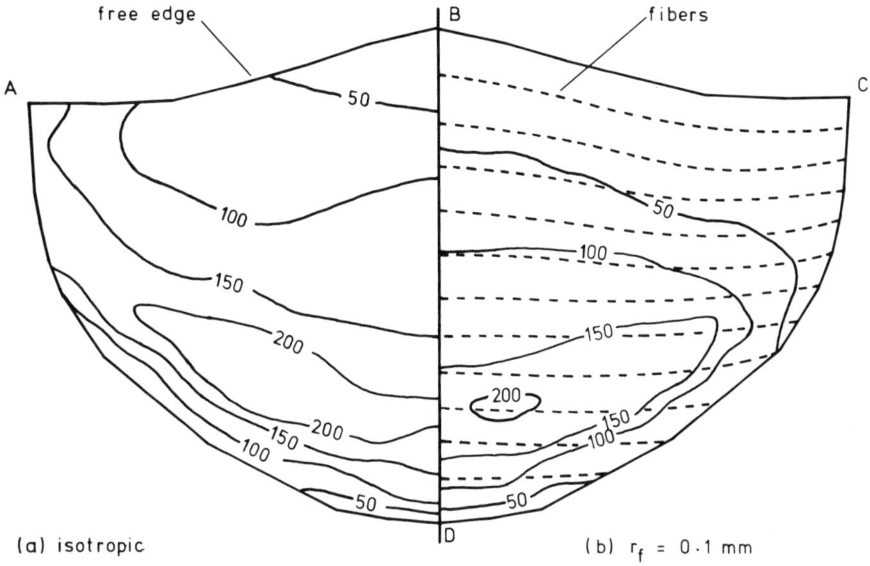

Figure 9.10 Membrane stress distribution for (a) an isotropic and (b) an anisotropic leaflet

9.8.2 Change in shape for anisotropic leaflets

As the fibre radii are increased from zero up to the maximum considered here of 3.0 mm, several major changes in the mechanical response of the leaflet surface to load were noted. Firstly, the downward displacement of all points in the leaflet was retarded. Secondly, there was a corresponding movement of the leaflet towards the coaptation interface which increased with increasing fibre radii. This situation contrasts sharply with that previously noted for isotropic leaflets where there was no noticeable movement towards the coaptation interface and hence, little dependence of coaptive area on applied pressure. Figure 9.11 shows the quasi-static motion of the leaflet under pressure (0.01 kPa to 16 kPa) for $r_f = 3.0$ mm.

Because the net movement of the leaflet is towards the coapating surface in the reinforced membrane, the area of coaptation was found to increase with both pressure and fibre radii as is shown in Figure 9.12. The discrete jumps in the area are the result of a new element coming into contact with the coaptive boundary while the gradual rise on the remaining portions of the curves is due to the stretching of elements, already in contact, under applied pressure.

In the isotropic case, no new elements came into contact with the coaptive boundary but with increasing degrees of circumferential reinforcement a substantial rise was found in the area of contact between adjacent leaflets. In fact, for the relaxed shape used in this model, the final coaptive area achieved appears

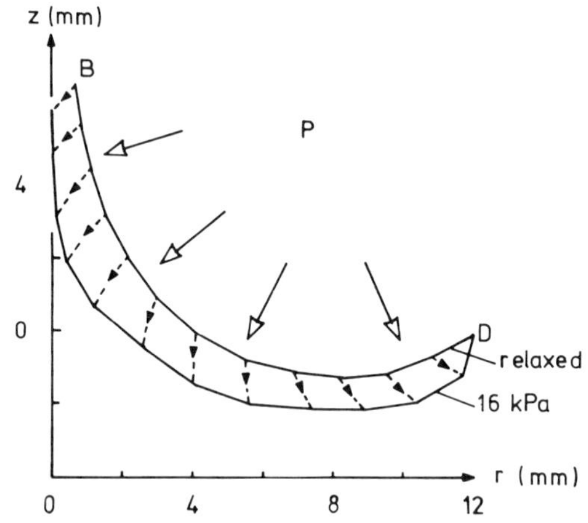

Figure 9.11 Displacement of line BD for an anisotropic leaflet ($r_f = 3.0$ mm) as a function of pressure

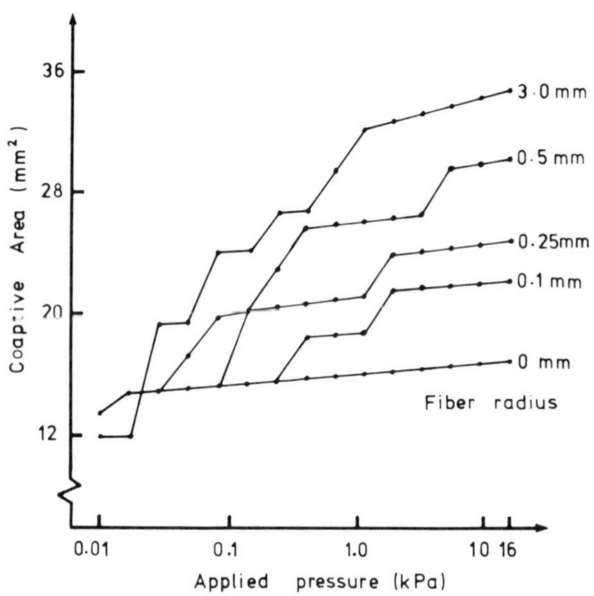

Figure 9.12 Coaptive area as a function of pressure

to level off at about a factor of two greater than that of the isotropic membrane. Note that some of the membrane is already in contact in its relaxed configuration and were it not for this, it appears reasonable to suppose that the isotropic membrane would have formed no coaptive surface.

This result lends strong credence to modelling natural aortic valve leaflets as a fibre-reinforced membrane structure. Leaflets with almost any degree of circumferential reinforcement can be expected to have significantly greater coaptive surface area than the corresponding isotropic leaflet.

9.8.3 Orientation of the stress field in anisotropic leaflets

From the previous model, the stress field in isotropic leaflets was found to be predominantly aligned in the circumferential direction at least at points not close to the leaflet boundary and, furthermore, this alignment was virtually independent of the applied pressure load.

For anisotropic leaflets, this simple pattern of behaviour changes progressively as the circumferential reinforcement increases. Figure 9.13 shows the orientation of the stress field in the membrane elements relative to the circumferential direction at the leaflet belly for a range of r_f values. For small values ($r_f \leqslant 0.1$), the alignment is almost the same as in the isotropic case. For intermediate values ($0.50 < r_f < 2.0$), there is a strong dependence of stress alignment on pressure, the alignment being radial at low pressures and circumferential at high pressures. For high values ($r_f \geqslant 3.0$ mm), the stress field remains radially aligned over the entire range of pressures considered.

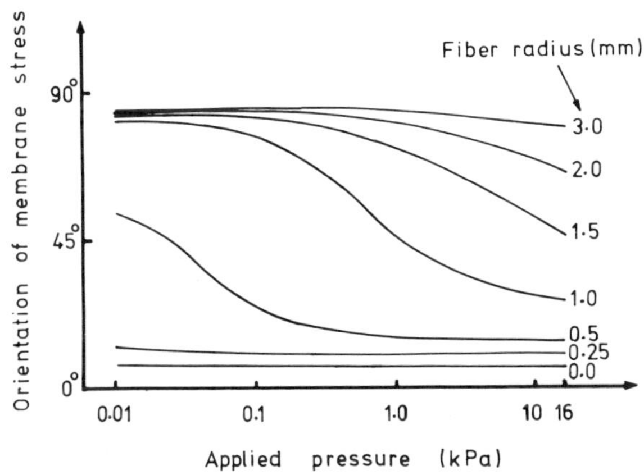

Figure 9.13 Orientation of the membrane stress field relative to the fibre direction in the leaflet belly as a function of pressure

Figure 9.14 gives a plot of membrane stress field alignment taken at a pressure of 16 kPa along the same fibre traversing the membrane from the central belly to the boundary at the base of the stent post. It may be observed that the change in alignment with fibre radii is greatest in the centre of the leaflet and becomes progressively less significant towards the rigid boundary.

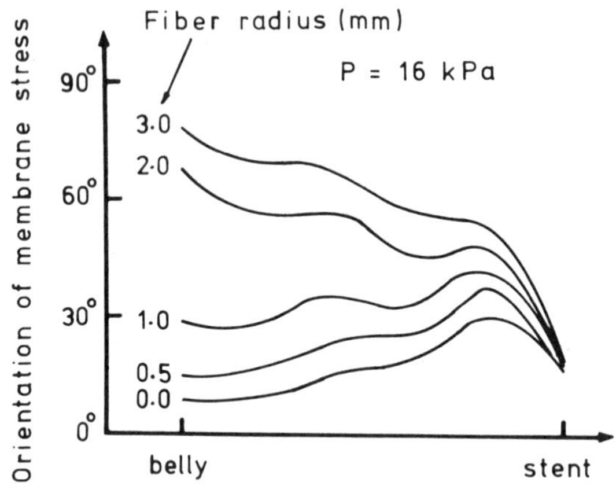

Figure 9.14 Orientation of the membrane stress field relative to the fibre direction across the leaflet at $P = 16$ kPa

9.8.4 Force transmission to the stent frame

The distribution of load which the structure will be required to sustain is very relevant to the design of stent frames. The leaflet transmits the applied pressure load exerted by the blood on to the supporting structure, be it a stent or the aortic root as in natural valves. In terms of the static global equilibrium of the bioprosthesis, all forces exerted on the valve leaflets must ultimately be exactly balanced by reaction forces exerted by the boundary. Although each leaflet exerts force on its neighbouring leaflets via the coapting surface, these are exactly balanced by the symmetry requirement so it will suffice to examine the reaction forces only at nodes on the stent boundary.

Figure 9.15 gives the magnitudes of the nodal force vector components on the stent boundary as a function of the global ϕ coordinate measured from $\phi = 0°$ on the leaflet centre-line to $\phi = 60°$ at the top of the stent post. The curve giving the force distribution for an isotropic leaflet shows that there is a significant concentration of force transmitted to the stent at the commissures. Most of the remaining load is transmitted to the basal ring at an angle centred at about 32° from the centre-line. For increasing values of r_f the concentration at the commissures ($\phi = 60°$) is removed and the force distributed further down the stent post, the change becoming more and more pronounced, until at $r_f = 3.0$ mm the two zones of concentration have virtually coalesced. Clearly, by appropriate adjustment of the fibre radii of particular fibres down the stent post, an almost uniform distribution of load on to the stent post could have been achieved.

The alignment of the reaction force vector at nodal points on the stent boundary plotted versus the global coordinate, ϕ, is shown in Figure 9.16 for the isotropic case and for a strongly reinforced anisotropic case. For each of the

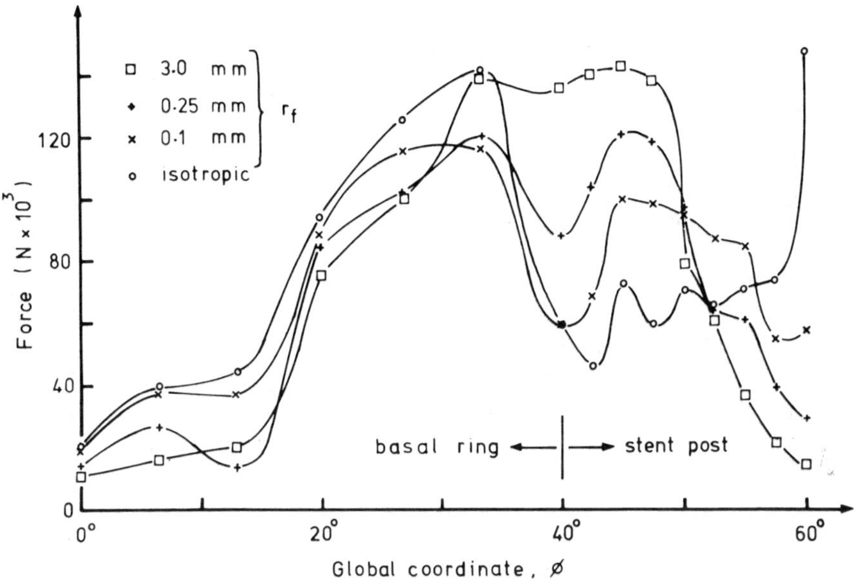

Figure 9.15 Reaction force magnitude at stent boundary (DC) as a function of global coordinate, ø

components r, ϕ, and z, the angle plotted is that subtended by the component to the positive branch of the corresponding global coordinate. Thus an angle of 90° shows that a component makes zero contribution to the magnitude of the vector.

In general, when the magnitude of the force is large, it is directed predominantly downward and when it is small, the force is predominantly radial.

9.9 CONCLUSION

A finite element model has been developed which has been used to explore the mechanics of tissue heart valves. The formulation described makes no *a priori* assumptions about the magnitudes of either the strains or the displacements and uses a new 'contact' boundary condition on the coapting edge of the leaflet. The constitutive law is specified in terms of a strain-energy function which is exponential in the strain invariants, allowing different tissues to be represented using a minimum number of elastic parameters.

Anatomical and morphological examinations of the aortic valve leaflets have shown that to a good level of approximation, the leaflets may be considered to be thin-membrane structures reinforced by thicker bundles of collagen fibres in the circumferential direction. Two distinct models were used in this study, the first approximating the leaflet as an isotropic leaflet and the second reinforcing the

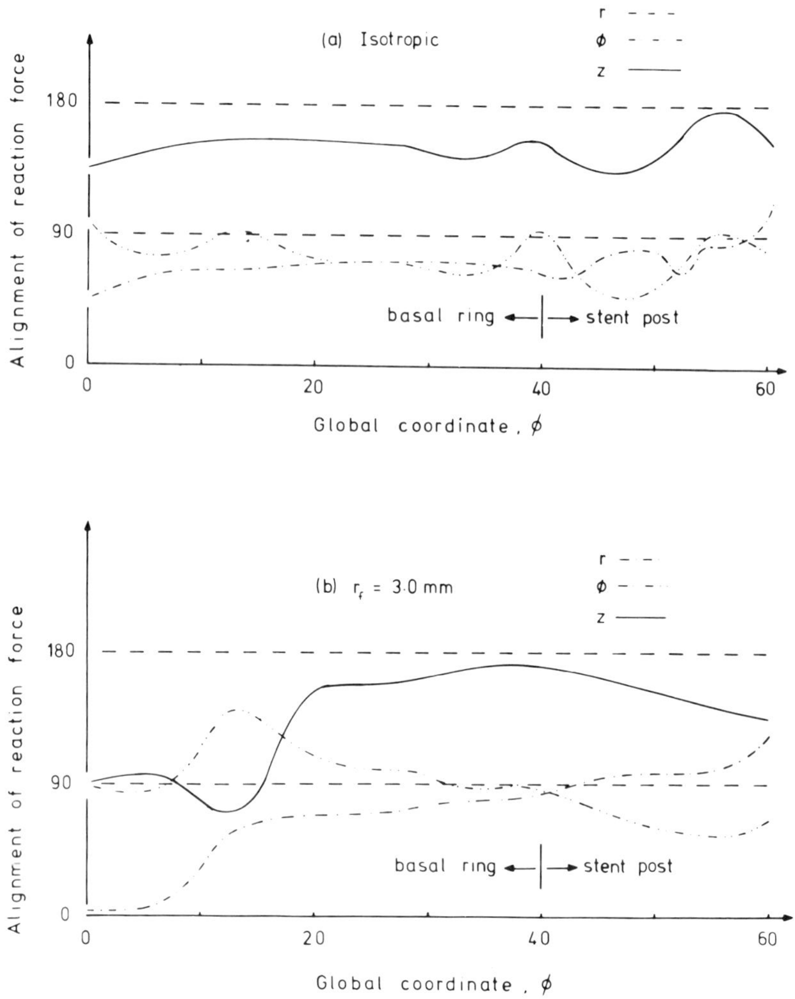

Figure 9.16 Alignment of reaction force at stent boundary as a function of global coordinate, ø

isotropic leaflet with circumferentially aligned fibres. Several important differences in the mechanics of the two models have been noted, the most significant being the increase in the interleaflet coaptive area with increasing tissue anisotropy and the elimination of the stress concentration at the commissures found in the isotropic model.

Although suitably detailed anatomical and elastic data are not yet available, it is nevertheless possible to explore how each particular parameter of the model affects the overall mechanics of bioprosthetic valves. Wherever possible, parameters have been systematically varied over limits set to encompass any physiological variations while holding all remaining parameters fixed. In this way numerical 'experiments' have been conducted which have led to an improved understanding of the relation between structure and function for bioprosthetic valves.

Many interesting aspects of this problem still remain to be explored using the static model, such as those relating to the effects of inadvertently distorting the leaflets during the mounting procedure and the influence of the choice of strain-energy function on the leaflet stresses. The incorporation of the dynamical terms in the equilibrium equations and the application of physiologically realistic loading is essential before any investigation of the effects of flexible stent posts, rapid fatigue cycling, and viscoelastic material properties can be attempted.

ACKNOWLEDGEMENT

The authors gratefully acknowledge the support of the Medical Research Council of New Zealand under Research Grant 133-MRC-188.

REFERENCES

Angell, W. W., Angell, J. D., and Kosek, J. C. (1979) Clinical and experimental comparisons establishing the gluteraldehyde treated xenograft as the standard for tissue heart valve replacement, *Tissue Heart Valves* (ed. M. I. Ionescu), Butterworths, Sevenoaks, pp. 89–126.

Bellhouse, B. J. and Talbot, L. (1969) Fluid mechanics of the aortic valve, *J. Fluid Mech.*, **35,** 721–735.

Binet, J. P., Duran, C. G., Carpentier, A., and Langlois, J. (1965) Heterologous aortic valve transplantation, *Lancet*, **2,** 1275.

Blatz, P. J., Chu, B. M., and Wayland, H. (1969) On the mechanical behaviour of elastic animal tissue, *Trans. Soc. Rheol*, **13,** 83–102.

Broom, N. D. (1977) The stress/strain and fatigue behaviour of gluteraldehyde preserved heart valve tissue, *J. Biomech.*, **10,** 707–724.

Broom, N. D. (1978) Simultaneous morphological and stress–strain studies of the fibrous components in wet heart-valve tissue, *J. Biomech.*, **10,** 707–724.

Broom, N. D. (1980) An 'in vitro' study of mechanical fatigue in gluteraldehyde-treated porcine aortic valve tissue, *Biomaterials*, **1,** 3–8.

Broom, N. D., and Thomson, F. J. (1979) Influence of fixation conditions on the performance of gluteraldehyde-treated porcine aortic valves: towards a more scientific basis, *Thorax*, **34,** 166–176.

Carpentier, A., Deloche, A., Relland, J., and Dubost, C. (1976) Valvular xenograft and valvular bioprosthesis: 1965–1975, *The Mitral Valve: a pluridisciplinary approach* (ed. D. Kalmanson), Edward Arnold, London, pp. 505–518.

Cataloglu, A., Clark, R. E., and Gould, P. L. (1977) Stress analysis of aortic valve leaflets with smoothed geometrical data, *J. Biomech.*, **10,** 153–158.

Chong, M., and Missirilis, Y. F. (1978) Aortic valve mechanics II. A stress analysis of the porcine aortic valve leaflets in diastole, *Biomaterials, Med. Dev. Artif. Organs*, **6,** 225–244.

Christie, G. W., and Medland, I. C. (1980) A static finite strain, finite element stress analysis of bioprosthetic heart valves, *Proc. Int. Conf. on Finite Elements in Biomechanics*, Feb. 18–20, 1980, University of Arizona, pp. 605–622.

Clark, R. E. (1973) Stress–strain characteristics of fresh and frozen human aortic and mitral leaflets and chordae tendineae, *J. Thorac. Cardiovasc. Surg.*, **66,** 202–208.

Clark, R. E., and Finke, E. H. (1974) Scanning and light microscopy of human aortic leaflets in stressed and relaxed states, *J. Thorac. Cardiovasc. Surg.*, **67,** 792–804.

Clark, R. E., Swanson, W. M., Kardos, J. L., Hagen, R. W., and Beauchamp, T. A. (1978) Durability of prosthetic heart valves, *Ann. Thorac. Surg.*, **26,** 373–335.

Cohn, L. H., and Collins, J. J. (1979) The gluteraldehyde-stabilized porcine xenograft valve, *Tissue Heart Valves* (ed. M. I. Ionescu), Butterworths, Sevenoaks, pp. 173–200.

Demiray, H. (1972) A note on the elasticity of soft tissues, *J. Biomech.*, **5,** 309–311.

Demiray, H., and Vito, R. P. (1976) Large deformation analysis of soft biomaterials, *Int. J. Engng. Sci.*, **14,** 789–793.

Ferrans, V. J., Spray, T. C., Billingham, M. E., and Roberts, W. C. (1978) Structural changes in gluteraldehyde-treated porcine heterografts used as substitute cardiac valves, *Am. J. Cardiol.*, **41,** 1159–1184.

Fung, Y. C. (1967) Elasticity of soft tissues in simple elongation *Am. J. Physiol.*, **213,** 1532–1544.

Gou, P. F. (1970) Strain energy function for biological tissues, *J. Biomech.*, **3,** 547–550.

Green, A. E., and Adkins, J. E. (1960) *Large Elastic Deformations (and non-linear continuum mechanics)*, Oxford University Press, Oxford.

Green, A. E. and Zerna, W. (1954) *Theoretical Elasticity*, Oxford University Press, Oxford.

Hamid, M. S., and Ghista, D. N. (1974) Finite element analysis of human cardiac structures, in: *Finite Element Methods in Engineering* (eds. V. A. Pulmano and A. P. Kabaila), Clarendon, Australia, pp. 337–348.

Hunter, P. J. (1975) Finite element analysis of cardiac muscle mechanics, D. Phil. Thesis, Oxford.

Ionescu, M. I., Pakrashi, B. C., Mary, D. A. S., Bartek, I. T., and Wooler, G. H. (1974) Long-term evaluation of tissue valves, *J. Cardiovasc. Thorac. Surg.*, **68,** 361–378.

Ionescu, M. I., and Tandon, A. P. (1979) The Ionescu–Shiley pericardial xenograft heart valve, *Tissue Heart Valves* (ed. M. I. Ionescu), Butterworths, Sevenoaks, pp. 201–252.

Karara, H. M. (1974) Aortic valve geometry, *Photogrammetric Engng*, **40,** 3193–4102.

Lee, C. S. F., and Talbot, L. (1979) A fluid-mechanical study of the closure of heart valves, *J. Fluid Mech.*, **91,** 41–63.

Oden, J. T. (1972) *Finite Elements of Non-linear Continua*, McGraw-Hill, New York.

Peskin, C. S. (1972) Flow patterns around heart valves: a numerical method, *J. Comp. Phys.*, **10,** 252–271.

Reis, R. L., Hancock, W. D., Yarbrough, J. W., Glancy, D. L., and Morrow, A. G. (1971) The flexible stent: a new concept in the fabrication of tissue valve prostheses, *J. Thorac. Cardiovasc. Surg.*, **62,** 683–689.

Roche, A. H. G., Barratt-Boyes, B. G., Heng, M. B., and Rutherford, J. D. (1977) Long-term results using antibiotic sterilized homograft valves for aortic, mitral or tricuspid replacement. *Proc. 2nd Henry Ford Hospital Symposium* (ed. J. Davita), Appleton-Century-Crofts, New York.

Rossiter, S. J., Miller, D. C., Stinson, E. B., Oyer, P. E., Reitz, B. A., and Shumway, N. E. (1979) Aortic and mitral prosthetic valve re-operations, *Arch. Surg.*, **114,** 1279–1283.

Snyder, R. W. (1972) Large deformation of isotropic biological tissue, *J. Biomech.*, **5,** 601–606.

Stinson, E. B., Griepp, R. B., Oyer, P. E., and Shumway, N. E. (1977) Long-term experience with porcine aortic valve xenografts, *J. Cardio-vasc. Thorac. Surg.*, **73,** 54–61.

Streenhoven, van A. A., and Dongen, van M. E. H. (1979) Model studies of the closing behaviour of the aortic valve, *J. Fluid Mech.*, **90,** 21–32.

Swanson, W. M., and Clark, R. E. (1973) Aortic valve motion during systole, *Circ. Res.*, **32,** 42–48.

Swanson, W. M., and Clark, R. E. (1974) Dimensions and geometric relationships of the human aortic valve as a function of pressure, *Circ. Res.*, **35,** 871–882.

Tan, A. J. K. (1974) The tensile behaviour of aortic heart valve leaflets and sterilization effects, Masters dissertation, University of Auckland.

Thomson, F. J. (1976) Progress report on the development of a flexible stent for heart valve homografts, *Report No. 132 (M7615)*, Dept. of Mech. Engin., University of Auckland.

Thomson, F. J., and Barratt-Boyes, B. G. (1977) The gluteraldehyde-treated heterograft valve: some engineering observations, *J. Thorac. Cardiovasc. Surg.*, **74,** 317–321.

Veronda, D. R., and Westmann, R. A. (1970) Mechanical characterization of skin-finite deformations, *J. Biomech.*, **3,** 111–124.

Vito, R. P. (1973) A note on arterial elasticity, *J. Biomech.*, **6,** 561–564.

Zuhdi, N., Hawley, W., Voehl, V., Hancock, W., Carey, J., and Greer, A. (1974) Porcine aortic valves as replacements for human heart valves, *Ann. Thorac. Surg.*, **17,** 479–491.

Finite Elements in Biomechanics
Edited by R. H. Gallagher, B. R. Simon, P. C. Johnson, and J. F. Gross

Chapter 10

The Role of Finite Element Models in Orthopaedics

Edmund F. Rybicki

10.1 INTRODUCTION

Current trends in orthopaedics are toward increased use of joint replacements and internal fixation devices. Advancing the state-of-the-art in design and utilization of these implants requires a multidisciplinary approach. Biological as well as certain metallurgical and material property aspects of implants and bones must be recognized. Examples are biocompatibility, stiffness, ultimate strength, failure modes, wear properties, and fatigue characteristics. The loading environment of healthy bones is substantial. Peak loadings and repeated stressing can cause failures of implant materials and interfaces. It has been found that stress levels can influence bone remodelling, bone healing, and resorption. Thus, a knowledge of the stress behaviour of orthopaedic implants and supporting bone tissue is an important link to understanding the success or failure of orthopaedic devices.

The process of determining stresses in any body, commonly called stress analysis, is a well-established discipline. Techniques for stress analysis can be classified into two categories. One is experimental stress analysis; the other is theoretical stress analysis. In experimental stress analysis, direct measurements are taken. Stresses are inferred from these measurements. There are a variety of theoretical stress analysis techniques. The most popular method is the finite element method. The method has its origins over 25 years ago. However, its heritage goes back over 200 years in the origins of mechanics. It is appropriate to note that Volterra and Gaines (1971) tell us that some of the early contributors in mechanics, such as Thomas Young (1773–1829), were also physicians. It was, however, a long time until mechanics contributed to an orthopaedics problem. Probably the first recorded application of mechanics to orthopaedics was in the work of Meyer (1867), who did a stress analysis of the upper portion of the human femur. This work was later extended first by Wolff (1870) and later by Koch (1917), who was both an engineer and a physician. Thus the origins of mechanics principles applied to orthopaedics problems go back over a century.

The finite element method, as we know it today, is a natural extension of the previous work in mechanics, the main distinction being that the finite element method is amenable to automated computation and it gained popularity as digital computers became more widely accessible. The popularity of the finite element method is also due to its ability to handle various geometries, loading conditions, and material properties. It has been used in many engineering problems including areas of civil, mechanical, petroleum, and structural engineering.

In orthopaedics, finite element analysis is most commonly associated with solid mechanics and stress analysis. However, two other applications have been noted. One is modelling electromagnetic fields to study their effect on bone healing as reported by Hassler, Rybicki, Rotaru, and Hughes (1974). The other area is the use of finite elements for temperature analysis. Huiskes and de Wijn (1978) used a finite element analysis to study the effects of exothermal reactions in PMMA used to set the stems of joint replacement devices. The main focus here will be on solid mechanics applications.

Numerous applications of the finite element method are found in the literature. This is testimony to the interest and value of the method as applied to orthopaedics-related problems. Before reviewing a representative segment of these papers, it is important to set the topic of finite element analysis in orthopaedics in proper persepctive by discussing the role of finite elements in orthopaedics. Thus, the following section is devoted to the role of finite element analysis in orthopaedics problems. Subsequent sections discuss applications for the finite element method in terms of selected papers cited in the literature. The last sections are devoted to some of the possible directions for finite element analysis in orthopaedics for the next decade.

10.2 ROLE OF FINITE ELEMENT ANALYSIS IN ORTHOPAEDICS

Applications of finite element stress analysis in orthopaedics includes understanding the stress behaviour of healthy bones, fixation and prosthetic devices, reconstructive surgery and failed implants. Unlike many other problems areas where finite element analysis has been applied, orthopaedics is different in two ways. First, because of the nature of the individual living tissues involved, there is an uncertainty with regards to material properties and mechanical behaviour which is not experienced, for example, in structural analysis. Second, the nature of the problems involved requires a multidisciplinary team. This team is comprised of professions that have not worked together until recently and often have different views on the values of computational results. Because of this, it is important to have clear, concise goals and expected results. Communications are an important ingredient in selecting an effective team. The emphasis in this chapter is on handling the first distinction, involving the inherent uncertainties in biologic material properties.

This aspect of uncertainty in orthopaedics is best addressed by using a coordinated approach involving finite element analysis, experiments, and clinical

Table 10.1 Three approaches to orthopaedics problems

Approach	Advantages	Disadvantages
Computational model	• 'Low' costs • 'Short' time	• Need to verify model with measurements • Time and cost required
Experimental model	• Measurements for stresses rather than computations	• Difficult to study a range of parameters • Difficult to extrapolate to new conditions
Coordinated computational and experimental	• Have measurements and a degree of confidence in computations • Reduced time and costs Can do parametric study • Can extrapolate to other conditions	• Requires more thought in planning and coordinating program

evaluations. Finite element analysis is an engineer's tool. Just like all tools, it must be used with other tools so that the results are directed toward meeting specified goals. To illustrate this approach, consider the following three possible approaches to orthopaedics-related problems and the advantages and disadvantages of each. The approaches are listed in the first column of Table 10.1 . The first approach is called a computational approach. This method involves using of the finite element method without experiments or clinical evaluations. The second approach involves experiments alone without finite element analysis or clinical evaluations. The third approach involves a coordinated combination of computational, experimental, and clinical evaluations. The advantages and disadvantages of each approach are listed in columns 2 and 3 of Table 10.1. The advantages of using the computational model alone are generally in terms of saving time and costs as compared with the experimental approach. This is particularly so when stresses are sought experimentally on the surfaces and through the thickness of the bones and implants. The disadvantage of the computational approach is that the important factor of verifying the capabilities of the computational model are missing. This could cause serious questions concerning the relation between the results of the finite element analyses and the orthopaedic problem.

The second approach is the experimental analogue of the computational approach. That is, only experiments are used. In this approach, the advantage is that the behaviour being studied is determined from measurements rather than computed values. There are, however, three disadvantages in this approach. As mentioned previously, the amount of time and costs to conduct experimental studies are generally higher than those for an analytical approach. The second disadvantage is that it is difficult to study a range of parameters and their effect on the behaviour being studied. This occurs because it is often more difficult to control parameters experimentally than with the computational model. The third disadvantage of the experimental approach is probably the most serious. This is that it is difficult to look at the experimental results and extrapolate to new conditions. The reason is that coupling of effects between variables is often too complex to extrapolate accurately to new conditions.

The third approach listed in Table 10.1 involves a coordinated combination of computations, experiments, and clinical evaluations. This approach is a combination of the first two approaches and retains the advantages of these two approaches, while removing the disadvantages. For example, this approach affords both measurements and computational values for the results. The experimental results provide an opportunity to verify the finite element model by comparing computed values with experimental data. A combination of finite element analysis and experiments can reduce the total number of experiments needed and thus reduce costs and time involved. The finite element model can be used to conduct parametric studies to determine the sensitivity of varying parameters on the results. Finally, another advantage of the combined approach is that the analyses and experiments can be combined to extrapolate to other conditions. The disadvantage of this approach is that it does require more thought and planning to coordinate the efforts of the various contributors. This can be a serious disadvantage because if the program is not planned properly, the tendency is for the combined approach to break down into the two independent approaches which are described in the top two rows of Table 10.1. Thus the disadvantages associated with each of these two approaches are possibilities.

Figure 10.1 illustrates the interactions between the finite element analysis, laboratory experiments, and clinical evaluations in the combined approach. The process starts with a definition of the problem and setting goals. This provides information to start the finite element analysis and experiments. Interactions between the experiments and finite element analyses are indicated by the arrows going both ways. This includes the verification of the finite element model and the selection of experiments. After enough information is obtained, the process can move toward clinical evaluations. The whole process is more complex than illustrated here. Success depends on iterations between the various disciplines involved and communications between the various contributors. After the clinical evaluations, the efforts have either met the goals or have not. If the goals are not met, then it may be necessary to redefine the goals based on what was learned from the first time through. It should be kept in mind that since most of the problems being addressed are outside of the scope of established design codes and procedures, it is unusual for something to work perfectly the first time. The ultimate success may depend on the degree of learning that goes on and the iterations between the contributors in Figure 10.1.

Figure 10.2 provides a more detailed description of the interactions between the various components involved, with focus on the role of the finite element model. In proceeding through the centre column of Figure 10.2, there are several questions that should be addressed. These questions pertain to the development

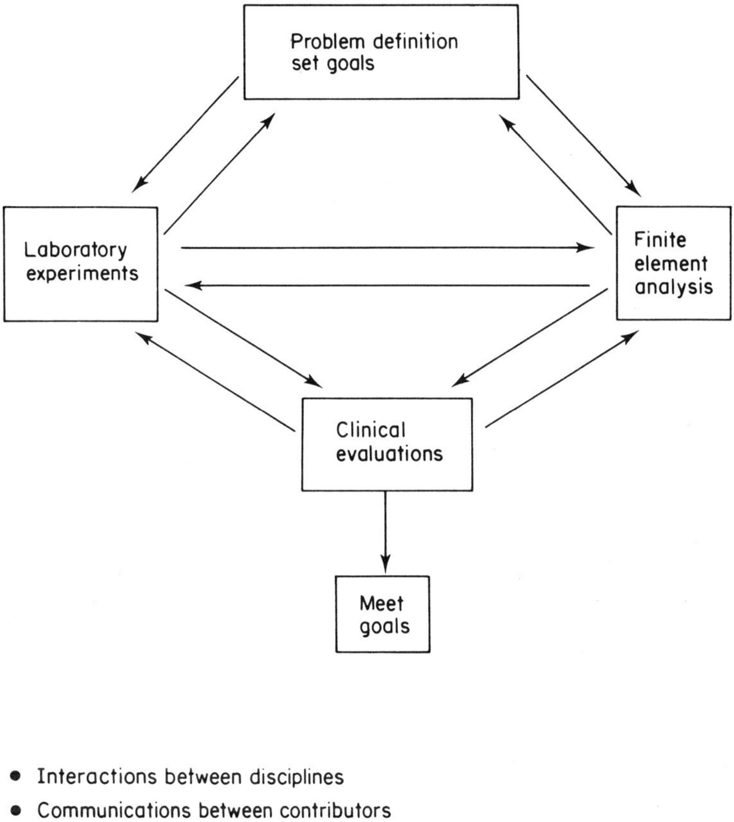

Figure 10.1 The role of finite element analysis in orthopaedics

and utilization of the finite element model. For example, in developing a representative model, questions concerning the type of loading conditions, material properties, geometries, and boundary conditions are important. The preliminary verification in Figure 10.2 gives an idea of how accurate the finite element model is in representing the laboratory experiments. After being satisfied that the model is appropriate, the next step is to use the finite element model to examine the effects of changes in the parameters and to determine the experimental test matrix. Often important experiments can be identified and others can be omitted in this step. There may be a reason to further modify the finite element model at this stage depending on the results from the experiments. The next step is to proceed to clinical evaluations when enough evidence has been obtained to support this action.

Several advantages of this approach can be seen. First, the approach has the potential to reduce the number of experiments that would be needed with an all-

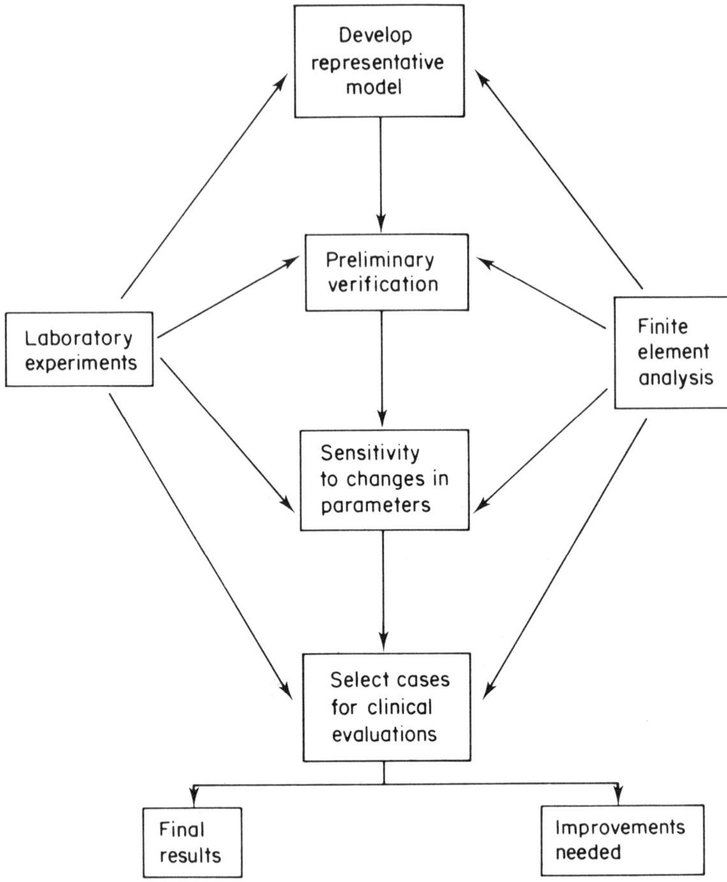

Figure 10.2 Interactions between finite element analysis, experiments, and clinical evaluations

experimental program. Second, as computer processing equipment becomes more sophisticated and available, the overall time and costs can be reduced. A third advantage is that the results contain a sensitivity study which can be used for other purposes. Finally, the results contain both experimental data and computed values and if enough confidence is gained in the model, it can be used to examine other cases of interest.

It is important to emphasize the value of iterations between the various steps of the combined computational and experimental approach described here. It is also important to verify the capabilities of the finite element models and to obtain representative data for verifying the model. Once this is accomplished, the approach described here provides a framework for a systematic and efficient method for utilizing the capabilities of the finite element analysis in orthopaedic problems.

10.3 SELECTION OF A REPRESENTATIVE FINITE ELEMENT MODEL

A few words are in order about evaluating how accurate the finite element models are. First, in developing the model, one should determine whether important mechanisms have been included. For example, if there is slipping at an interface, or a different range of stiffnesses involved, how important is it to include these factors? Another question is, 'Do the models agree with experimental data and clinical observations?' If so, how good is the agreement? If not, it shoud be determined why not. Here one should not preclude asking how accurate are the data and the observations. It is important to maintain the applications of the finite element model within the scope of verification. That is, if the model was verified for static conditions, it is not guaranteed to be useful for dynamic loadings. Because of the complexity and uncertainty of material properties, geometries, and loading conditions involved in orthopaedics problems, an important part of the finite element modelling is a healthy amount of verification of the model.

With this as background information and a description of the role of finite element analysis in orthopaedics, the next section highlights some of the applications of finite element analysis in orthopaedics.

10.4 RECENT APPLICATIONS OF FINITE ELEMENT ANALYSIS

Finite element analysis in orthopaedics-related problems increased in popularity during the last five years of the 1970s. The numerous papers on this subject are testimony to the interest and value of finite elements in orthopaedics. There are, however, too many papers to cover all of them here. Instead, selected papers and applications have been pointed out to illustrate current trends and specific points.

Applications of finite elements in orthopaedics have primarily been stress analysis. However, other applications have included electromagnetic fields for enhanced bone healing, temperature analysis for PMMA, and joint lubrication. The primary interest here is on the stress analysis applications, but recognition is also given to other categories. Applications of finite element analysis in orthopaedics fall into several categories. These include:

(a) stresses in intact bones;
(b) fixation devices;

(c) prosthetic devices; and
(d) other applications

The following sections discuss each of these categories..

10.4.1 Stress in intact bones

Understanding stresses in intact bones has been a topic of longstanding interest as evidenced by the work of Meyer (1867), Wolff (1870), and Koch (1917). These studies, of course, were not done with finite elements. However, they did provide a certain amount of motivation for the finite element studies of the femur. The results of these studies suggested an interesting relation between stresses and bone architecture. There have been many two- and three-dimensional finite element studies of the femur. Rybicki, Simonen, and Weiss (1972) carried out two-dimensional finite element studies of the proximal femur and took into account the effects of muscle forces. There was agreement between the trabecular structure and the lines of principal stresses from the finite element analysis. Several three-dimensional finite element stress analyses of the femur have been conducted, including studies by Wood (1975), Valliappan, Svensson, and Wood (1977), Harris, Chao, Block, and Weingarten (1978); and Brown and Ferguson (1978), Rohlmann, Bergmann, and Kolbel (1979) have compared finite element stresses with laboratory measurements to study the stresses in bones. These studies showed results for out-of-plane loadings and geometries that could not be represented in two-dimensional studies.

Finite element analyses of the tibia and knee joint have been done. For example, Hayes, Swenson, and Schurman (1978) have modelled the tibial plateau in terms of an axisymmetric finite element model. Chand, Huang, and Rim (1976) have investigated stresses in the human knee joint. These two references shed some light on how loads are transferred through the knee joint. Simon, Stanley, Kamel, and Peltier (1974) have obtained a better understanding of the relation between joint function and joint stresses with the aid of a finite element model. Rybicki, Glaeser, Strankowski,and Tamm (1979) and Rybicki, Strankowski, Tamm and Glaser (1978) have developed a finite element based model for joint stresses including the effect of lubrication. Here the model was first verified by comparisons with experimental data. Then it was used to show that the synovial fluid properties and cartilage stiffness effect the lubrication behaviour. Another anatomical feature of the leg receiving attention is the patella or knee cap. Among others, Hayes (1977) has incorporated structural features of the patella in a finite element representation and showed their effect on distributing forces.

Remodelling and healing of bones are topics that have been studied with the aid of finite element methods. Hassler, Rybicki, and Mills (1975) have investigated stresses during remodelling of a rabbit calvarium. They found a relation between applied mechanical stress and the level of remodelling activity for this particular bone. Woo *et al.*, (1979) have investigated remodelling in long bones in terms of Wolff's Law. The results suggested trends in the application of Wolff's law to remodelling. Quantitative studies on the effects of compressive mechanical stresses on bone healing were carried out with the aid of finite element stress analysis by Hassler, Rybicki, Simonen, and Weis (1974). A range of stresses was identified for the maximum level of healing. Enhanced bone healing due to electrical stimulation was examined with the aid of a finite element model for the electrical current distributions by Hassler, Rybicki, Diegle, and Clark (1977). A correlation between current density and healing was suggested by the data.

Intervertebral discs have been modelled as a combination of fluid and solid material using finite element analysis. Studies have been reported by Belytschko, Kulak, Schultz, and Galante (1974) and Kulak, Belytschko, Schultz, and Galante (1976). These studies show some insight into the interactions between solids and fluids for transferring loads.

10.4.2 Fixation devices

Finite elements have proven to be very helpful in investigations of fixation devices. The device receiving most attention has been the bone plate. Internal fixation by bone plates has been investigated by finite element and experimental programs. Studies by Plant and Bartel (1974), Rybicki *et al.* (1974), Simon *et al.* (1977), Woo *et al.*, (1977), and others have shown good agreement between experimental measurements and finite element results. Rybicki and Simonen (1977) have examined the mechanics of an oblique fracture held together with a compression plate using a two-dimensional finite element model. The initial tension in the plate was found to have a beneficial effect on closing the fracture. External fixation devices have been studied by Chao, Briggs, and McCoy (1979) using a combination of finite element analysis and experimental results. This approach provides data to supplement the understanding of the mechanics of the device as well as evaluate the results of the finite element analysis.

10.4.3 Prosthetic devices

The prosthetic device receiving the most attention through finite element modelling has been the various hip prostheses. In such studies the finite element method has proven to be a valuable tool for evaluating the effects of different factors on stresses in hip prostheses. Some of the papers are mentioned in the following. McNeice (1974) investigated stresses in the femur due to a hip arthroplasty. Bartel and Ulsoy (1975) investigated the effect of stem length and stiffness on the stresses associated with a bone prosthesis. Andriacchi, Galante, and Belytschko (1976) have evaluated stresses in the stem of a total hip-replacement device. Questions about stresses in hip reconstruction were addressed by McNeice, Eng, and Amstutz (1975). The effect of the prosthesis cross-sectional geometry was examined by Bartel and Desormeaux (1976). McNeice and Amstutz (1976) and Bartel and Samahyek (1976) have investigated the stresses in the cement used to hold prostheses in place.

Combinations of finite element and clinical studies of the total hip replacement have been reported by McNiece, Ayres, and Raso (1977). Other investigators of the mechanics of the total hip prosthesis have been Svensson, Valliapan, and Wood (1977), Crowninshield and Branch (1978), Scholten, Röhrle and Sollbach (1978), Tarr *et al.* (1979), Huiskes (1978), Yettram and Wright (1979), Kwak, Lim, Kim, and Rim (1979), Hampton, Andriacchi, Galante, Belytschko Belytscho (1976), and Crowninshield, Brand, and Johnston (1979).

The topic of finite element analysis in the design of prostheses has been discussed by Lewis (1977) and Huiskes, Elangovan, Banens and Sloof (1977). Knee prostheses are receiving increased attention in terms of finite element analyses, as evidenced by the works of Lewis, Jaycox, and Wang (1977) and Askew, Lewis, and Kerr (1979).

10.4.4 Other applications

The category of other applications is selected to mention some of the newer areas of orthopaedics that are presently being investigated with the finite element analyses. These areas include the effects of heating due to the setting of PMMA cement (Huiskes, 1979), and the effects of a porous material as a means of prosthesis stabilization (Rybicki *et al.*, 1975; Rybicki, 1977; and Klawitter *et al.*, 1979). Other applications such as joint lubrication and electrical stimulation for bone healing were mentioned earlier.

10.5 SUMMARY AND FUTURE OF FINITE ELEMENTS IN ORTHOPAEDICS

The number publications and people applying finite element analysis to orthopaedics problems is evidence of the utility and potential of the method for orthopaedics applications. This paper described a role for finite element analysis as part of an effort coordinated with experiments and clinical evaluations. This role is advocated because the state-of-the-art in design of orthopaedic devices has not been developed to the point where a handbook for design is available. Thus new discoveries are being made with the design and use of new and modified orthopaedic devices. In time, all of the available information will be sufficient to develop design guides.

Concerning the future, the direction seems to be toward more coordinated studies involving engineers, materials scientists, and physicians. Prosthesis design and bone-healing techniques will become more sophisticated. The interactions of stresses and healing and stress and remodelling will probably be recognized and used to advantage in orthopaedics. Orthopaedic fixation devices and joint replacements could be tailored to meet individual needs.

Better understandings of the geometry effects in implant design and the forces and stresses acting on joints and bones will help to establish a data base in design. More experience will enable people to establish design criteria. Bone architecture has been of interest for over one hundred years and will most likely continue to be popular in the future. Research toward quantifying Wolff's law is likely, along with further investigations of function and structure relations. Finally, the effects of electrical stimulation on bone remodelling and bone healing will probably be refined and other factors effecting healing, such as temperatures, will most likely be investigated. In all, the decade of the 1980s appears to hold much promise for many finite element–related advances in orthopaedics.

REFERENCES

Andriacchi, T. P., Galante, J. O., Belytschko, T. B., and Hampton, S. (1976) A stress analysis of the formal stem in total hip prostheses, *J. Bone Joint Surg.*, **58**(A), No. 5.

Askew, M. J., Lewis, J. L., and Kerr, L. M. (1979) The effect of post geometry, material and location on interface stress levels in tibial components of total knee prostheses, *25th Annual Meeting Orthopaedic Research Society (AMORS)*, San Franciso, Cal., February.

Bartel, D. L., and Desormeaux, S. G. (1976) Stresses in prosthesis stems as a function of cross-sectional shape, *22nd AMORS*, New Orleans.

Bartel, D. L., and Samahyek, E. (1976) The effect of cement modulus and thickness on stresses in bone-prosthesis systems, *22nd AMORS*, New Orleans.

Bartel, D. L., and Ulsoy, G. A. (1975) The effect of stem length and stem material on stresses in bone-prosthesis systems, *21st AMORS*, San Francisco.

Belytschko, T., Kulak, R., Schultz, A., and Galante, J. (1974) Finite-element stress analysis of an intervertebral disc, *J. Biomech.*, **7**, 277–285.

Brown, T. D., and Ferguson, A. B., Jr. (1978) The development of a computational stress analysis of the femoral headmapping tensile, compressive and shear stress for the varus and valgus positions, *J. Bone Joint Surg.*, **60**(A), 619–629.

Chand, R., Huag, E., and Rim, K. (1976) Stresses in the human knee joint. *J. Biomech.* **9**, 417–422.

Chao, E. Y., Briggs, B. T., and McCoy, M. T. (1979) Theoretical and experimental analysis of Hoffmann–Vidal external fixation system, *External Fixation – The Current State of the Art* (eds. A. F. Brooker and C. C. Edwards), The Williams and Wilkins Co., Baltimoro, pp. 345–370.

Crowninshield, R. D., and Branch, A. J. (1978) The effect of stem cross-sectional shape on the load transmission from total hip prostheses, *24th AMORS*, Dallas, Tex., February.

Crowninshield, R. D., Brand, R. A., and Johnston, R. C. (1979) An analysis of femoral stem design in total hip arthroplasty, *25th AMORS*, San Francisco, Cal., February.

Hampton, S. J., Andriacchi, T. P., Galante, J. O., and Belytschko, T. B. (1976) Analytical approach to the study of stresses in the femoral stem of total hip prostheses, Paper 32.1, *29th ACEMB*, Boston, Mass., November.

Harris, L. J., Chao, R., Bloch, R., and Weingarten, V. (1978) A three-dimensional finite element analysis of the proximal third of the femur, In: *Trans. Orthop. Res. Soc.*, **3,** 16.

Hassler, C. R., Rybicki, E. F., Diegle, R. B., and Clark, L. C., (1977) Studies of enhanced bone healing via electrical stimuli: Comparative effectiveness of various parameters, *Clinical Orthopaedics*, **24,** 9–19.
Hassler, C. R., Rybicki, E. F., and Mills, E. J., (1975) Quantification of bone stresses during remodeling, *28th ACEMB*, New Orleans, LA.
Hassler, C. R., Rybicki, E. F., Rotaru, J. H., and Hughes, K. E. (1974) The effects of electrode configuration and calculated current density uponelectrically augmented bone healing in rabbit calvarium, *Proceedings of the 27th ACEMB*, Phildelphia, PA., p. 297.
Hassler, C. R., Rybicki, E. F., Simonen, F. A., and Weis, E. B., Jr. (1974) Measurements of healing at an osteotomy in a rabbit calvarium: The influence of applied compressive stress on collagen synthesis and calcification, *J. Biomech.*, **7,** 545–550.
Hayes, W. C. (1977) Structural analysis of the patella, *Bull. Hosp. Joint Diseases*, **38,** 96–98.
Hayes, W. C., Swenson, L. M., and Schurman, D. J. (1978) An axisymmetric model of the laterial tibial plateau, *J. Biomech.*, **11,** 21–23.
Hayes, W. C., Swenson, L. W., and Scherman, D. J. (1978) Axisymmetric finite element analysis of the lateral tibial plateau, *J. Biomech.*, **11,** 21–33.
Huiskes, R. (1978) Stress analyses of implanted orthopaedic joint prostheses for optimal design and fixation, *First Conf. Europ. Soc. of Biomechanics*, Brussels, Belgium, May.
Huiskes, R. (1979) Some fundamental aspects of human joint replacement. Analysis of stresses and heat conduction in bone-prosthesis structures, Thesis, Eindhoven University, Nijmegen.
Huiskes, R., and de Wijn, J. R. (1978) Heat conduction in acrylic bone cement, a thermodynamic analysis, *Trans. J. Bone and Joint Surg.*, **2**(1).
Huiskes, R., Elangovan, P. T., Banens, J. P. A., and Slooff, T. J. J. H. (1977) Finite element computer methods for design and fixation problems of orthopaedic implants; In: *Biomechanics VI B*, Int. Series on Biomechanics, (eds. E. Asmussen and K. Jørgensen), University Park Press, Baltimore, MD.
Klawitter, J. J., Cook, S. D., Weinstein, A. M., Clemow, A. J. T., and Lunceford, E. M. (1979) Finite element stress analysis of bone ingrowth stabilized hip prostheses, *Trans. of the 11th International Biomaterials Symposium Clemson University*, Clemson, South Carolina.
Koch, J. C. (1917) The laws of bone architecture, *Amer. J. Anat.*, **21,** 177–298.
Kulak, R., Belytschko, T., Schultz, A., and Galante, J. (1976) Nonlinear behavior of the human intervertebral disc under axial load, *J. Biomech.*, **9,** 377–386.
Kwak, B. M., Lim, O. K., Kim, Y. Y., and Rim, K. (1979) An investigation of the effect of cement thickness on an implant by finite element stress analysis, *International Orthopaedics*, Springer-Verlag, 351–319.
Lewis, J. L. (1977) Analytical methods in prosthesis design, *Proc. Workshop on Internal Joint Replacement*, Northwestern Univ. Chicago, pp. 127–134.
Lewis, J. L., Jaycox, O., and Wang, O. (1977) Stress analysis of some features of knee prostheses by finite element, *23rd AMROS*, **2,** 55, Las Vegas, Nevada.
McNeice, G. M. (1974) Effect of hip arthroplasty on stress fields in the proximal femur–A guide to prosthesis design, *Proc. 5th Canadian Medical and Biological Engineering Conference*, Montreal.
McNeice, G. M., and Amstutz, H. C. (1976) Stresses in prostheses stems and supporting acrylic–A finite element study of hip replacement, *22nd AMORS*, New Orleans, January.
McNeice, G. M., Ayres, R. K., and Raso, V. J. (1977) The total hip – finite element and clinical studies, In *Mechanics in Engineering*, University of Waterloo Press, Waterloo, pp. 83–110.

McNeice, G. M., Eng, P. and Amstutz, H. C. (1975), Finite element studies in hip reconstruction; *5th Int. Conf. of Biomechanics*, Jyväskylä, Finland; Published (1976) in: *Biomechanics VA*, Int. Series on Biomechanics, (ed. P. V. Komi, Univ. Park Press, Baltimore, M. D.

Meyer, H. (1867), Die Architectur der Spongiosa, *Archiv. f. Anat. Phys. und Wissensch Medizin*, p. 613.

Plant, R. E., and Bartel, D. L. (1974) Finite element analysis of bone–plate screw system, *Adv. Bioeng. Proc. ASME-WAM*, pp. 85–95.

Rohlmann, A., Bergmann, G., and Kobel, R. (1979) A comparative study of finite element and experimental stress analysis of a bone, *25th AMORS*, San Francisco, California.

Rybicki, E. F. (1977) Studies on the effect of mechanical stress on bone healing, bone ingrowth and remodelling, *Bulletin of the Hospital for joint Diseases*, **37**(1).

Rybicki, E. F., *et al.*, (1974) Mathematical and experimental studies on the mechanics of plated transverse fractures, *J. Biomech*, **7**, 377–387.

Rybicki, E. F., Glaser, W. A., Strankowski, J. S., and Tamm, M. A. (1979) Effects of cartilage stiffness and viscosity on a nonporous compliant bearing lubrication model for living joints, *J. Biomech.*, **12**, 403–409.

Rybicki, E. F., and Simonen, F. A. (1977) Mechanics of oblique fracture fixation using finite element model, *J. Biomechanics*, **10**, 141–148.

Rybicki, E. F. Simonen, F. A., and Weiss, E. B. (1972) On the mathematical analysis of stress in the human femur, *J. Biomech.*, **5**, 203–215.

Rybicki, E. F., Strankowski, J. S., Tamm, M. A., and Glaser, W. A. (1978) A finite element model for complaint bearing lubrication using a minimization algorithm, *WEAR*, **47**, 279–292.

Rybicki, E.F., Wheeler, K. R., Karagianes, M. T., Simonen, F. A., Hassler, and Hubert, L. E., (1975) On the design philosophy for a porous metal orthopaedic implant, *Pro. 28th ACEMB*, New Orleans, LA.

Scholten, R., Röhrle, H., and Sollbach, W. (1978) Analysis of stress distribution in natural and artificial hip joints, using the finite–element-method; *The South Afr. Mech. Eng.*, **28**, 220.

Simon, B. R., Stanley, G. M., Kamel, H. A., and Peltier, L. F. (1974) Relation of subchondral bone and articular cartilage stresses to joint function, Paper 33, *Proc. 5th Intl. Conf. Exp. Stress Analysis*, Udine, Italy.

Simon, B. R., Woo, S. L. Y., Stanley, G. M., Olmstead, S. R., McCarthy, M. P., Hemmott, G. F., and Akeson, W. H. (1977) Evaluation of one-, two-, and three-dimensional finite element and experimental models of internal fixation plates, *J. Biomech.*, **10**, 79–86.

Svensson, N. L., Valliappan, S., and Wood, R. D. (1977) Stress analysis of human femur with implanted Charnley prosthesis; *J. Biomech.*, **10**, 581–588.

Tarr, R. R., Lewis, J. L., Jaycox, D., Sarmiento, A., Schmidt, J., and Latta, L. L. (1979) Effect of materials, stem geometry and collar–calcar contact on stress distribution in the proximal femur with total hip, *25th AMORS*, San Francisco, Cal.

Valliappan, S., Svesnsson, N. L., and Wood, R. D. (1977) Three-dimensional stress analysis of the human femur, *Comp. in Biol. and Med.*, **7**, 253–264.

Volterra, E., and Gaines, J. H. (1971) *Advanced Strength of Materials*, Prentice-Hall, Inc., Englewood, N. J.

Wolff, J. (1870), Über die Innere Architectur der Knochen und ihie Bedeutung für die Frage vom Knochenwachtum, *Virchow's Arch. f. path. Anat. u. Phys.*, **50**, 389.

Woo, S. L. Y., Kuei, S. C., Dillon, W. A., White, F. C., and Akeson, W. H. (1979) The effects of stresses on long bone remodelling: A study of Wolff's law, *Trans. 25th AMOR*, **4**, 93.

Woo, S. L. Y., Simon, B. R., Akeson, W. H., and McCarty, M. P. (1977) An interdisciplinary approach to evaluate the effect of internal fixation plates on long bone remodeling, *J. Biomech*, **10,** 87–95.

Wood, R. D. (1975) Stress analysis of the femur; Thesis, Univ. of New South Wales, Australia.

Yettram, A. L., and Wright, K. W. J. (1979) Biomechanics of the femoral component with particular reference to the stress in the bone cement; *J. Biomed, Eng.*, **1**(4), 281.

Finite Elements in Biomechanics
Edited by R. H. Gallagher, B. R. Simon, P. C. Johnson, and J. F. Gross

Chapter 11

Biomechanical Analysis of External Fixation Devices for the Treatment of Open Bone Fractures

E. Y. S. Chao and K.-N. An

11.1 INTRODUCTION

The management of trauma involving long-bone fractures requires many forms of treatment methods and devices. In the course of developing these devices, there is usually a lack of quantitative information to justify their design and application. The recent advances in orthopaedic sciences have greatly assisted surgeons in selecting ideal treatment modalities for improved results. In addition, the acceptance of biomechanical engineering principles in orthopaedics has greatly enhanced the development and application of many innovative appliances, ranging from artificial joints for the treatment of arthritis to metal plates and screws for bone fracture fixation. The recent advancement in external fracture reduction devices is another outstanding spin-off application of modern biomechanics in orthopaedic surgery.

The increasing frequency of severely compounded and infected long-bone fractures with substantial soft-tissue involvement due to high-speed travel and heavy machinery usage has stimulated interest in utilizing rigid external skeletal fixation devices. The use of such devices has many advantages, in addition to the ease of facilitating soft-tissue management. The treatment of infected non-union, stabilizing bone fragments in fusion, osteotomy, and limb lengthening are only a few primary indications. Because of their versatility in adapting to different skeletal locations and the ability of providing varying degrees of rigidity, the external fixators have been utilized in tibial, femoral, pelvic, and forearm cases. As in most other orthopaedic appliances, numerous different types of external fixation apparatus were developed, but few provide quantitative data describing their mechanical behaviour under different loading conditions. Such deficiency makes successful application of these devices difficult.

Preliminary biomechanical analyses of different fixators and their structural elements were performed by a few investigators on cadaver bones and artificial models (Brinker *et al.*, 1975; Hierholzer *et al.*, 1978; Jernberger, 1970; Jorgensen, 1972; Klip and Bosma, 1978; Gunterberg *et al.*, 1978; McCoy *et al.*, 1980). The data generated from these studies provides some useful clinical information. They also found that the mechanical behaviour of each device was highly variable and sensitive to the manner of its application. Vidal *et al.*, (1970) evaluated the resistance to compression of different geometric configurations of the Hoffmann–Vidal apparatus (Figure 11.2) based on experimental models. These studies all pointed out the need of more quantitative analyses to establish the basic guidelines and precautions in order to fully utilize the advantages provided by such apparatus in patient care.

The most commonly used external fracture fixation devices can be classified according to five geometrical configurations based on their specific frame structure. As shown in Figure 11.1, these five configurations can be classified as; (1) unilateral, (2) bilateral, (3) triangular, (4) quadrilateral, and (5) circular. The triangular design can have all half pins (penetrating the skin and bone only on one side). In the circular configuration, the outer frames can either be closed or open, depending on the specific design preference. Figure 11.2 depicts a number of popular external fixators used for long-bone fracture treatment which match the defined frame configurations.

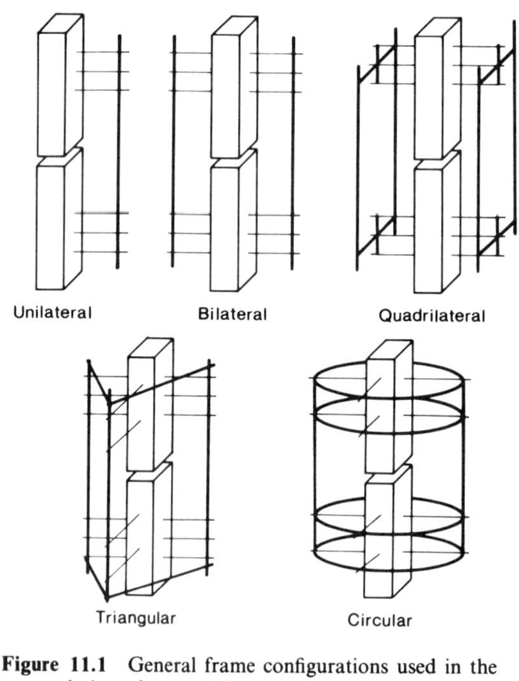

Figure 11.1 General frame configurations used in the design of external fracture fixation devices

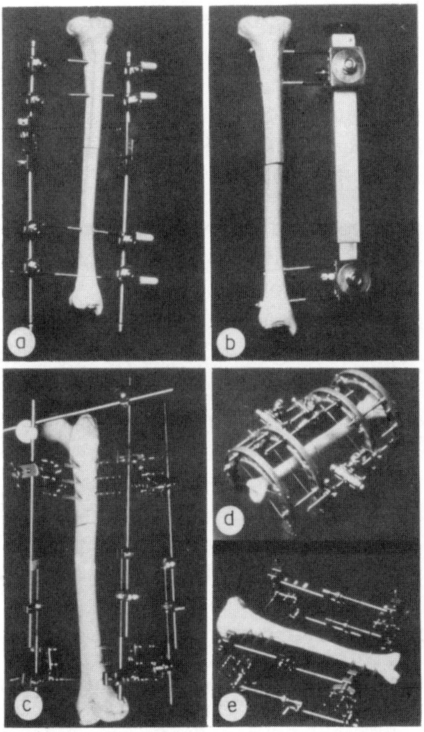

Figure 11.2 Common external fixation apparatus for long-bone fracture treatment
(a) ASIF tubular system – bilateral
(b) Wagner device – unilateral
(c) Hoffmann femur apparatus – triangular
(d) Volkov–Oganesian device – circular
(e) Hoffmann-Vidal apparatus – quadrilateral

Rigidity of the fracture fragments after fixation has been regarded as one of the most important biomechanical factors in judging the performance of an external fixator. An experimental method has been used to evaluate several devices (Briggs, 1979; McCoy, 1979). Axial, bending, and torsional stiffness values were used as the main assessment criterion. A universal testing machine was used to obtain these stiffness values based on different application configurations. Although such analysis was effective in providing overall information concerning the mechanical characteristics of each device studied, it has the following limitations:

1. The testing procedure is extremely time-consuming to accommodate a wide range of variation among multiple parameters and loading modes.

2. No information is available concerning the internal stresses for each component of the fixator, as well as the crucial pin–bone interface, to predict various potential failure modes.
3. Changing the basic geometric design and material property of the apparatus on an experimental model is infeasible to provide the needed information for future modification of the device.

For these reasons, theoretical models were developed based on concepts of structural analysis so that not only the system geometry can be changed at will, but the internal stresses for each component and interface can also be determined with great accuracy. Besides, the pin and frame material property and the variation of all application configurations can easily be simulated for thorough investigation. Experimental results were used to validate the theoretical predictions. In this chapter we present the modelling procedure and the analysis results for idealized frame configurations, which can adequately describe the mechanical behaviour of several popular external fixators currently in use.

11.2 STRESS AND RIGIDITY ANALYSIS

11.2.1 Structural model development

All external fixators, when applied to fractured bone fragments, can be modelled as frame structures (see Figure 11.1). The pins are threaded to provide rigid fixation to the bone. These pins are fastened to side bars or circular frames through tightly fitted set screws to prevent sliding or rotation. Therefore, the entire structure can be considered as a rigid frame, since all members are connected by rigid joints. When the bone is loaded three-dimensionally, the reactions at each joint of the entire frame structure cannot be determined by the principles of statics. Such statically indeterminate structures require special methods (Wang, 1953) to obtain the complete solution for bending moments and deflections.

However, when the device becomes more complex and involves numerous geometry and material changes, the available structural analysis programs may not be suitable for direct application. Modification of the existing programs must be performed and the use of the digital computer would also be necessary. The use of finite element analysis becomes more attractive, since it can handle the indeterminate problem easily. The ability to provide systematic formulation, graphical display of the results, and frame deformation plots are some of the additional advantages. For the present problem, the finite element method was a natural choice to obtain both the stiffness property of the system and the internal stresses in the key structure elements.

A simple frame structure, as shown in Figure 11.3, was chosen for the basic model. This model simulates both the unilateral and bilateral frame

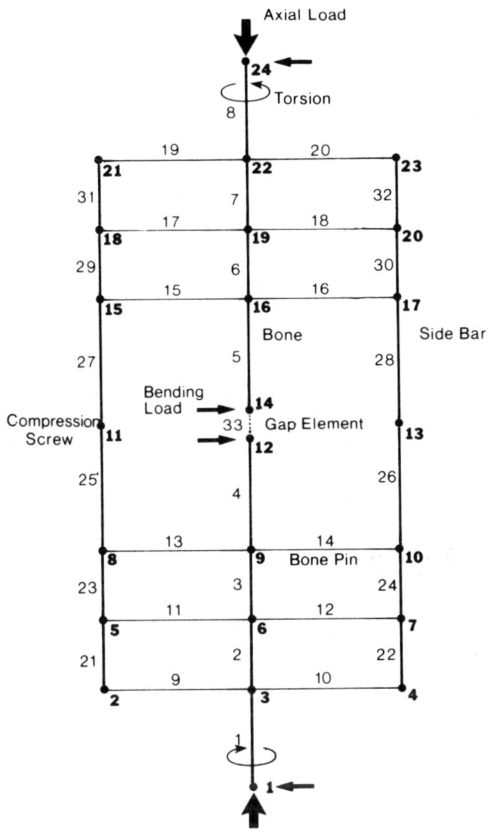

Figure 11.3 Finite element model using three-dimensional beam elements to analyse the structural rigidity and internal stress of a bilateral external fixator

configurations. For the quadrilateral configuration, this model was also considered adequate, since the rigidity of the side bar is at least an order of magnitude higher than that of the transfixing bone pins. Three pins on either side of the bone fracture site were used in the standard model. Triangular configurations were also modelled with moderate changes in the fundamental geometry of the frame structure. All variations on the pin–bone orientation, pin size and number used, pin material, and loading modes were easily implemented through the finite element program.

Simple two-dimensional elastic beam elements were used for the plane load case. The cross-sectional shape of the element is symmetric with respect to the neutral axis so that the maximum bending stresses are calculated at one-half of the height of the beam. These elements must remain in the x–y plane throughout the deformation process and all elements must not have zero length or area.

Under transverse bending loads and torsion applied through the bone ends, three-dimensional elastic beam elements were used. This element has six degrees of freedom at each node along the three orthogonal nodal axes, as seen in Figure 11.4. Cubic polynomials are used for the displacement function in order to calculate rotations.

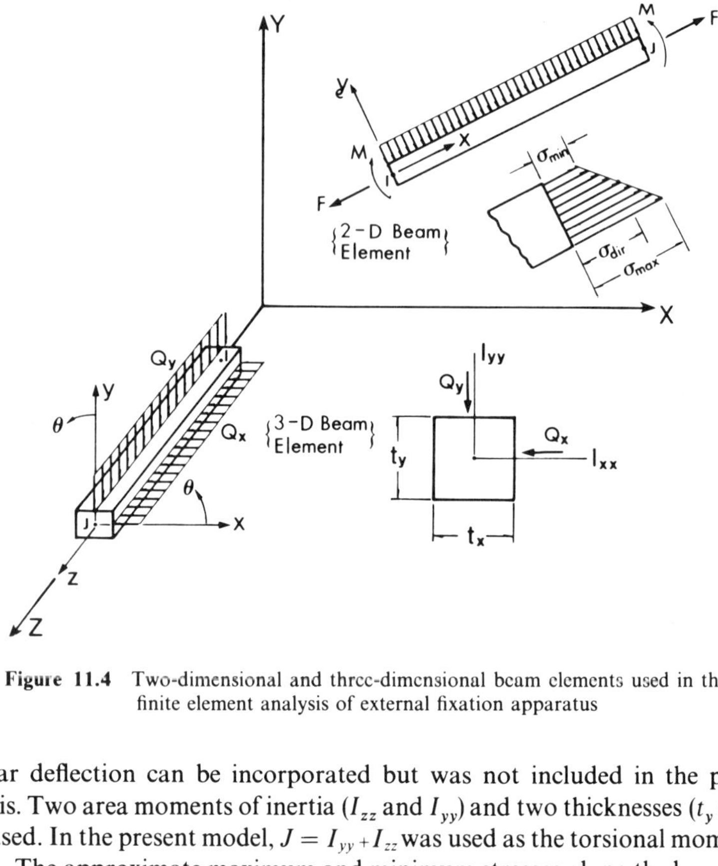

Figure 11.4 Two-dimensional and three-dimensional beam elements used in the finite element analysis of external fixation apparatus

Shear deflection can be incorporated but was not included in the present analysis. Two area moments of inertia (I_{zz} and I_{yy}) and two thicknesses (t_y and t_z) were used. In the present model, $J = I_{yy} + I_{zz}$ was used as the torsional moment of inertia. The approximate maximum and minimum stresses along the beam were calculated according to the diagrams shown in Figure 11.4.

Under specific cases, bone fracture gap (element no. 33 in Figure 11.3) was modelled as either a two- or a three-dimensional interface element (gap element, Figure 11.5), which helps to predict fracture-fragment contact and displacement after compressive load or bending was applied. If there was no gap between bone fragments to begin with, the bone fracture site was modelled by a single node. Symmetric boundary loading conditions were used at the bone ends and the fracture site in order to maintain the structure in static equilibrium. Displacements in non-loading planes for the bone were properly constrained in order to

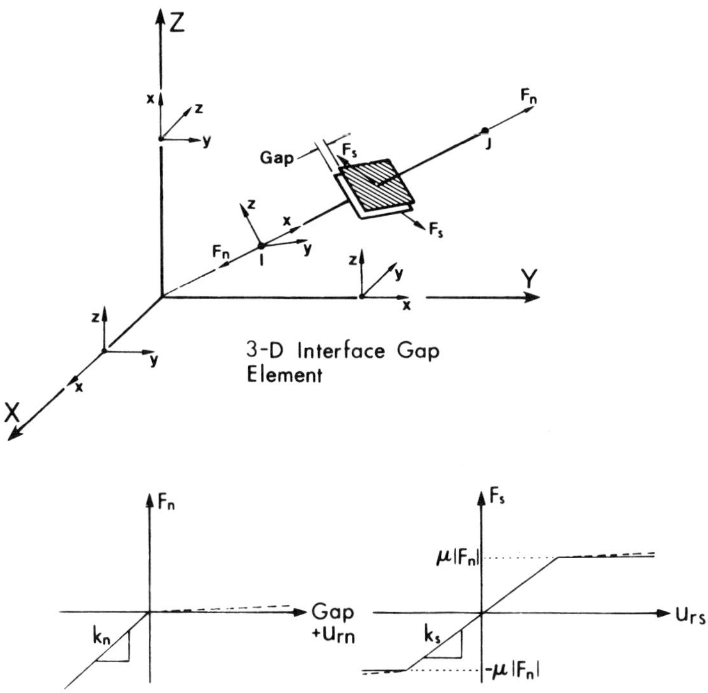

Figure 11.5 Three-dimensional gap element used to model the fracture interface in the finite element analysis

avoid a singular matrix. The key input data describing the material property and geometry of the basic model is summarized in Table 11.1. Various parameters were subsequently changed in the model to provide the parametric analysis.

Two finite-element programs were used to solve the present problem. The ANSYS program (Swanson Analysis System, Inc.) was used for its superior pre- and post-processor capabilities. A convenient grid generation scheme and a graphic package for frame deformation plots were used to facilitate easy presentation of the model and the results. The computer output consists of the direct (axial) stress, the bending stresses at each end of a beam element, and the maximum and minimum stresses at the nodes. The maximum stress is computed as the sum of the direct stress and the absolute values of both bending stresses (in the *y* and *z* directions; see Figure 11.4). The twelve-member forces and moments (six at each nodal point) are also available through an option. The SAP-IV program (Earthquake Engineering Research Center, University of California, Berkeley) was used to determine bone fracture stiffness property and bending moments at the critical element. Due to its simpler format and limited data-processing capability, substantial computer time can be saved.

Table 11.1 Structural parameters of the 2-D finite element model of an external fracture fixation apparatus

Structural component	Element no.	Length (in)	Diameter (in)	Area (in^2)	I_{zz}, I_{yy} (in^4)	I_{xx} (in^4)	E (10^6 psi)	K_n (lb/in)	v
3-D bone element	1, 8 2, 3, 6, 7 4, 5	3.321 0.875 2.0	1.0	0.785	0.049	0.098	2.2	–	0.3
Bottom pins	9–14	2.5	0.16	0.02	0.32×10^{-6}	0.64×10^{-6}	29	–	0.3
Upper pins	15–20	2.5	0.16	0.02	0.32×10^{-6}	0.64×10^{-6}	29	–	0.3
Left side bar	21, 22, 25, 26, 23, 24	0.875 2.125	0.315	0.08	4.8×10^{-4}	9.6×10^{-4}	29	–	0.3
Right side bar	27, 28, 31, 32, 29, 30	0.875 2.125	0.315	0.08	4.8×10^{-4}	9.6×10^{-4}	29	–	0.3
Fracture gap	33	0.25	–	–	–	–	–	10^6	–

I_{xx} = Polar area moment of inertia, J
K_n = Axial stiffness of the 3-D gap element.

11.2.2 Pin– bone interface model

In order to analyse bone stress near the pin, a refined finite element model was developed using eight-node three-dimensional isoparametric solid elements. A simplified doubly symmetric model was used (Figure 11.6) in which the element mesh was generated for only one-quarter of the entire system. The mesh density for two sections of the model for bone and bone–pin interface is shown in Figure 11.7. The nodes on the surfaces of the quarter model were constrained to move only along the plane of symmetry. No out-of-plane displacement was allowed. This permitted a realistic response of the cylindrical bone segment supported by a transverse pin against axial load. The complex geometry of the element grid system for bone and pin was developed with the aid of the powerful UNISTRUC II program (Control Data Corporation, Graphic Division), which is available through the CYBERNET Services.

To further simplify the system, the remaining section of the bone pin not in contact with the bone was modelled by three-dimensional beam elements (Figure

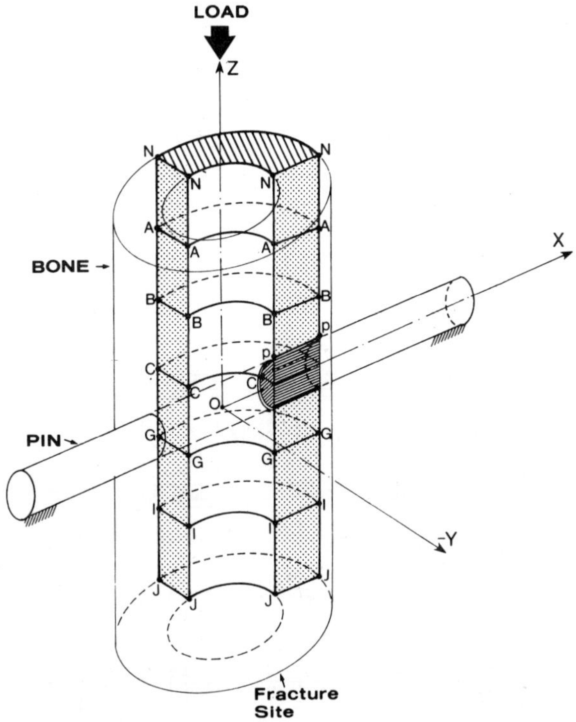

Figure 11.6 The doubly symmetric structure used to develop the finite element model for pin–bone interface system. Only a quarter of the system is required for the element mesh generation

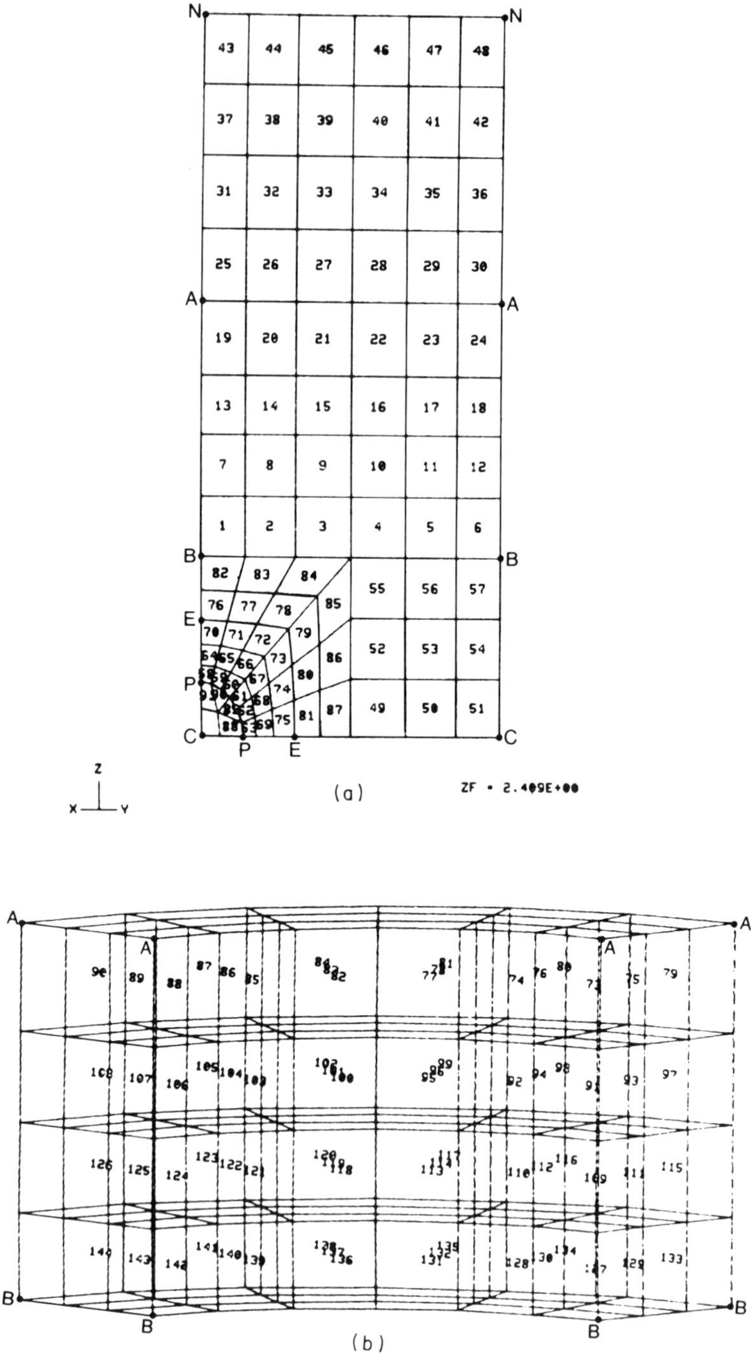
N
N
A
A
B
B
E
P
C
C
P
E
Z
X
Y
(a)
ZF • 2.409E+00
A
A
A
A
B
B
B
B
(b)

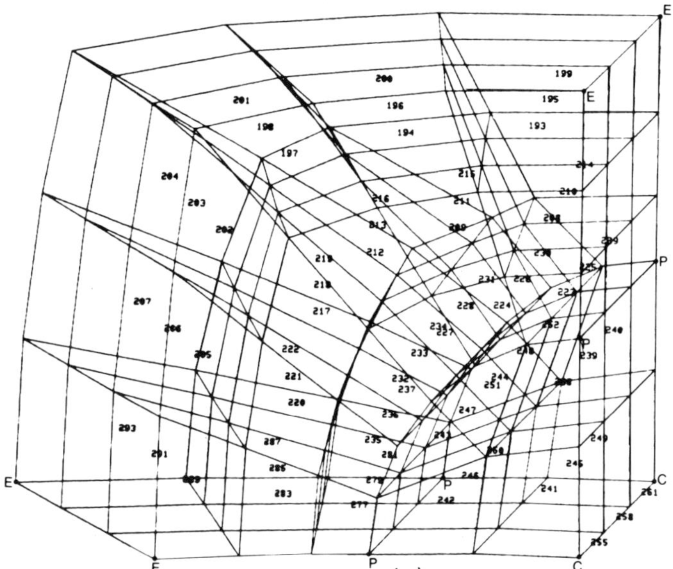

(c)

Figure 11.7 Mesh density and element orientation of the bone-pin finite element model
(a) Plane geometry of the grid used to generate the three-dimensional model with different key blocks indicated
(b) Element distribution of a section of the bone (90° segment) away from the pin
(c) A close-up section of the bone and pin (P–P section) interface

11.8). The nodes on the pin surface were assumed to couple with the nodes of the bone on the interface. Thus, no separation was allowed. The nodes on the side faces of the pin were assumed to have free displacement along the x-direction (direction of the pin). Such simplification should not affect the overall response of the system under the present loading condition. The pin was rigidly fixed on its ends and axial load was applied only at one end of the bone, leaving the other end free. Based on this model, critical stress field in the bone near the bone–pin interface can be analysed adequately.

The present model can also be used to study bone–pin interface stress subjected to other types of loads. From the previous structural model, bending moments in the pin elements were obtained. Such moments can be used as the loading conditions for the refined bone–pin model to determine the bone stresses. In this case, bone ends must be properly fixed to ensure the static equilibrium requirements.

The bone–pin model consisted of 552 three-dimensional, eight-node iso-parametric elements and 9 three-dimensional beam elements with a total of 1218 nodes. The ANSYS finite element analysis program was used to calculate

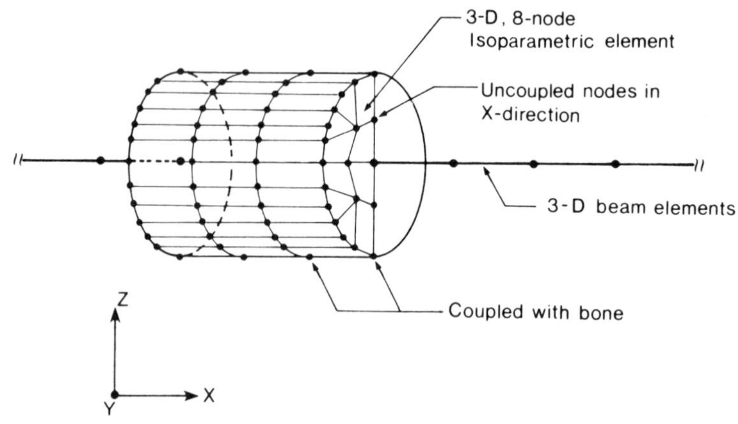

Figure 11.8 The bone pin was modelled by three-dimensional 8-node isoparametric elements for the section interfaced with the bone. The remaining portions of the pin were modelled with three-dimensional beam elements

element and node stresses. A post-processor was used to obtain stress contour plots to better visualize stress field in the bone adjacent to the metal pin. Each run for a selected loading case consumed 478 seconds on a CDC-7600 computer.

11.2.3 Experimental model testing

In order to quantitate the fracture stiffness properties of various types of external fixators, physical models were tested on an MTS material testing machine. Different loading conditions were applied through the bone fragments and the displacements at the fracture site were recorded to derive the stiffness moduli. For the convenience of providing uniform bone property, synthetic bone models made of Swedish body putty* material was used. This material was tested against human cortical bone and found that they were compatible in mechanical properties.

Load cells, LVDT, RVDT, and micrometers were used to indicate the applied load and the corresponding displacements and the results were recorded on an x–y plotter. For the convenience of documenting the amount of fracture site compression, a flat load washer was placed at the fracture gap. Several external fixators are capable of introducing fracture compression by turning the adjustment screws on the side bars. Such a load washer can help to calibrate the amount of compression introduced based on the number of turns of the screws. The testing apparatus as mounted on the MTS machine. The flat load washer and the micrometers are illustrated in Figure 11.9.

* Produced by Aktiebolaget Plastic Co., Götenberg, Sweden

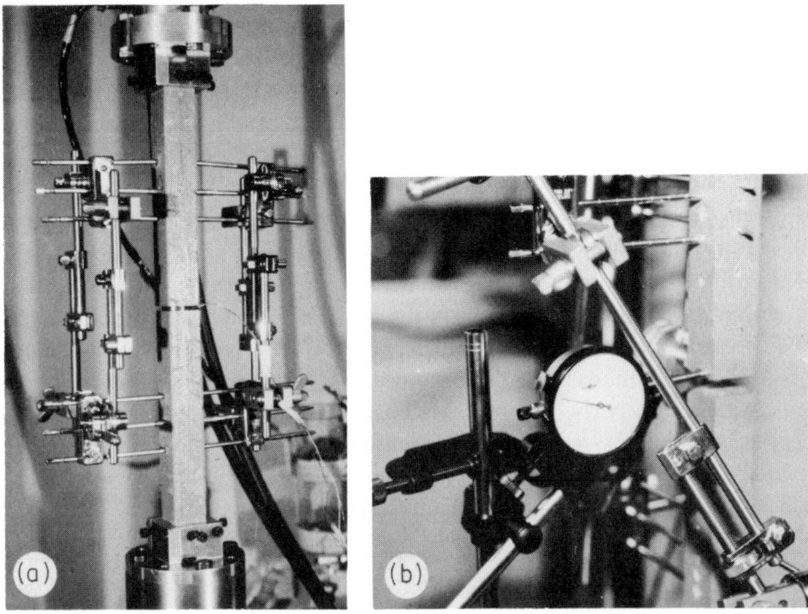

Figure 11.9 Experimental set-up utilized to measure bone compressive force and fracture stiffness
(a) A flat load washer placed at the fracture gap
(b) Micrometers used to measure bone fragment displacement at the fracture site

It is interesting to compare the fracture stiffness property of the internal fixation methods (the use of bone plate-screws and intramedullary rods) and the external fixation techniques. Paired bone specimens were fractured experimentally and fixed with eight-hole plates, Küntscher nails, and external fixators. Their respective stiffness values were recorded. The objective of this test was to examine whether external fixation devices are able to duplicate the stiffness property provided by the well-established internal devices to achieve bone union.

11.2.4 Testing criteria

The main performance criterion for the external fracture fixation device is the ability to control fracture site displacement under applied load exerted through the bone fragments. Based on physiological observation and experimental testing, there are three types of loads applied through the bone ends. These are: (1) axial load, (2) bending in both anteroposterior (AP) and lateral directions, and (3) torsion. The corresponding displacement that occurred at the bone fracture site reflects the stiffness property of the fixation apparatus. If the applied load is plotted against the fracture displacement, three stiffness values can be determined:

1. *Axial stiffness*, K_1 – Axial load versus longitudinal displacement of the bone fragment at the fracture site.
2. *Bending stiffness* (both AP and lateral), K_2, K_3 – Bending load versus AP or lateral displacement of the fracture site.
3. *Torsional stiffness*, K_4 – Torque applied through the bone ends versus the axial rotation of the bone fragments occurring at the fracture site.

Since the structural model is symmetric and the external loads are uniformly applied through the bone ends, the cross-coupling effect in displacement will be minimal and thus ignored. The mechanical behaviour of the frame structure was kept within its elastic range, therefore the linear superposition principle can be applied to combine the stiffness values for overall evaluation. It is difficult to evaluate the performance of a device and its varying configurations based on a specific stiffness property. A combination of all stiffness values must be utilized. However, each stiffness value has a different range of magnitude and dimensions. It would be more convenient to develop an equivalent stiffness value (or modulus) which is the weighted sum of all the different individual stiffness values. The equivalent stiffness modulus, κ, can be expressed as:

$$\kappa = \sqrt{\sum_{i=1}^{4} \lambda_i K_i^2}$$

λ_i are the relative rating scores to be determined. For the present investigation, each K_i value was rated equally with $\lambda_i = 1$. The magnitude of external loads was derived from the experimental measurements of patients performing level walking. The axial load, bending moments, and torque were varied to accommodate body weight variation among the population.

In order to perform parametric analysis, the following parameters were varied systematically:

1. number of bone pins used;
2. symmetric versus non-symmetric frame configurations;
3. pin diameters;
4. presence and absence of bone fragment compression;
5. pin group separation;
6. half-pin (unilateral configuration) versus full-pin (bilateral or quadrilateral configuration) application; and
7. pin and frame material property.

In addition to the stiffness property of the bone fragments, pin and frame stresses were also determined in the analysis. Based on these results, the mechanical behaviour of the external fixation system was evaluated. In order to assess the accuracy of the theoretical model developed, experimental tests were conducted on the stiffness property of the pin, frame, joints, and the entire fixator–bone system.

11.3 RESULTS

11.3.1 Structural stiffness

The stiffness values for different loading modes were found to be linear with respect to the magnitude of load applied. This made it easy to study the combined loading case, since the superposition principle could be used. The use of the equivalent stiffness value (κ) has greatly facilitated relative comparison of different devices, as well as the variation of geometric configurations.

In the standard configuration, as shown in Figure 11.10, the side-bar separation was chosen to be 5 inches ($a = b = 2.5$ inches) with the two inner pins on either side of the fracture spaced at a distance of 4.25 inches ($L = 2.125$ inches). The individual pin separation (h) within each group was placed at 0.875 inches apart. In this case, the material for both pins and the side bars were assumed to be 316 L stainless steel ($E \simeq 29 \times 10^6$ psi). The stiffness values for each loading mode, as well as the equivalent stiffness, are presented in Table 11.2. The effects of structural geometry on the stiffness values were investigated by varying the parameters shown in Figure 11.10. It was found that the overall stiffness of the structure can be increased significantly by increasing the pin number, the pin diameter, and decreasing the side-bar separation. These effects are not linear with respect to the incremental change of the parameters. More specifically, the effects of pin number to the relative change in stiffness values of the structure are plotted in Figure 11.11. If only one pin was used in one side of the bone, the axial stiffness would be further reduced. However, one pin cannot maintain the bone stability under AP bending load; such a configuration would be clinically infeasible. The increase of overall stiffness was significant when the total pin number was increased from six to eight. The precise effects of pin diameter and side-bar separation on the axial stiffness are summarized in Table 11.3.

Table 11.2 Stiffness value of the standard external fixation configuration

Loading mode	Stiffness value
Axial	2059 lb/in
AP bending	56 lb/in
Lateral bending	636 lb/in
Torsion	20 in lb/deg
Combined (κ)	2156

The effects of material property are equally significant on the overall structure behaviour. Since the stiffness property of the side-bar elements are more than ten times that of the pin elements, the material effect will be more pronounced in the pin–bone interface, thereby affecting the structural stiffness value of the fixation system substantially. Two types of material currently being used to manufacture

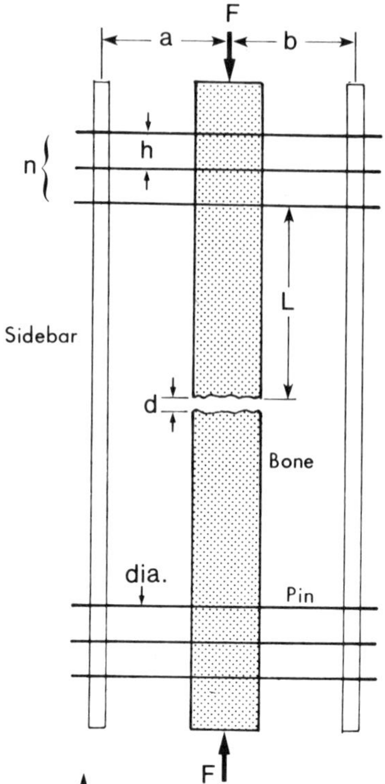

Figure 11.10 Geometric parameters used in the standard model for variation analysis

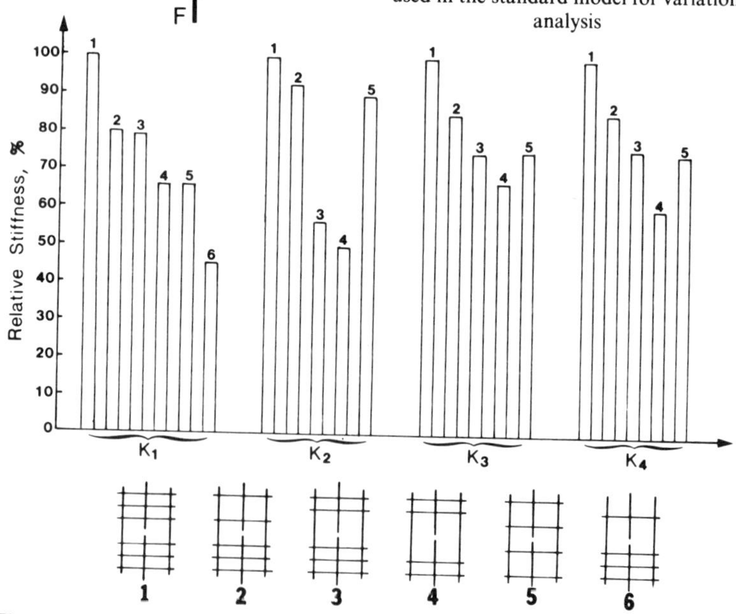

Figure 11.11 The effects of pin number on the reduction of fracture stiffness under different loading conditions

Table 11.3 Variation of axial stiffness based on pin diameter and side bar separation

	Parameter	K_1	K_2	K_3	K_4	κ
Pin diameter	$D = 0.16''$ (standard)	2059	56	636	20	2156
	$D = 0.20''$	4854	94	1180	39	4996
	$D = 0.24''$	9457	127	1963	60	9660
Side bar separation	$L = 2''$	3954	76	735	23	4022
	$L = 2.5''$ (standard)	2059	55	636	20	2156
	$L = 3''$	1201	42	567	18	1329

the external fixators were investigated. They were 316 L stainless steel ($E \simeq 29 \times 10^6$ psi) and the commercially available pure titanium ($E \simeq 14.5 \times 10^6$ psi). Four material combinations were assumed: (1) stainless steel pins and side bars; (2) stainless steel side bars and titanium pins; (3) titanium side bars and stainless steel pins; and (4) titanium side bars and pins. Under various loading conditions, the corresponding stiffness values for each material combination are presented in Table 11.4.

Table 11.4 Material properties versus structural stiffness

Material combination	Structural stiffness K_1	K_2	K_3	K_4	κ
S. S. pin and bar	2059	56	636	20	2156
Ti pin and S. S. bar	1041	32	419	11	1123
S. S. pin and Ti bar	2029	43	546	18	2102
Ti pin and Ti bar	1033	28	331	10	1058

It is obvious that if the bone pins were made of materials with a small modulus, the fracture stiffness will be reduced substantially. However, if only the outer frame was made of a more elastic material, the reduction in stiffness value would become insignificant. The advantage of using titanium is the considerable reduction in weight, which makes patient application more attractive. Using the combination of titanium bars and stainless steel pins will satisfy this functional requirement without sacrificing the structural rigidity. However, stainless steel pins are easier to deform due to the low yield stress of the material. In pin removal, this may be an important clinical factor to be considered.

One of the common forms of external fixators used clinically involves half-pins under the unilateral configuration shown previously in Figure 11.1. A dramatic reduction of fracture stiffness is expected, which was quantified based on the current theoretical model. A relative comparison of the stiffness values between the standard bilateral and the unilateral half-pin configurations is summarized in Table 11.5. It is interesting to note that the reduction of the axial and AP bending stiffness values is much more significant than other loading modes when half pins

Table 11.5 Relative comparison of fracture stiffness between bilateral full-pin configuration and the unilateral half-pin configuration

Configurations	Relative stiffness, lb/in or in lb/deg K_1	K_2	K_3	K_4
Full-pin bilateral	2059	55	636	20
Half-pin unilateral	318	28	104	8
Percent reduction	84	49	84	60

were used. This defect can be remedied by applying half pins in two perpendicular planes, thus forming the triangular configuration. In the regular bilateral design with full pins, the weakness in AP bending can be improved by using additional half pins in the anteroposterior direction. The overall stiffness will also be increased.

The effects of other parameters, such as the individual pin separation, h; pin group separation, L; and asymmetry of pin and frame placement on the overall rigidity of the bone fragments, were less significant. However, in AP and lateral bending loads the parameters h and L have more effect on the bone fragment stiffness, which must be recognized. A summary diagram illustrating the effect of four major parameters to the stiffness property of an external fixator is shown in Figure 11.12. Only relative effects were presented, since each parameter has a different influence on the individual stiffness value.

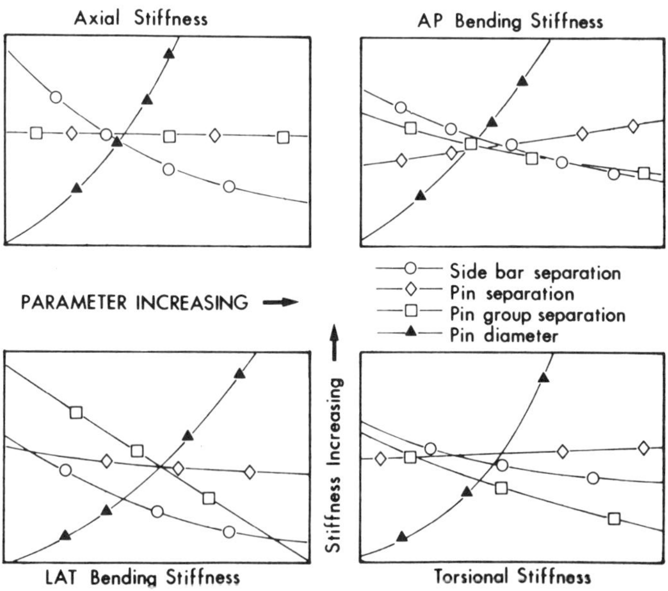

Figure 11.12 A parametric presentation of the effects of fixator geometry to fracture stiffness property

11.3.2 Pin and bar stresses

Due to the large difference in cross-sectional area between the pins and the bars, only the pin stresses were significant to render further investigation. The bone stresses near the bone–pin interface were grouped separately and will be discussed in the next section. The axial (direct) and shear stresses in the pin due to axial and shearing loads were found to be substantially smaller than the bending stresses and thus ignored.

The maximum bending stress in each pin varies linearly with the applied load. Under axial load, the bending stresses among the pins in each group vary only minimally with the middle pin carrying slightly higher stress. The inner two pins carry tensile load and the outer pin is in compression. However, the axial stress is substantially lower and combined stress analysis was not performed. Under different loading conditions, the maximum pin stresses and their location are summarized in Table 11.6. The magnitudes for axial, torsional, and bending loads assumed are higher than the clinically expected values. For the convenience of providing relative comparison, magnitudes of 10 and 100 were used.

Table 11.6 Maximum pin stress and location under various loading conditions

Loading condition	Magnitude	Pin location	Pin stress (psi)
Axial	100 lb	Middle pins	105 500
AP bending	10 lb	Inner pins	71 000
Lat. bending	10 lb	Outer pins	22 750
Torsion	10 in lb	Inner pins	14 075

These stress data were based on the condition that there was no bony opposition at the fracture site. However, when bone fragments are directly in contact, the maximum pin stress is reduced by 97 % of the original value. Under regular weight bearing in a patient fitted with such a device, the pin stress can reach beyond the yield limit of the material. Permanent pin deformation is quite possible. When fewer pins were used, the pin stresses could easily have reached the yield or fracture point, even under moderate load. This explains why pin deformation occurred in certain clinical applications, especially with stainless steel pins.

The effects of other geometric parameters on pin stresses were found to be much less significant when compared with the results of the stiffness values. This finding is not surprising since the beam element deflection is proportional to the cube of the length, while the bending moment (or stress) is only proportional to the length itself. Increasing the pin diameter and pin material modulus resulted in less dramatic reductions in pin stress. However, when the half pins (unilateral configuration) were used under axial loading and AP bending, the maximum pin bending stress was increased nearly 100 % as compared to the standard bilateral full-pin configuration. On the other hand, the pin stress was substantially smaller under torsion and lateral bending.

11.3.3 Bone stresses near pin holes

Based on an axial compression through the bone ends on a standard external fixator, each of the three pins on the same bone fragment carries about the same amount of load. Therefore, the net force applied to the pin through the bone–pin interface should be about one-third of the total end force. Since the axial rigidity for the bone is quite large as compared to the flexural modulus of the pins, the same amount of bending moment would be shared by the pins too. However, for both torsion and bending loads, each pin carries different moments. These moments for all the pins were available from the previous frame structural analysis. For the sake of presenting results easily, a dimensionless unit compressive load was applied through the bone–pin system, as shown in Figure 11.6.

Based on the three-dimensional finite element model developed, the maximum compressive stress was found at the outer portion of the bone cortex immediately above the pin. Another region of high compressive stress is located at the inner cortex below the pin. Since the model and loading conditions are symmetric, the maximum compressive stresses are nearly along the z-direction (σ_z) in the global coordinate system. When stainless steel pins with a diameter of 0.15625 inch (4 mm) are used, the maximum compressive stress has a magnitude of $114F$ psi, where F is the dimensionless unit load applied through the bone end. The stresses in the pin are much higher with values of $326F$ psi at the pin–bone interface and $330F$ psi at the ends fixed to the side bars. Checking with a simplified beam model with both ends fixed, pin stresses agree quite well with the finite element solution.

For the convenience of visualizing the stress field in the critical regions of the bone, contour stress plots for the Von Mises stress were processed. Two sections of bone were examined and both were directly adjacent to the pin–bone interface. One section is a longitudinal cut through the bone in the x–z plane directly above the pin (Figure 11.13). The other section is a transverse cut in the x–y plane immediately adjacent to the surface of the pin (Figure 11.14). Both plots provide a clear description of stress distribution near the pin–bone interface. It is quite obvious that high bone stresses are concentrated close to the pin, as one would anticipate intuitively. The tensile stress in bone does exist based on the current model, but its magnitude slightly less than the compressive stress generated.

In addition to stainless steel pins, titanium material was also tried for the pin based on identical geometry. It was interesting to find out that the bone stress was increased by about 17% when titanium pins were used. No geometric variation was tried, since the model is too expensive to regenerate for the purpose of

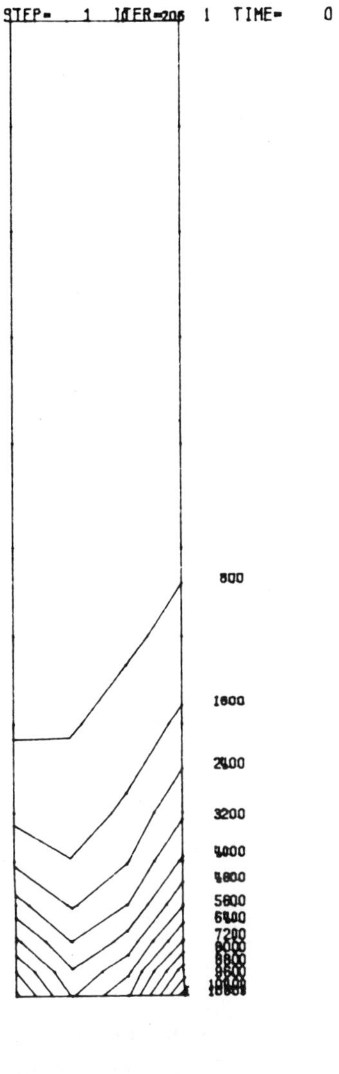

Figure 11.13 Contour plot of the Von Mises stress through a longitudinal section of the bone immediately above the pin. The stress magnitude has been magnified by a factor of 100

parametric analysis. However, different loading conditions were tried by applying the bending moments determined from the structural analysis to the pin–bone interface as input forcing function without changing the present finite element model geometry and mesh density. This portion of the results will not be included here. In general, axial force is worst in generating high bone stress, especially when there is no direct contact at the fracture gap. For various magnitudes of load, the current results can be extrapolated, since a linear system was assumed. This will also allow combined stress analysis using the superposition principle.

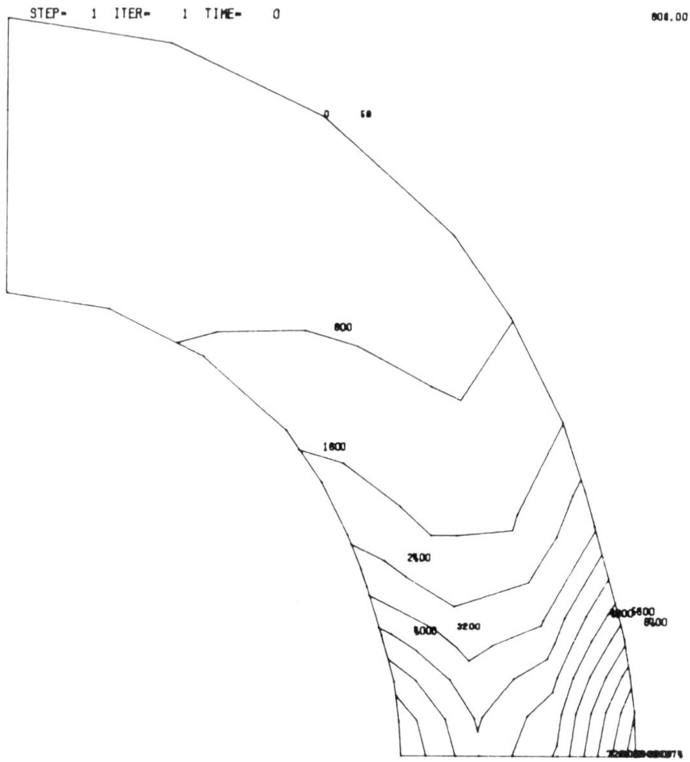

Figure 11.14 Contour plot of the Von Mises stress through a transverse section of the bone at a distance of 0.08 in. (2 mm) above the pin. Only one-quarter of the plot is shown and the rest is symmetric with respect to the two orthogonal end lines. The stress magnitude has been multiplied by a factor of 100

11.3.4 Experimental test results

For the purpose of validating the finite element model developed, experimental tests were conducted. Under similar geometry and loading conditions, the comparisons of stiffness values between the theoretical and the experimental models are presented in Table 11.7. In general, the stiffness values obtained from the experimental model are lower. In the actual model (quadrilateral configuration of the Hoffmann–Vidal type) more mechanical joints are involved, while their stiffness characteristics are difficult to define. These joints usually are not absolutely rigid, and they will undoubtedly result in a slightly less stiff structure. The same difference occurs at the bone–pin junctions. This explains why the stiffness values between the two models are at variance. However, in the comparison of bone fragment compression measured by a transducer with that

Table 11.7 Comparison of stiffness values between the theoretical and the experimental models

Load mode	Stiffness values Theoretical model	Experimental model
Axial	2059 lb/in	1347 lb/in
AP Bending	56 lb/in	41 lb/in
Lateral Bending	637 lb/in	440 lb/in
Torsion	20 in-lb/deg	14 in lb/deg

predicted by the theoretical model, the data agreed closely. This result suggests that the theoretical model can simulate the mechanical structure adequately.

In the actual application of the present analysis, emphasis was placed on the relative comparison of the results as influenced by the geometric and material changes of the structure. As long as the theoretical model behaves similarly to the experimental model, the exact match of the stiffness property is immaterial. Extensive experimental tests were conducted to investigate the mechanical behaviour of different devices following similar parametric analysis presented here (Chao, Briggs, and McCoy, 1979). The experimental results hold a parallel trend with the theoretical predictions. Consequently, the current finite element model was considered adequate and the results generated should provide relevant information for apparatus design and patient application.

In testing bone specimens stabilized with internal and external fixation devices, the relative stiffness comparison is shown in Figure 11.15. As one would expect,

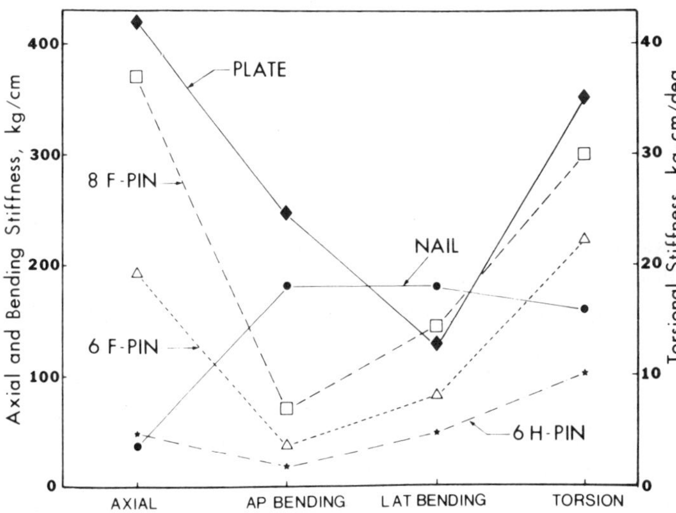

Figure 11.15 Relative comparison of fracture stiffness of bone specimens fixed with both internal and external fixation devices

each type of fracture fixation apparatus provides drastically different stiffness properties in each loading mode. The stiffness patterns for plates and nails are completely opposite. This means that bone fracture healing can take different mechanical environments provided by either the rigid plate fixation or by less rigid intramedullary nail fixation. The stiffness properties among the external fixators with six full pins, six half pins, or eight full pins are different in magnitude, but they do follow a similar pattern. If one wishes to use an external fixator to achieve bone fracture healing, it would be logical to attempt to duplicate the stiffness characterstics inherent to the internal devices. Unfortunately, none of the devices studied were able to achieve this without significant geometric modification.

11.4 DISCUSSION

Certainly all the data produced by the analytical model could be obtained through experimental testing, but the number of tests one would have to perform based on the combination of multiple variables makes it nearly impossible to accomplish. Certain information cannot be produced easily by experimental means. However, a fair amount of experimental testing is necessary in order to assess the reliability of the theoretical model developed. Close correlation between the theoretical and the experimental data has been obtained, which greatly enhances the applicability of the present methodology.

Although the structural model was considered quite simple by engineering standards, it is statically indeterminate and involves three-dimensional loading conditions. The use of finite element analysis is a logical choice which can produce an enormous amount of information with only limited computer time and analysis effort. Many different configurations encountered in realistic clinical situations can easily be simulated by the same model to obtain further information. This effort has well exemplified the power of the analytical method in biomechanical investigations.

The mechanical behaviour of the standard external fixator for bone fracture fixation depends greatly on the geometry and material property of the device. With such knowledge, better design can be conceptualized to achieve more ideal results. More importantly, the present data can help surgeons achieve optimal treatment results if certain apparatus is given. The effects of various configuration changes on the rigidity of the fixator have been quantified to establish proper guidelines for patient application. Such variability can be utilized to duplicate the stiffness characteristics of internal devices. It is, therefore, reasonable to conclude that external fracture-fixation apparatus can and should be used as a means to treat fractures and obtain bone union similar to any other form of internal fixation.

Pin stress developed in the system during application of an external fixator can be extremely high, which may cause pins to deform permanently or even fracture.

If less pins or half-pins are used, such an effect would be even more dramatic. However, large pin deformation will also cause substantial bone fragment motion which will irritate soft tissue and cause pain. The pain can serve as a control reflex to reduce load bearing, which will in turn decrease pin stress and deformation. Such feedback mechanism will serve as a safety measure to avoid overloading on bone pins. When the fracture is beginning to unite, bone fragment opposition will be present which can significantly reduce pin stress, even under large axial load. This explains why, at a later stage of fixator application, the patient can begin load bearing without pain. However, if the fracture is not healing, weight bearing should be discouraged to prevent potential failure of the bone pins.

Two possible factors can cause pin deformation clinically. Firstly, the pins can be permanently deflected under high axial compressive loads transmitted through the bone while the bone ends do not maintain contact at the fracture site, and secondly, while the bone fragments are being forced together to carry high compressive loads by contracting the side bars. In this case, the pins are under a bending load in the reverse direction, similar to the case when bone fragments are subjected to distraction forces. Although such mechanical failure is infrequent in clinical practice, biological failure pin–bone loosening and infection may occur more often and could be related to interface stress. The demonstrated high bone stress at the pin–bone interface identifies one of the sources of this potential complication. During AP bending, gravity and inertial effects can easily introduce bone stress based on the same principle. Under this situation, there is the additional component of axial rotation along the pin axis which can stimulate more loosening effects, since the resistance between the bone and pin under such a loading mode would be small, especially when smooth pins are used.

Under certain circumstances, the amount of fracture compression and bone stress near the pin holes induced by the fixation apparatus can be estimated by measuring the deflectiin of the bone pins on X-ray, provided that the pins are still within their elastic limit. The presence of fracture site compression will not only increase structural stiffness of the system and reduce pin stresses; it may also enhance bone healing. However, if the bone compressive stress becomes too large, necrotic reactions may occur which can cause adverse effects. Application of bone fragment compression through adjustment screws on the side bars is not a simple task. First, the screws must be turned symmetrically to prevent distortion being introduced to the frame rather than generating uniform compression to the bone fracture ends. On the other hand, each turn of the screw can produce a significant amount of compression when the bone ends are already in contact with each other. Using mechanical aids to turn the screw is definitely not recommended.

Other commonly used external fixators bear the forms of triangular and circular configurations (Figure 11.2). In the bilateral and quadrilateral configurations, additional half-pins can be used in other planes to enhance bone fracture rigidity. These types of application are required to create specific stiffness properties for the external fixators that can provide similar rigidity obtained through the use of internal fixation devices. Therefore, it is quite possible that ideal stiffness matching between external and internal devices can be achieved, as shown in Figure 11.16. To do that, optimal design concepts must be utilized.

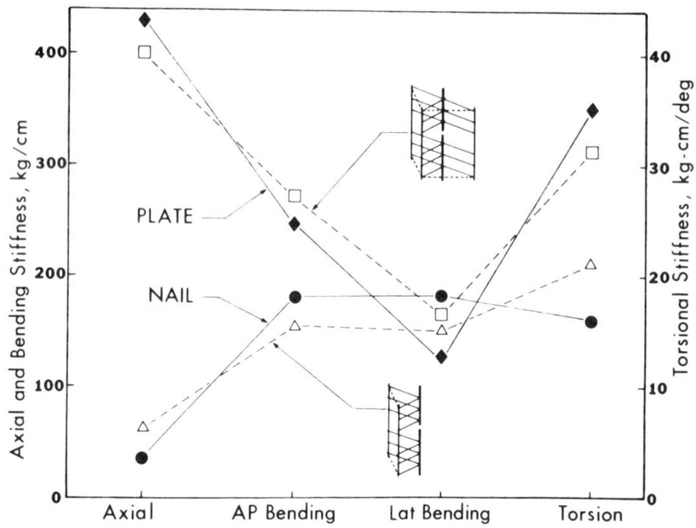

Figure 11.16 Fracture stiffness matching between the internal and external fixation devices. Triangular configurations were used to obtain proper stiffness characteristics required

In modelling the pin–bone system, a coupled boundary condition was assumed for the mating nodal points between the pin and the bone surfaces. Such an assumption allows the transmission of tensile stress through the interface which is unlikely to occur under realistic situations. In order to model the interface correctly, gap elements must be used which require improved knowledge concerning interface mechanical properties. Besides, a proper iterative scheme must be used in the finite element solution process in order to maintain the compatability and equilibrium conditions when gaps are forming at the interface. This process is extremely time consuming especially on a sizable model. Other more efficient methods must be utilized to overcome this problem. As a result of the current assumption, the compressive stresses obtained at the interface would probably be lower than the actual value. Therefore, the present data on bone stress near the pin holes represents the lower-bound results. For the purpose of demonstrating the potentially hazardous condition in the bone due to high pin loads, the simplified model used should be adequate.

Finally, no biological knowledge is currently available concerning the ideal amount of stiffness required for optimal bone healing. It is difficult to establish rational guidelines for patient application when the type of bone fracture varies. Bone fractures can heal via different mechanisms, regardless of the form of treatment. Under certain conditions, a rigid frame may be indicated while others may require less rigid fixation. Carefully designed biomechanical–histological studies using animal models may help to provide the essential knowledge to enhance fracture treatment results. Until this is accomplished, the analysis of the mechanical system can only be used based on established clinical experience.

11.5 SUMMARY

One of the common forms of bone-fracture treatment involves the application of an external fixation system that can maintain bone-fragment stability via the use of transcortical pins. These pins can be applied remotely from the trauma site which provides the distinctive advantage of soft-tissue management. The stiffness property of the apparatus has a significant influence on the treatment results which must be quantified according to biomechanical methods. The entire structure is a statically indeterminate frame system which can be best studied using the finite element method. A more detailed three-dimensional model was also developed to quantify the bone stress at the regions near pin insertion sites. The structural stiffness and pin stresses were analysed under different geometric and material variations. The results indicate that the bone fracture stiffness can be improved by increasing the pin numbers, pin diameter, pin material modulus, and decreasing side-bar separation. Other parameters have less effect on the overall performance of the system. The weakest loading mode is AP bending, which is a realistic condition and must be carefully controlled in patient management. Pin stress can reach a dangerous level causing permanent deformation and fracture. Even at lower loads, the bone stress at the pin–bone interface can become high enough to cause pressure necrosis leading to pin-track infection and loosening. However, if bony opposition can be achieved at the fracture site, such stresses will be reduced to a less damaging level. Experimental tests were conducted to verify the theoretical model and close correlation was found. The present results can be used to improve the design of current devices, but most importantly, they can help to establish guidelines for proper surgical application. Further studies are necessary in order to identify the ideal biological conditions for fracture healing. When such information becomes available, the optimal treatment modalities for various long-bone fractures using external fixation apparatus can then be rationalized.

ACKNOWLEDGEMENT

The experimental study of various types of external fracture fixation devices was performed by Drs. B. T. Briggs and M. T. McCoy. The authors are indebted to their contribution.

REFERENCES

Briggs, B. T. (1979) Biomechanical analysis of the Hoffmann–Vidal external fixation apparatus, M. S. thesis, Dept. of Orthopaedic Surgery, Mayo Clinic and Graduate School of Medicine.

Brinker, W. O., and Gretchen, L. F. (1975) Principles and application of external skeletal fixation, *Vet. Clin. No. Am.*, **5,** 197–208.

Chao, E. Y., Briggs, B. T. and McCoy, M. T. (1979) Theoretical and experimental analysis of Hoffmann–Vidal external fixation system, in: *External Fixation – The Current State of the Art* (eds. A. F. Brooker and C. C. Edwards, The Williams and Wilkins Co., Baltimore, pp. 345–370.

Gunterberg, B., Goldie, I., and Slatis, P. (1978) Fixation of pelvic fractures and dislocations and sacro-iliac dislocation after external compression fixation, *Acta Orthop. Scand.*, **49,** 278–286.

Hierholzer, G., Kleining, R., Horster, G., and Zemenides, P. (1978) External fixation – Classification and indications, *Arch. Orthop. Traumat. Surg.*, **92,** 175–182.

Jernberger, A. (1970) Measurement of stability of tibial fractures – A mechanical method, *Acta Orthop. Scand.*, Suppl. 135.

Jorgensen, T. E. (1972) Measurement of stability of crural fractures treated with Hoffmann osteotaxis. 1. Method and measurements of deflection of autopsy crura, *Acta Orthop. Scand.*, **43,** 188–206.

Klip, E. J., and Bosma, R. (1978) Investigations into the mechanical behaviour of bone-pin connections, *Engineering in Medicine* (The Institution of Mechanical Engineers, London), **7,** 43–46.

McCoy, M. T. (1979) A biomechanical study of different external fixation devices based on bone fracture stiffness, M. S. thesis, Dept. of Orthopaedic Surgery, Mayo Clinic and Mayo Graduate School of Medicine.

McCoy, M. T., Briggs, B. T., and Chao, E. Y. (1980) A comparative study of external fixators based on bone fracture stiffness, *Trans. 26th Ann. Meeting, Orthopaedic Research Society*, **5,** 173, February.

Vidal, M. M., Rabischong, P., Bonnel, F., and Adrey, J. (1970) Etudé Biomechanique du fixator externe d'Hoffmann dans les fractures de jambe, *Societe de Chirurgie de Montpellier*, Seance du 20 Janvin, pp. 45–52.

Wang, C. K. (1953) *Statically Indeterminate Structures*, McGraw-Hill Book Co., New York.

Finite Elements in Biomechanics
Edited by R. H. Gallagher, B. R. Simon, P. C. Johnson, and J. F. Gross

Chapter 12

Stress–Morphology Relationships in Trabecular Bone of the Patella

Wilson C. Hayes, B. Snyder, B. M. Levine, and S. Ramaswamy

12.1 INTRODUCTION

The adaption of trabecular bone to alterations in its mechanical environment has long been a subject of interest in musculoskeletal physiology. Reductions in bone density are observed after periods of disuse such as prolonged immobilization or exposure to a reduced gravitational field. Increases in bone density and changes in cortical width have also been reported in athletes and experimental animals exposed to heavy exercise (Jones *et al.*, 1977; Woo *et al.*, 1979). In the orthopaedic literature, this adaptive response is generally referred to as 'Wolff's law' after the German anatomist who first called attention to it (Wolff, 1870a). It is surprising to note, however, that a quantitative formulation of this law does not exist. Thus, it is not currently possible to specify the characteristics or the magnitudes of a multiaxial stress field which relate to bone formation or resorption.

Trabecular bone also plays a central role in the normal functioning of human synovial joints since it serves to transmit the large contact stresses applied at the articular surfaces to the compact bone of the cortical shaft (Hayes *et al.*, 1978). Radin and co-workers (1972) advanced the hypothesis that subchondral trabecular stiffening is an initiating factor in the development of degenerative arthritis. Marked reductions in trabecular bone density are also observed in rheumatoid arthritis and the associated reductions in bone stiffness and strength can markedly influence the prognosis of prosthetic replacement of the joint.

Previous studies of the adaptive changes in the trabecular bone of synovial joints have focused on complex regions such as the proximal femur (Simon *et al.*, 1974) and the tibiofemoral articulations of the knee (Hayes *et al.*, 1978). These regions are characterized both by uncertain loading conditions and by their complex geometries. The human patella, by contrast, is a comparatively simple structure which is subjected to a statically determinant force system. It has a well-developed trabecular architecture which has been analysed in some detail (Raux *et al.*, 1975; Townsend *et al.*, 1975). The patella is thus an ideal region for studying the functional adaptation of trabecular bone and the aetiology of degenerative processes such as chondromalacia and osteoarthritis.

The purpose of the work described here is to develop mathematical models for the internal stresses induced in the patella and then to use these models: (1) to examine relationships between the trabecular architecture and the predicted stress distributions; and (2) to study the effects on trabecular stresses of variations in apparent density, subchondral stiffness, patello-femoral contact area, and trabecular anistropy. The eventual goals are to identify the most significant parameters controlling the mechanical behaviour of normal and diseased synovial joints and to provide a quantitative formulation for 'Wolff's law'.

In this chapter, a literature review is provided on the numerous hypotheses that have been generated to explain the stress-related remodelling response in trabecular bone. Previous investigations directed toward predicting stresses in trabecular structures and in quantitatively describing trabecular architecture are also summarized. This review is followed by a description of the materials and methods used in the present investigation of stress–morphology relations in trabecular bone of the human patella. The resulting quantitative correlations between predicted stresses and the observed architecture are then described and discussed and used to generate a hypothesis on the stress parameters controlling trabecular architecture.

12.2 BACKGROUND

It is widely assumed that mechanical stresses influence the form and structure of bone. The concept of stress- or strain-induced bone remodelling was popularized by the German anatomist Julius Wolff (1870a, b; 1892) and is often referred to as Wolff's law. Prior to Wolff a number of other authors recognized the adaptability of bone (Galileo, 1638; Wyman, 1849), and Meyer (1867), working with the Swiss engineer Culmann, first suggested that trabecular architecture is organized along principal stress trajectories. The voluminous early literature on this subject is summarized by D'Arcy Thompson (1917) in his remarkable book, *On Growth and Form.* While qualitative support for the 'Law of bone transformation' (Wolff, 1892) has been noted in many instances, it should be emphasized that the mathematical laws which Wolff assumed to govern bone remodelling are not available. The characteristics or magnitudes of the stress or strain fields which control bone remodelling are not known.

12.2.1 Hypotheses of Wolff's law

In one view of Wolff's law, the remodelling mechanism is viewed as a 'black box' and the focus of interest is the architectural change and its relation to stress. The assumption is that a feedback loop can be described in which the input is a changing loading pattern and the output an altered bone architecture. This process is directed presumably toward some equilibrium state. Wolff adopted this view in stating:

> Every change in the form and the function of a bone or of their function alone is followed by certain definite changes in their internal architecture, and equally definite secondary alterations in their external conformation, in accordance with mathematical laws (quoted by Keith, 1918).

Wolff is reported to have been considerably influenced by Culmann's hypothesis that the trabeculae of the proximal femur were organized along the lines of principal stress (Koch, 1917). Roux (1895) also proposed a 'trajectorial architecture' to describe the geometry of trabecular bone and to explain how it is organized to provide maximum strength with a minimum of material. According to Murray (1936) the fundamental idea in the trajectorial theory is that trabeculae follow the stress trajectories in a homogeneous body of the same form. Pauwels (1948) conducted photoelastic experiments directed toward this theory, and Kummer (1966) demonstrated qualitatively the similarity between a three-dimensional trajectorial model of the femur and the trabecular architecture (Kummer, 1972).

Koch (1917) conducted a detailed analysis of the strength of materials of the human femur with particular attention to the architecture of trabecular bone. While based on a number of simplifying assumptions, he concluded that: (1) the trabeculae of the proximal femur are arranged in compressive and tensile systems corresponding to the principal stress directions; (2) the thickness and spacing of trabeculae vary with the stress magnitudes; (3) the density of trabecular bone is related to the shear stress magnitudes; and (4) the trabecular arrangement results in a structure of maximum strength with minimal material.

Several authors have challenged these views, refuting Koch's use of static loads and suggesting that trabecular alignment is controlled by compressive stresses (Carey, 1929; Frost, 1964; Hassler *et al.*, 1977). Throughout, there has been a lively debate in the orthopaedic and anatomic literature on the mechanical factors controlling the architecture of trabecular bone (Dixon, 1910; Jansen, 1920; Milch, 1940; Scott, 1957; Tobin, 1968; Oxnard, 1972; Takechi, 1977). More recent stress analysis of the proximal femur, based largely on the finite element method, have tended to support the trajectorial theory (Toridis, 1969; Rybicki *et al.*, 1972; Simon *et al.*, 1974; Brown *et al.*, 1980). Clinical interest has also focused on changes in the trabecular architecture of the proximal femur as an index of osteoporosis (Singh, 1970; Jowsey, 1977).

A number of papers have dealt primarily with cortical bone. Frost (1973) suggested that bone form may change in response to intermittent surface strain, supporting the theory that bone drifts toward the load-induced concavity, thereby neutralizing bending and putting the whole bone in compression. Lanyon *et al.* (1976) refuted this theory with experimental evidence of tensile strain occuring on sheep bones *in vivo*. Lanyon (1974) also showed that in the sheep calcaneous, the trabeculae are aligned approximately along principal strain directions, and later suggested (Lanyon *et al.*, 1976) that this alignment may be advantageous for reducing shear strains.

12.2.2 Mechanistic hypotheses for Wolff's law

Cell-based remodelling systems have been described in elegant detail by Frost (1964a, b) and these have served as the basis for systems of histomorphometric characterization of bone remodelling. Excellent reviews of remodelling dynamics have recently been provided by Jowsey (1977) and Albright and Brand (1979). In Frost's view, remodelling activity is divided into internal remodelling, which is the process of deposition and resorption of bone within already established osteons, and surface remodelling, which is the process of deposition of new osteons on periosteal or endosteal surfaces. Continuum models for internal remodelling where the remodelling rate is a function of strain have been proposed (Martin, 1972; Cowin and Hegedus, 1976; Cowin and van Buskirk, 1978). A surface remodelling hypothesis was suggested by Gjelsvick (1973) in which the remodelling rate is linearly proportional to surface stress. Pauwels (1948) suggested an optimal strain level as the basis for bone remodelling.

For trabecular bone, Radin *et al.* (1972) and Pugh *et al.* (1974) suggested a damage-accumulation hypothesis for Wolff's law and suggested a central role for trabecular microfractures in the remodelling of trabecular bone. The observed microfractures in various stages of healing trabecular bone in both humans and rabbits. It was also demonstrated that healing increases the mass and stiffness of trabeculae. Similar microfractures have been observed in degenerative femoral heads (Todd *et al.*, 1972).

Several hypotheses are based upon evidence that bone demonstrates a piezoelectric effect, i.e. that bone converts mechanical stress to an electrical potential that in turn influences the activity of osteoblasts and osteoclasts. This piezoelectric property was first discovered by Fukada and Yasuda (1957) and by Bassett and Becker (1962). Since then, many other investigators have demonstrated the existence of a piezoelectric effect in dry and wet bone *in vitro* (Shamos *et al.*, 1963; Reinish and Nowick, 1975) and in bone under physiologic conditions (Cochran *et al.*, 1968; Black and Korostoff, 1974; Lanyon and Hartman, 1977). There is also ample evidence that application of direct current (approximately 1 to 40 μA) results in the formation of new bone and also accelerates the fracture-healing process (Friedenberg *et al.*, 1974; Brighton *et al.*, 1977; Spadaro, 1977; Becker *et al.*, 1977).

The electrical properties of bone and cartilage have recently been summarized in a conference proceedings (Brighton *et al.*, 1979). The above considerations have led many workers to postulate that the piezoelectric properties of bone play an important role in the remodelling of the skeletal system (Marino and Becker,

1970; Bassett, 1971; Gjelsvik, 1973; Eriksson, 1976). However, no direct experimental evidence that this is the case is yet available since it is not currently possible to monitor the piezoelectric character of bone non-invasively, and invasive methods are difficult to maintain for long-term experiments. Saha and Lakes (1979) have recently reported a non-contacting electromagnetic device which may be useful for these purposes. In order to use this piezoelectric effect as a theoretical basis for bone remodelling, a number of workers have provided a general, theoretical description of bone as a piezoelectric material (Korostoff, 1979; Martin, 1979). Direct mechano-chemical transduction methods, mediated through strain-induced diffusion changes, have also been proposed (Justus and Luft, 1970), as have direct vascular effects (Trueta, 1962; 1968).

12.2.3 Adaptive remodelling of bone

Both *in vitro* and *in vivo* experiments have been conducted to investigate the stress-related adaptive response of bone. Tissue (Gluckmann, 1938) and organ culture methods have been used to study the effects of cyclic (Hall, 1968) and static (McMaster and Weinert, 1970) loads on foetal bone explants (Bassett, 1968; Rodan *et al.*, 1975). An increase in the rate of calcium loss during space flight, as compared to pre-flight values, has been measured consistently in many astronauts and used as evidence in support of Wolff's law. Rambaut and Johnson (1979) recently reviewed current understanding of calcium loss in humans in low-gravity environments and compared space-flight date with predicted results based on comparable periods of bedrest. The mechanisms involved in bone mineral loss are not clear. Yagodovsky *et al.* (1976) reported inhibition of growth in rats following a 22-day Soviet space flight. In another Soviet study, a 45% decrease in periosteal bone growth was measured, also in rats. Morey and Baylink (1978) found flight animals to have experienced complete cessation of bone formation as exhibited by arrest lines. Much of this experimental work indicates that mineral balance may take considerable time to return to normal if it does so at all, suggesting that permanent damage to calcium control mechanisms may result from prolonged lack of exposure to stress.

'Plate-induced osteopenia' during the late course of fracture treatment with plate fixation is also often cited as evidence for the adaptive response of bone. The regional loss of bone mass due to the application of a plate to the canine femur has been studied experimentally by several investigators (Tonino *et al.*, 1976; Akeson *et al.*, 1976; Moyen *et al*, 1978; Daniels *et al.*, 1978). Unfortunately, the change in bone stresses caused by plate application is not known for these studies. Carter *et al.* (1980) recently used *in vivo* strain gauges and beam theory to calculate these stresses, and found a strong correlation between changes in the stress sense (i.e., from tension to compression) and metabolic activity. This study is a landmark in that it represents one of the very few attempts to compare stress predictions with *in vivo* remodelling activity.

Prolonged physical exercise has also been shown to result in significant changes in the structure of bone. Radiographic studies of the humeri of professional tennis players showed a pronounced hypertrophy of bone on the playing side. The cortical thickness increased 34.9 % in men and 28.4 % in women in response to this training (Jones *et al.*, 1977). Using non-invasive methods, bone density changes have also been observed in althletes (Nilsson and Westlin, 1971; Dalen and Olsson, 1974). Similar effects of exercise have been demonstrated in experimental animals (Saville and Whyte, 1969; King and Pengelly, 1973; Kiiskinen, 1977; Kiiskinen and Heikkinen, 1978; Chavapil *et al.*, 1973). Animal experiments have also suggested that the effects of exercise may depend on its intensity and on the age and size of the experimental animal (Booth and Gould, 1975; Tipton *et al.*, 1972; Rosenfeld *et al.*, 1973). Recently, Woo *et al.* (1980) have subjected swine to 12 months of chronic exercise with non-exercised matched swine as controls. From the cortical cross-sections obtained at sacrifice, increases in cortical thickness of 17 % were demonstrated. Making use of their material, we have further demonstrated changes in cross-sectional area, area moments of inertia, and direction of the principal moments using special computer programs developed for this purpose (Nagurka and Hayes, 1980).

Attempts have also been made in experimental animals to apply controlled loads to investigate the adaptive response of cortical bone. Chamay and Tschantz (1972) resected portions of the radius in adult dogs and reported marked thickening of ulna after a few weeks. They interpreted their results in terms of loading levels resulting in fatigue damage accumulation, a process which has also been studied *in vitro* (Carter and Hayes, 1976; Carter, Hayes, and Schurman, 1976). Hert *et al.* (1969) applied static bending loads to growing rabits and showed remodelling effects which were not observed in adult animals. Liskova and Hert (1971) extended the work to cyclic bending loads and demonstrated bone deposition in highly stressed areas in both young and mature animals, and resorption in lightly stressed areas. Recently, Churches *et al.* (1979) have extended these methods to the ovine metacarpus and demonstrated adaptive remodelling in the mid-diaphyseal region resulting in approximately 17 % increase in cross-sectional area. These studies are all compromised by the lack of information on the stress fields active to cause remodelling.

12.2.4 Stress analysis in trabecular bone

Recent trends in total joint replacement and clinical failures in some early designs have provided a strong impetus for the development of accurate stress analyses in normal and prosthetic joints. Two trends have emerged in attempts to model stresses in trabecular bone. Most workers have assumed that trabecular bone is a homogenous material of heterogeneous modulus (Lewis, 1977; Hayes *et al.*, 1978; Bartel *et al.*, 1980; Brown *et al.*, 1980a; 1980b; Hayes *et al.*, 1980). These studies have generally emphasized the importance of modulus variations and loading assumptions, and de-emphasized the effect of anisotropy in governing the predicted stress distributions. There is also an increasing trend, however, toward microstructural modelling with attempts made to predict individual trabecular stresses in either idealized structures (Pugh *et al.*, 1974) or in structures based on direct tracings of trabecular morphology (Williams and Lewis, 1979; Cook *et al.*, 1980; Lavernia *et al.*, 1980). The Pugh model has been criticized as unrepresentative of actual trabecular architecture, and adjustments were necessary in trabecular spacing and width to match macroscopic modulus predictions. The morphological approach has been criticised as being so specialized to the particular architecture modelled as to be of little predictive value. Despite these criticisms, there is little doubt that the finite element method provides a powerful technique for stress analyses in trabecular bone and that there will be an increasing trend toward coupling of models based on macroscopic homogeneity and those directed toward microstructural predictions.

12.2.5 Trabecular bone morphology

The basis for most quantitative studies of trabecular architecture is stereology, a technique for the description of microscopic structural parameters (Weibel and Elias, 1967; Dehoff and Rhines, 1968; Underwood, 1970; Weibel, 1973). Major contributions to the application of these techniques to trabecular bone were provided by Merz (1967), who combined area measurements from point counting with profile measurements from linear intersection counts. Merz and Schenk (1970a; 1970b) used these methods in a study of aging and observed a continuous decrease with age, of volume and surface density. The specific surface (surface density/volume) separated into distinct populations after age 50 with one group exhibiting a thickening and reduction in number of trabeculae. Similar techniques were used by others to study the iliac crest (Wakamatsu and Sisson, 1969). Radin *et al.* (1972; 1973; Pugh *et al.*, 1974) used stereology in conjunction with modulus measurements in support of their microstructural model for trabecular bone. They proposed the ‘contiguity ratio’ as a means for specifying the extent of lateral bracing in trabecular bone. Raux *et al.* (1975) later applied these techniques to the patella. There have also been a few attempts to automate stereologic measurements (Lloud *et al.*, 1978).

Whitehouse (1974) generalized stereological techniques to a consideration of anisotropy and these were applied to the patella by Townsend *et al.* (1975). Whitehouse's studies in three dimensions suggested that trabecular bone is orthotropic and this has recently been supported by Williams and Lewis (1979) using stereologic techniques. Recently, Hayes *et al.* (1980) have combined quantitative stereology with stress predictions using finite element analysis of the human patella, and have provided support for the trajectorial theory of trabecular architecture.

Whitehouse and others have consistently emphasized that the prediction of three-dimensional structure from observations of plane sections is the most difficult problem of stereology. Scanning electron microscope observations have thus been used to describe three-dimensional trabecular architecture (Whitehouse and Dyson, 1974; Whitehouse, 1975; 1977). Singh (1978) has recently summarized the controversy over the architecture of trabecular bone and proposed a classification into types based on: (a) interconnecting rods; (b) interconnecting rods and plates; and (c) interconnecting plates. Observations by Williams and Lewis (1979) confirmed this scheme in the proximal tibia. Our laboratory has recently proposed in addition an open-celled porous foam model for the subchondral region where highly random hydrostatic stress states are to be expected.

12.2.6 Summary

From the above, it is clear that the question of Wolff's law has been an issue of long-standing interest in musculoskeletal physiology. Examples of the stress-related adaptive response include space flight, internal fixation of fractures, and chronic exercise. Many hypotheses have been proposed for the characteristic stress states leading to bone remodelling and a number of microstructural mechanisms have been proposed as mediators of this stress-related adaptive response. Powerful numerical techniques are now available for the quantitative description of trabecular architecture. Yet, to our knowledge, except for our recent study in the patella (Hayes *et al.*, 1980), these methods have never been combined in an attempt to elucidate the mechanical factors controlling trabecular architecture.

12.3 MATERIALS AND METHODS

Human synovial joints are characterized by complex geometries, non-linear and heterogeneous constitutive properties, large deformations, and highly indeterminant loading conditions. Faced with these complexities, numerical techniques such as the finite element method become the only viable approach to the analysis of synovial joint stresses. To create a finite element model, input data must be provided to describe the geometry of the structure, its material properties, and the applied loads.

To characterize the morphology of trabecular bone in the patella, specimens were embedded, sectioned, and polished using conventional metallographic techniques. Reflected-light photomicrographs were then used to conduct strerological investigations of trabecular morphology. The accuracy of point-counting techniques was verified using computerized planimetry. Special test patterns consisting of oriented ellipses were used to determine the accuracy of the directed secant method used for characterizing trabecular orientation and anisotropy. With this background information on the accuracy of the stereologic techniques, the methods were used to measure areal density, contour density, average trabecular width, trabecular orientation angle, and per cent orientation in regions of the patellar photomicrograph corresponding to the element subdivisions of the finite element models. These stereological measurements and the finite element stress predictions were the basis for stress–morphology comparisons that were the principal objective of this investigation.

12.3.1 Finite element analysis

12.3.1.1 Microstructural analysis of patellar geometry

To provide input data on patellar geometry for the finite element models, both patellae of the 20-year-old autopsy specimen were embedded in methylmethacrylate and sectioned using a diamond saw. After polishing using standard metallographic techniques, reflected-light photomicrographs were taken at 41 × and montaged to create the mid-sagittal cross-section shown in Figure 12.1a. This photomicrograph was also used to determine the location of tendonous insertions which were then used to apply loads to the finite element model (Figure 12.1b). Contact microradiographs were also made of the same mid-sagittal slice and scanned with a photodensitometer to provide approximations for the trabecualr bone density and thus its stiffness and strength. Independent estimates of trabecular areal density were also obtained from the photomicrograph using the stereologic techniques described below.

12.3.1.2 Constitutive relations

Articular cartilage is a dense, connective tissue which provides the low-friction articulating surface in human synovial joints. It is a two-phase gel consisting of a permeable, viscoelastic solid matrix and an incompressible fluid which is free to flow in response to matrix consolidation (Mow and Lai, 1979). Most previous studies of the mechanical properties of cartilage have provided only lumped parameters based on the assumption that cartilage is a single-phase, homogeneous material. On the basis of these lumped parameters, it has not been possible to separate changes in matrix viscoelastic properties from those due to changes in flow characteristics. Thus, new techniques were developed for measuring the dynamic viscoelastic properties of the cartilage matrix phase uncoupled from the effects of interstitial fluid flow, and these measured biomechanical parameters were correlated with biochemical assessments of the tissue (Hayes and Bodine, 1978; Kaplan *et al.*, 1979). Although cartilage behaves in bulk as a bi-phasic, viscoelastic material, for purposes of the static, linear elastic finite element models developed here, the cartilage was assumed to be linear, elastic and nearly incompressible with elastic modulus, $E = 10$ MPa and Poisson ratio, $\nu = 0.47$.

To determine the compressive strength and stiffness of trabecular bone spanning a large density range, both human and bovine subchondral trabecular

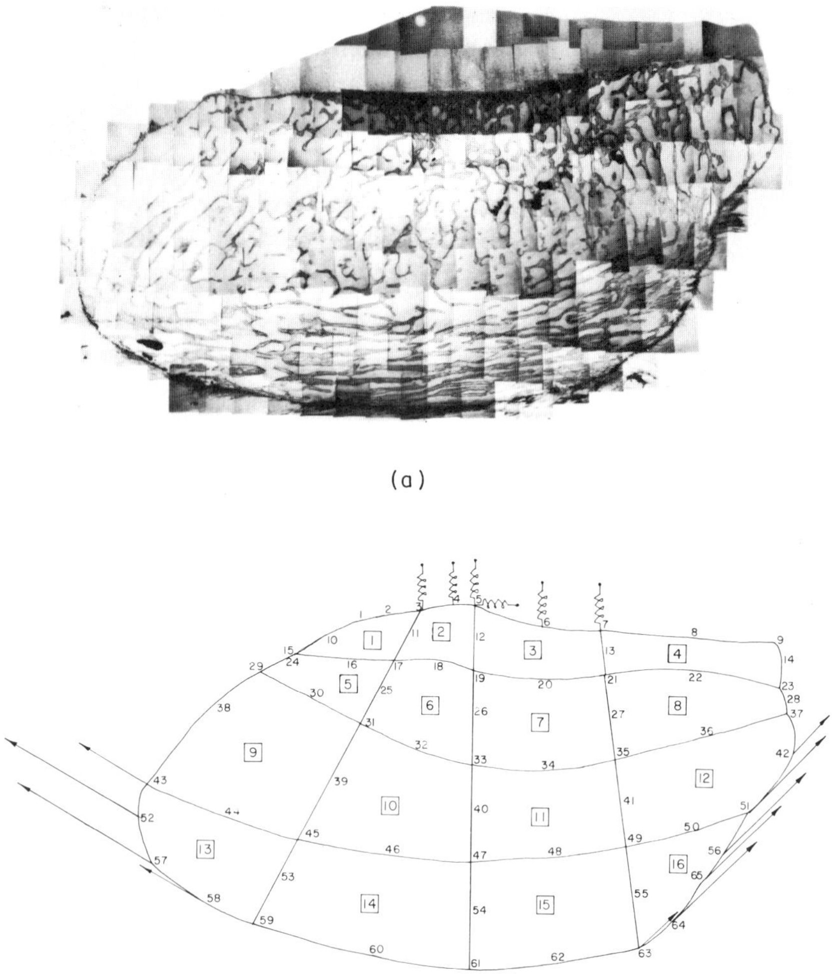

(b)

Figure 12.1 Mis-sagittal cross-section of the patella: (a) reflected-light photomicrograph (41 ×); (b) plane-strain finite element model (MARC)

bone was tested (Carter and Hayes, 1976a, b; 1977). In addition to specimen density, the effect of strain rate and the effect of marrow *in situ* were examined. The effect of specimen apparent density on compressive strength was found to be approximately proportional to the square of the apparent density. The stiffness varied as the cube of the apparent density (Figure 12.2). Here this power-law

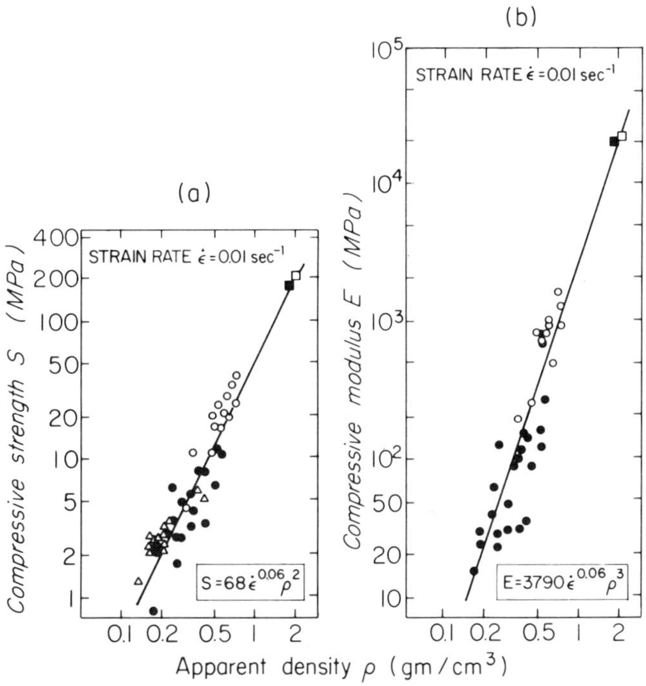

Figure 12.2 Mechanical properties of trabecular bone as a function of apparent density: (a) compressive strength; (b) compressive modulus

relationship between the apparent density of trabecular bone and its elastic modulus was used to estimate modulus both from photodensitometry and from direct sterologic measurements of areal density. A point-counting technique described below was used to determine areal density and thus to calculate an approximate value for volumetric apparent density (Snyder and Hayes, (1980). The modulus was then taken as proportional to the cube of the apparent density using the relationships established by Carter and Hayes (1976a, b, 1977).

These results provided input data for the finite element models and also much-needed clinical guidelines for predicting bone strength and stiffness in patients with decreased skeletal mass. Osteoporosis is present in many pathological conditions and is a common finding in bedridden or elderly patients and in some arthritic patients in need of total joint replacement. Photon absorption densitometry studies have shown that the ulnar bone densities in osteoporotic patients may be reduced to as little as one-third of normal. The results of these power-law relations between strength and density suggest that in these patients, the bone compressive strength is reduced to one-ninth of normal.

12.3.1.3 Boundary conditions and loads

The articular cartilage surfaces of the patella articulate with articular cartilage present in the intercondylar region of the femur. The cartilage thickness is not small compared with the dimensions of the patella, and this layer was also included in the finite element model. Femoral articular cartilage was modelled by means of grounded springs with spring stiffnesses equal to the cartilage modulus (Figure 12.1b). In order to prevent rigid-body motion of the fiite element model, simple constraints on patella movement in the proximal-distal direction were also introduced.

The mechanics of patello-femoral articulation of the human knee has been analysed extensively (Lieb and Perry, 1968; Morrison, 1968; Reilly and Martens, 1972; Smidt, 1973; Maquet, 1976). Considering the loading experienced by the patella, it should be noted that the force system is statically determinant. The quadriceps force, the patellar tendon force, and the patello-femoral joint reaction force form a closed triangle and specification of any two of the forces fixes the third. In the present two-dimensional analysis, the small force that is experienced by the patella in the lateral direction (due to the Q angle between the quadriceps muscle force and the patellar tendon force) was ignored. For these analyses, a representative force system corresponding to slow, stair climbing was used since this represents a relatively severe, but quite frequent loading situation. For the normal patella, the quadriceps force was taken to be 2690 N acting at 48° to the horizontal axis used as a reference system for the finite element analysis (Figure 12.1b). For equilibrium, the patellar ligament force was chosen such that the resultant force along the horizontal axis was zero and the moment due to the quadriceps force about the centre point of the contact region was equal and opposite to the moment due to the patellar ligament force about the same point. It should be noted that these two conditions uniquely define the magnitude and direction of the patellar ligament force. For the normal patella, the patellar ligament force was calculated to be 1920 N acting at an angle of 20° to the horizontal.

12.3.1.4 Finite element models

An enlarged tracing of the mid-sagittal slice of Figure 12.1a was used to construct the two-dimensional finite element model of the patella (Figure 12.1b). Two finite element models were generated. For the first model, a general-purpose finite element program (MARC Analyses Research Corp) with 16 isoparametric, plane-strain quadrilateral elements was used. Both articular cartilage and trabecular bone were assumed homogeneous and trabecular bone was assumed anisotropic with its modulus varying according to the photodensitometry scans of the mid-sagittal microradiograph. The effects of variations in patello-femoral contact area and subchondral trabecular stiffness were studied parametrically.

The complete test matrix for the MARC patellar finite element model allowed parametric investigations of: (1) density variations; (2) subchondral stiffness variations; (3) trabecular bone anisotropy; and (4) patello-femoral contact area reductions. Results were obtained as contour plots of global stress and strain components, deformed geometries, and von Mises equivalent stresses.

As part of the development of a laboratory-based finite element modelling package, a second plane-strain finite element model was created using GIFTS (Graphics Oriented Interactive Finite Element Time-sharing System) developed at the University of Arizona. GIFTS is a static and dynamic, linear, elastic structural analysis program with extensive pre- and post-processing capabilities. The program is resident on a VAX 11/780 at the Joint Computing Facility at Massachusetts Institute of Technology. All program input and output were transmitted via telephone lines to and from the Orthopaedic Biomechanics Laboratory and output on a TEKTRONIX 4010 graphics terminal with hard copy provided by a TEKTRONIX 4631 hard-copy unit.

The second finite element model consisted of a 168, 4-node, isoparametric plane-strain elements. In this case, material properties were assigned on the basis of the quantitative morphologic analyses of the mid-sagittal slice and previously described power-law relationships between trabecular density and modulus. The major difference between the two finite element models was a marked increase in the trabecualr density in the region just beneath the articular cartilage. This difference is attributed to inaccuracies inherent in scanning photodensitometry used to estimate trabecular density in the first model.

12.3.2 Quantitative stereology

Stereology is the technique for the quantitative microscopic analysis of structural parameters. The method has been used primarily in the material sciences and in cytology. Here, stereologic techniques were applied to the study of trabecular bone after the preparation of reflected-light photomicrographs. The main objectives in these studies were to determine accurate measures of areal density and average trabecualr width and to provide parameters which could characterize the orientation and anisotropy of trabecular bone. To determine areal density, a random-grid point-counting method was used after first comparing results with independent measurements using computerized planimetry. To characterize trabecular orientation and anisotropy, the method of directed secants was used. In this case, measurement accuracy was investigated using test pattern arrays consisting of ellipses of known geometry and orientation.

12.3.2.1 Accuracy comparisons with computerized planimetry

Quantitative morphological parameters were determined for twelve regions of trabecular bone corresponding to the planar elements of the finite element model shown in Figure 12.1. Computerized planimetry was used to measure bone perimeter length, L_A, and areal bone density, A_A. Stereological methods were used to confirm these measurements. A 300-point random grid was used to determine A_A by point counting and a Merz grid (Merz and Shenk, 1970a, b) was used to determine L_A. The parameters measured both by planimetry and stereology were used to calculate: (1) volume density, V_B; (2) interface length, B_A; (3) specific interface length, B/A; and (4) average trabecular width, W_B. Calculations were performed in four quadrants of each element and averaged to provide centroidal values. Correlation analyses showed the independently determined parameters from planimetry and stereology to agree with coefficients of determination, R^2 of 0.98 for A_A (Figure 12.3a) and 0.94 for B_A. The resulting calculation of average trabecular width, W_B, also exhibited good agreement between the stereological results from the Merz grid and computerized planimetry ($R^2 = 0.854$) (Figure 12.3b).

12.3.2.2 Method of directed secants

Morphologic parameters were also measured using the method of 'directed secants' (Saltykov, 1958). To use this technique, an array of parallel lines of constant separation was laid across the pattern under investigation at 10° intervals between 0° and 180° (Figure 12.4). The number of intersections, I, between the parallel line grid and the boundaries between bone and marrow on the plane surface were counted by hand. From the number of intersections between the grid lines and the cross-sectional profiles, and from the total length of the grid lines, the total projected length of the boundary perimeter on a line at right angles to the scan was estimated.

The total number of intersections per unit grid line length in a given direction, θ, is indicated by $I_L(\theta) = I_L(\theta)$ (oriented) $+ I_L(\theta)$ (isotropic). $I_L(\theta)$ is a minimum when the grid lines are parallel to the oriented portion of the bone pattern. Therefore, the direction of orientation was specified by the angle at which the minimum number of intersections per unit test length occurred. When the test sęcants were directed perpendicular to the orientation axis of the test ellipses, the mean number of intersections per unit test line length was a maximum. The departure from isotropy was described by plots of the mean of the number of intersections per unit grid line length at 10° intervals between 0° and 180°.

The method of directed secants was first tested for accuracy by analysing the angle of orientation, the extent of anisotropy, and B_A, the perimeter length density, of idealized structures on a plane surface for which these parameters could be simply specified. Arrays of ellipses of varying dimensions were analysed since three-dimensional trabecular bone structures have been modelled as networks of prolate spheroids formed by the rotation of an ellipse about its major axis (Whitehouse, 1974). In five of the test arrays, the degree of anisotropy was

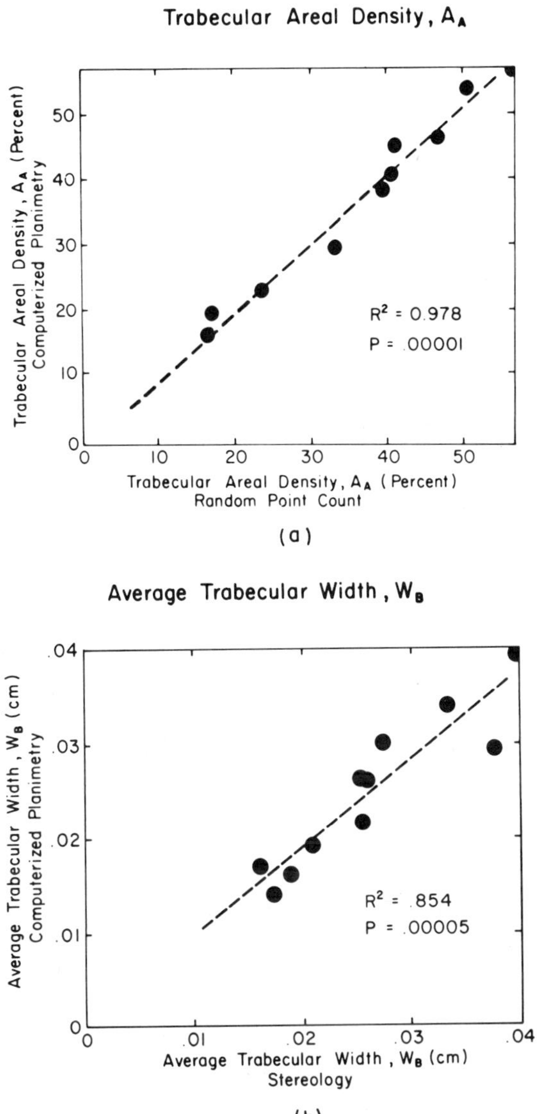

Figure 12.3 Accuracy of stereological parameters for trabecular bone of the patella: (a) areal density; (b) average trabecular width

progressively decreased as the shape of the ellipses approached that of a circle. In all cases the angle of orientation remained constant. In one of the test patterns, the size, orientation, and ratio of major axis to minor axis of the ellipses were randomly selected with the aid of a random numbers table (Figure 12.5b).

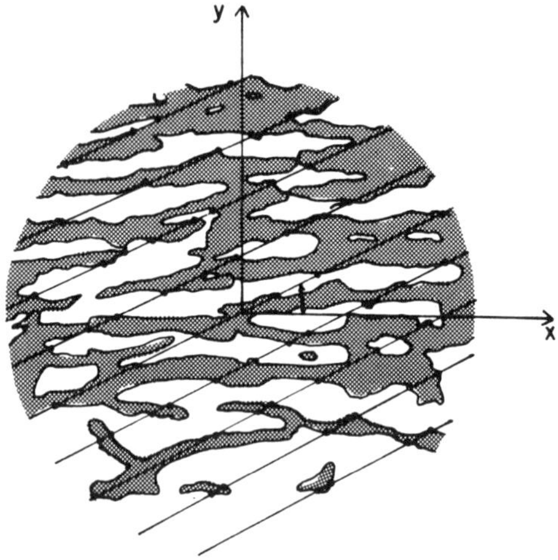

Figure 12.4 Grid lines for determination of morphologic parameters

Figure 12.5 demonstrates that the resultant mean of the number of intersections per unit grid line length approximates a sinusoid for a very oriented test pattern (Figure 12.5a), and a horizontal line for randomly oriented test patterns (Figure 12.5b). As the eccentricity of the ellipses progressively decreases and the ellipses become circles, the amplitude of the resultant sinusoid curve decreases and the resultant mean of the number of intersections curve approaches a horizontal line.

The method for measuring trabecular orientation for two elements corresponding to the finite element analysis is shown in Figure 12.6. Element 14, representing a highly anisotropic anterior region, yielded an orientation of approximately 78 % with an orientation angle of approximately 90°. Element 7, representing a high-density, nearly isotropic subchondral trabecular region, yielded an orientation of only 8 %.

12.4 RESULTS

Two plane-strain finite element models were developed based on a mid-sagittal slice of a normal human patella. The first model (MARC) was used to determine the most significant parameters influencing stress distributions in trabecular bone and specially to determine the relative importance of: (1) trabecular density variations; (2) bone anisotropy; (3) subchondral stiffness variations; and (4) patellofemoral contact area reductions. The latter two studies were conducted to explore hypotheses on the aetiology of degenerative arthritis and chondromalacia.

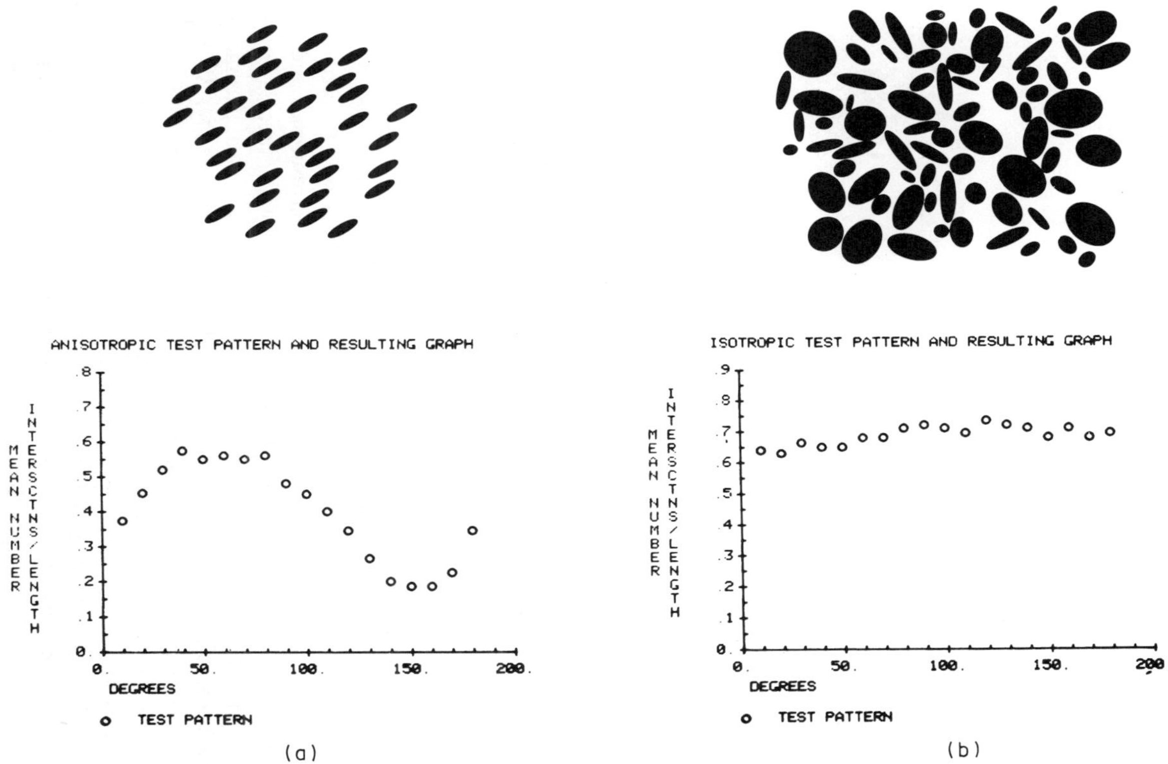

Figure 12.5 Orientation of elliptical test arrays using the method of directed secants: (a) anisotropic pattern; (b) isotropic pattern

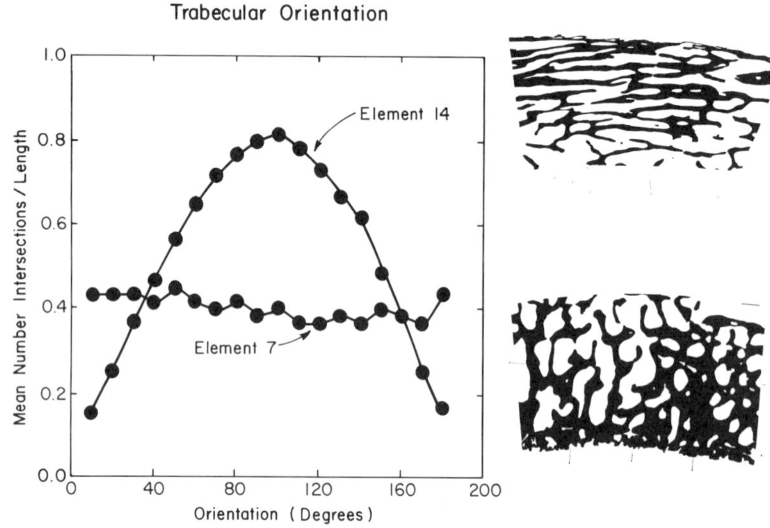

Figure 12.6 Specification of anisotrophy and orientation angle for two regions of trabecular bone from the patella

The second model (GIFTS) was used to provide more refined estimates of the stress distributions in patella trabecular bone. More accurate data on trabecular density was available from direct stereological measurements of areal density and were used in the GIFTS model. Because of the post-processing options of the GIFTS program, additional output describing the principal stress magnitudes and their direction was also available.

The ultimate goal of the finite element modelling effort was to provide a quantitative comparison between the predicted stress distributions and the morphological analysis of trabecular bone. For this purpose, trabecular stress measurements were systematically compared against morphological parameters describing trabecular density, orientation, and percent anisotropy. Various stress measures were used based on previous hypotheses for the stress components controlling the formation and resorption of trabecular bone. These included maximum principal stress (tensile), minimum principal stress (compressive), and von Mises effective stress (which is proportional to the difference in principal stress). The observed trabecular orientation and anisotropy, determined using the method of directed secants, were also compared against the finite element predictions for the magnitudes and direction of the principal stresses. These comparisons then formed the basis for a hypothesis on a stress parameters controlling trabecular architecture.

12.4.1 Finite element results: MARC model

The MARC finite element model was used to study parametrically the most important variables controlling stress distributions in the normal patella. The effects of non-uniform density, anisotropy, and variations in subchondral stiffness and patello-femoral contact area were investigated. The objective was to determine the sensitivity of stress predictions to the various parameters and to explore hypotheses on the aetiologies of osteoarthritis and chondromalacia.

12.4.1.1 Non-uniform density and anisotropy

Graphic output of von Mises effective stress (Figure 12.7) allows an assessment of the relative importance of density variations and anisotropy. In Figure 12.7a, where trabecular bone was assumed to be isotropic and of uniform density (resulting in a uniform modulus of 100 MPa), there is a gradual and uniform increase in trabecular effective stress proceeding from the subchondral region to the anterior surface of the patella. The assumption of uniform trabecular density results in effective stress values of about 1.8 MPa in the dense trabecular bone of the anterior patella with a highly concentrated region of 2.6 MPa near the distal attachment of the patellar tendon. The central patella is subjected to effective stresses in the range of 0.7 to 1.2 MPa with the subchondral region exhibiting contour values of 0.4 MPa.

In Figure 12.7b, where density variations based on photodensitometry scans were used to assign a non-uniform modulus, there are concentrated regions of effective stress corresponding to the regions of increased density in the patella (Figure 12.1a). Maximum contours corresponding to effective stresses in the range of 3.0 to 3.6 MPa occur in the dense, oriented anterior region. Minimum contour values of about 0.5 MPa occur in the central patella. Subchondral trabecular bone exhibits contour values of about 1.0 MPa.

A comparison of uniform and non-uniform density models thus reveals an approximately two-fold stress increase in the subchondral region under the assumption of non-uniform density. This corresponds to significantly increased concentrations of effective stresses in the anterior and subchondral regions where trabecular bone is dense and an unloading of the low-density central patella. The correspondence between the effective stress contour plot of Figure 12.7b and the trabecular architecture shown in Figure 12.1a is immediately apparent.

The effect of the additional assumption of trabecular anisotropy is shown in Figure 12.7c. In this case, trabecular bone was again assumed to be of non-uniform density and modulus and, in addition, modulus values were assumed to be anisotropic, based on the experimental findings of Townsend *et al.*, (1975). These workers found maximum modulus ratios of about 3 : 1 in small cubes of patellar trabecular bone. For this parametric investigation, this maximum value was assumed throughout the model and used to adjust modulus values in the principal material directions.

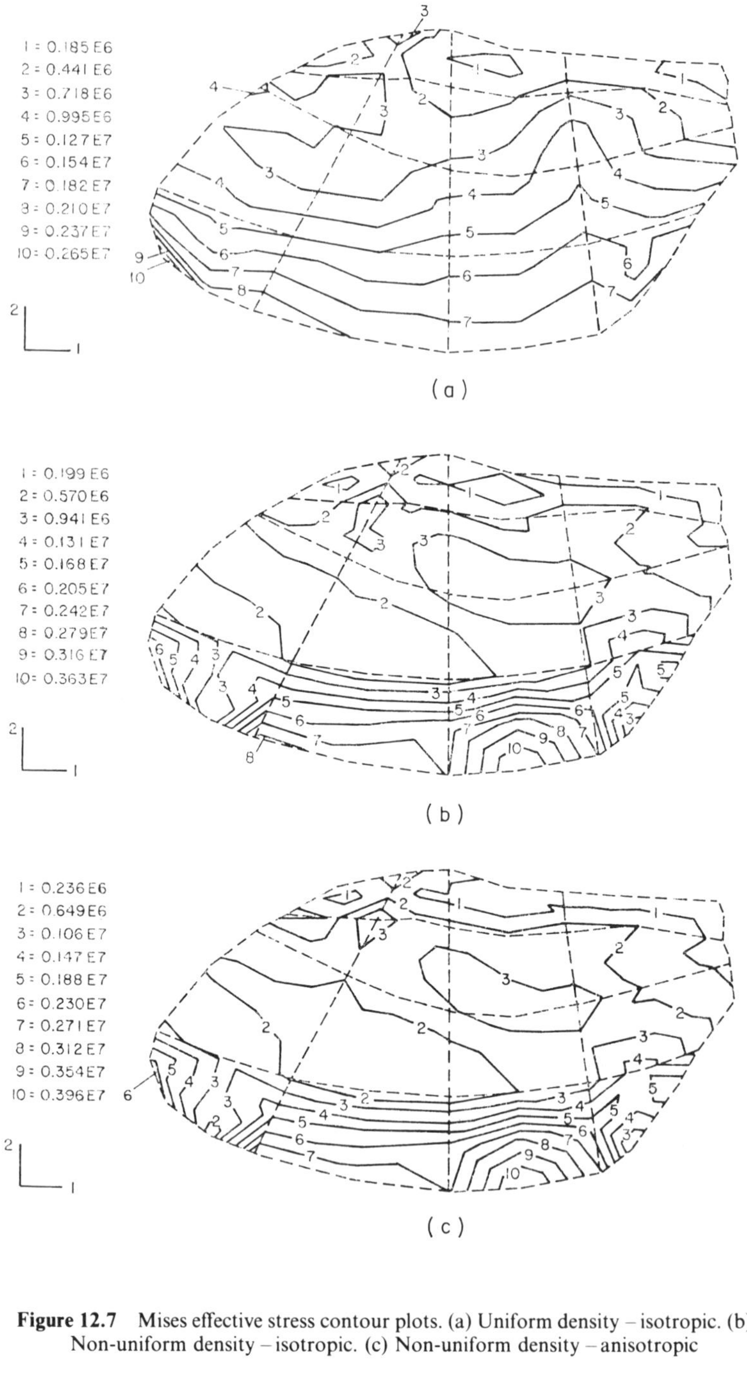

Figure 12.7 Mises effective stress contour plots. (a) Uniform density – isotropic. (b) Non-uniform density – isotropic. (c) Non-uniform density – anisotropic

A comparison of the anisotropic (Figure 12.7c) and isotropic (Figure 12.7b) effective stresses shows that the additional assumption of anisotropy results in approximately an 8 % increase in the interior, and an 11 % increase in the subchondral region. These differences are far less than the 200 % differences observed between the uniform and non-uniform density models. This greater sensitivity to density variations is due to the large modulus differences (which span two orders of magnitude) which result when a non-uniform density is assumed. By contrast, the assumption of trabecular anisotropy only results in a 2- to 3-fold variations in modulus. The obvious conclusion is that the assumption of non-uniform density is a necessity in finite element analysis of trabecular structures whereas the assumption of anisotropy is considerably less important.

12.4.1.2 Aetiology of osteoarthritis and chondromalacia

To investigate the hypothesis that subchondral trabecular stiffening is a contributing factor in the aetiology of osteoarthritis, the effect of wide variations in subchondral modulus was investigated. To examine this effect, the non-uniform density, isotropic model was used as a baseline. In one finite element run, the density of the subchondral trabecular bone was increased by 50 % from baseline values. In a second run, the density was decreased by 50 %. Note that the cubic relationship between trabecular apparent density and stiffness (Carter and Hayes, 1977) indicates that this is equivalent to 8-fold increases and decreases in subchondral trabecular modulus, a variation that spans nearly two orders of magnitude.

Figure 12.8 shows the effective stress contour plots for cases of decreased and increased subchondral density. Comparison of Figure 12.8a and 12.8b shows that this large variation in subchondral stiffness markedly affects the local subchondral trabecular stresses but does not significantly affect articular cartilage stresses. Beneath the centre of the patello-femoral contact region, this density increase results in an increase in trabecular effective stress from approximately 0.7 MPa to approximately 1.8 MPa, an increase by a factor of 2.5. Cartilage stresses change only by a few percent.

Comparison of the increased subchondral stiffness case (Figure 12.8b) with the normal non-uniform density case (Figure 12.7b) provides the most reasonable test of the hypothesis that subchondral stiffening increases articular cartilage stresses. With this 8-fold increase in modulus, the subchondral effective stresses increase by a factor of 1.6 and again the articular cartilage stresses do not change by more than a few percent.

Thus, these results do not support the hypothesis that generalized subchondral stiffening is a contributing factor in the aetiology of degenerative arthritis. Instead, the results suggest that cartilage stresses are nearly insensitive to wide variations in trabecular modulus. The reason for this insensitivity is the fact that trabecular modulus ranges between approximately 15 MPa and 1000 MPa with

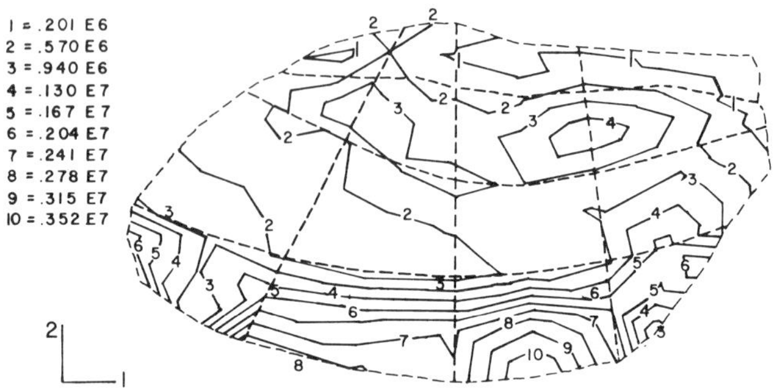

a. NONUNIFORM DENSITY - ISOTROPIC - 50% DECREASE

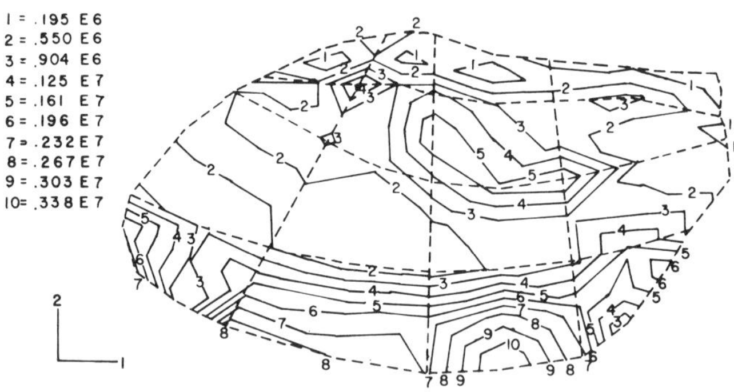

b. NONUNIFORM DENSITY - ISOTROPIC - 50% INCREASE

Figure 12.8 Mises effective stress contour showing effect of density changes in subchondral trabecular bone

the dense subchondral region approaching the upper limit. This represents an increase of two orders of magnitude over the cartilage modulus of about 10 MPa. Thus, since subchondral trabecular bone is initially so much stiffer than cartilage, an increase of even an order of magnitude in bone stiffness would be expected to have little effect on cartilage stresses.

The finite element models were also used to investigate the Outerbridge (1961) hypothesis that chondromalacia patellae is associated with localized increases in articular cartilage stresses due to contact with a medial femoral condylar ridge. Although this hypothesis has been questioned by many workers (see Abernathy *et al.*, 1978), it seemed useful to explore the importance of variations in patello-femoral contact area on cartilage and trabecular stresses. For this purpose, the number of support springs used to represent patello-femoral contact was reduced from 5 to 3. In this case, there were highly significant increases in effective stress in both subchondral trabecular bone and in articular cartilage when compared with the baseline case of figure 12.7b. with the reduced contact area, cartilage stresses increased from about 0.4 MPa to 1.34 MPa and subchondral trabecular stresses increased from 1.1 MPa to 1.9 MPa. These results are particularly striking for the articular cartilage stresses and indicate that reductions in patello-femoral contact area are far more influential on cartilage stresses than are increases in subchondral stiffness.

12.4.1.3 Effect of cortical shell

To investigate the effect on the predicted stresses of including a thin cortical shell in the finite element model, beam elements were added at the outer perimeters of the two-dimensional, plane-strain elements representing trabecular bone. Direct measurements were made of shell thickness (Figure 12.9a) using the reflected-light microscope and the polished mid-sagittal slice shown in Figure 12.1a. Although highly non-uniform local shell thicknesses were observed, values were averaged and used to assign beam thickness. The thickness varied from approximately 0.5mm in the immediate subchondral region to about 0.15 mm in the anterior patella. Inter-element compatibility was not maintained between the plane-strain and beam elements and thus the model underestimated the structural stiffness of the shell. The beam elements were assumed to have a uniform modulus of 1000 MPa.

The effect on von Mises effective stresses of the inclusion of a set of beam elements representing the cortical shell (Figure 12.9b) demonstrates only slight increases in stress levels when compared with the baseline model of Figure 12.7b. However, when the loading in the shell itself is examined (Figure 12.10), it is observed that the shell is loaded almost exclusively by in-plane membrane forces with the shell bending stresses being vanishingly small in all regions except the thickened subchondral area (Figure 12.10b).

This result suggests that the cortical shell exhibits structural efficiencies much like engineering shells which resist applied loads primarily by in-plane stressing. In the skeleton, the oblique insertion of liagaments and tendons (which seems to be a characteristic feature throughout the skeleton) ensures that the shell is loaded primarily by membrane forces. In the subchondral region, where out-of-plane bending moments are induced by articular contact forces, marked thickening of the trabecular bone is observed which helps to transmit these forces to other regions in the shell.

12.4.2 Finite element results: GIFTS model

Results from the plane-strain finite element model of the patella generated using the GIFTS finite element analysis package is shown in Figure 12.11. Note that for this model the orientation has been reversed with the anterior patella directed

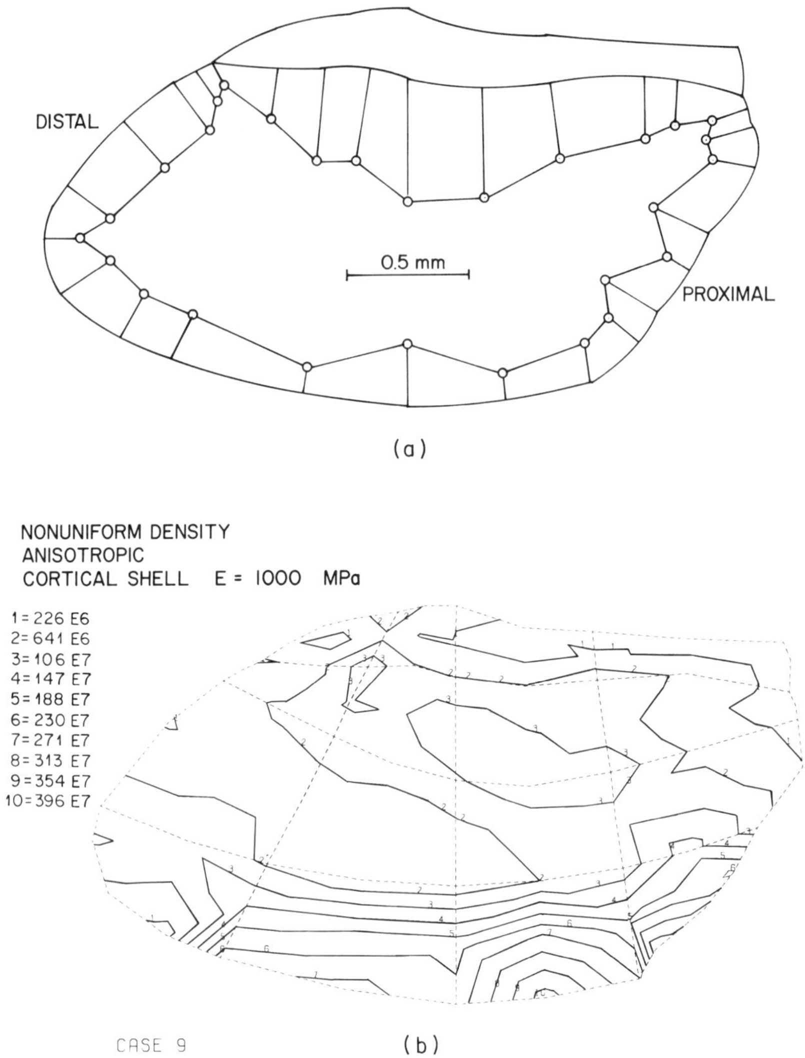

Figure 12.9 Cortical shell model: (a) shell thickness; (b) equivalent Mises stress contours with addition of cortical shell

upward and the proximal attachment of the quadriceps to the right. The number of plane-strain elements has been increased from 16 to 168 and the element type has been changed from an 8-noded isoparametric element in the MARC model to a 4-noded isoparametric element in the GIFTS model.

The maximum principal stresses are completely tensile (Figure 12.11a) and it is apparent that the combined action of the quadriceps mechanism and the patellar

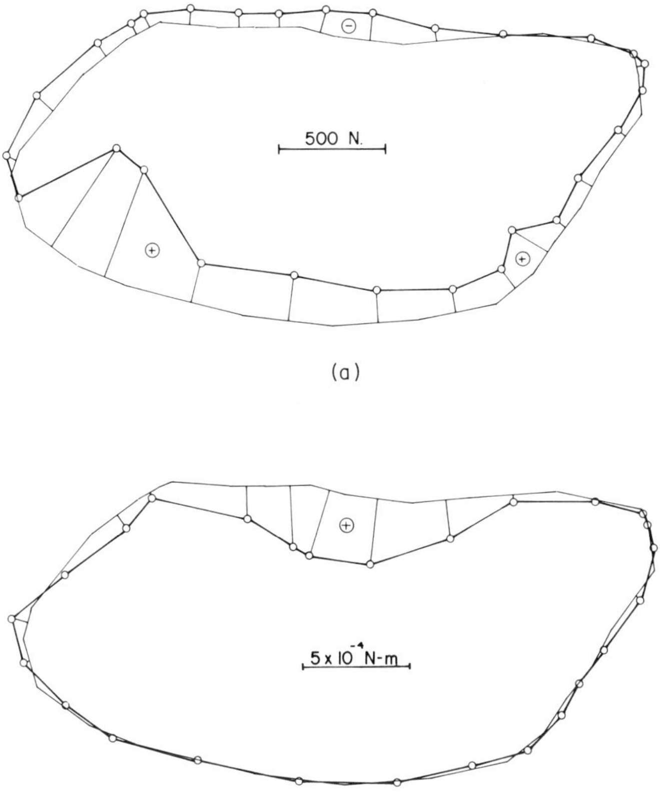

Figure 12.10 Loading of cortical shell showing primarily in-plane membrane forces (a) Membrane force. (b) Bending moment

tendon induce large tensile stresses in the anterior patella. The maximum tensile stress is approximately 8.5 MPa and occurs near the distal pole at the site attachment of the patellar tendon. Moving proximally, the tensile stresses remain constant at approximately 7 MPa.

The compressive principal stresses are shown in Figure 12.11b. The largest compressive stress of about 3 MPa occurs in the subchondral region with a bi-axial compressive stress state predicted in some subchondral elements. The articular cartilage is also seen to be in bi-axial compression, a result which is in agreement with the hydrostatic stress state predicted in cartilage in finite element models of the proximal tibia (Hayes *et al.*, 1978).

The predicted directions of the principal stresses are also shown in Figure 12.11. The tensile principal stresses are strongly oriented in an approximately

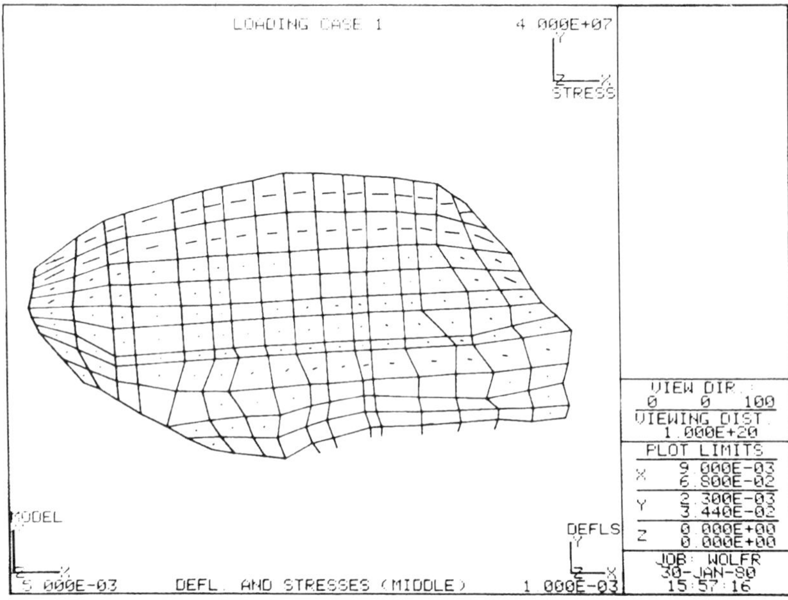

(a)

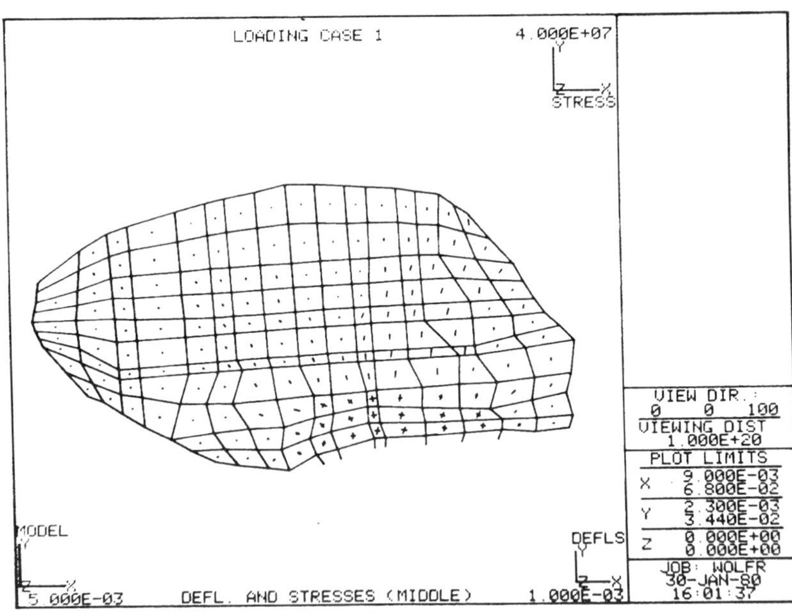

(b)

Figure 12.11 Maximum and minimum principal stresses and their directions from GIFTS model

proximal-to-distal direction in the anterior patella, a direction which corresponds closely to the trabecular orientation in the anterior patella shown in Figure 12.1a. The minimum principal stresses are oriented radially, eminating from the contact region and fanning out toward the anterior patella.

The predicted stress distributions indicate that the patellar loading and support conditions result in stresses similar to those in a short beam subjected to three-point bending with superimposed axial tension. There is tension in the anterior patella and compression in the subchondral region near the applied bending load. The correspondence between the predicted stress distributions and the trabecular architecture shown in Figure 12.1a is immediately apparent. The regions of maximum trabecular density are located in the anterior patella and in the subchondral bone. These regions correspond to areas of maximum tensile and compressive principal stresses in the finite element models. A correspondence between the principal stress directions and trabecular orientation also appears possible from a comparison of Figure 12.11 with the mid-sagittal photomicrograph shown in Figure 12.1a.

12.4.3 Stress – morphology relationships

The principal objective of this investigation was to provide quantitative comparisons between predicted stresses from the finite element models and the descriptors of trabecular architecture obtained from quantitative stereology. The basis for this comparison was not obvious due to the many conflicting hypotheses on the characteristics of stress fields which lead to the formation or resorption of bone.

Comparisons between the predicted stress distributions from the GIFTS finite element model and the stereological measurements of trabecular architecture are shown in Figures 12.12 and 12.13. Figure 12.12 is a plot of von Mises effective stress against trabecular areal density. A highly significant correlation ($R^2 = 0.561$, $P = 0.001$) is evident. This result indicates that bone density is greatest in regions where the principal stress differences are a maximum. It should also be noted that these correlations did not hold when either the maximum principal stress (tension) or the minimum principal stress (compression) were compared against areal density. This tends to refute the hypothesis that either tensile or compressive stresses alone govern trabecular architecture. Instead, the result suggests that a combined stress parameter (such as von Mises or maximum shear stress) best relates to regions of increased bone density.

The orientation results shown in Figure 12.13 further suggest that the trabecular architecture is aligned along the principal stress directions predicted by the finite element model. The predicted principal stress directions calculated from the finite element model are plotted against trabecular orientation as computed from the method of directed secants. A highly significant positive

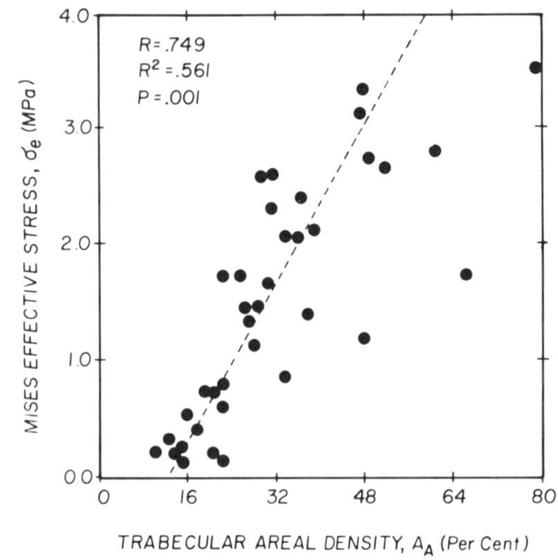

Figure 12.12 Effective stress versus trabecular areal density showing highly significant correlation between trabecular density and principal stress difference

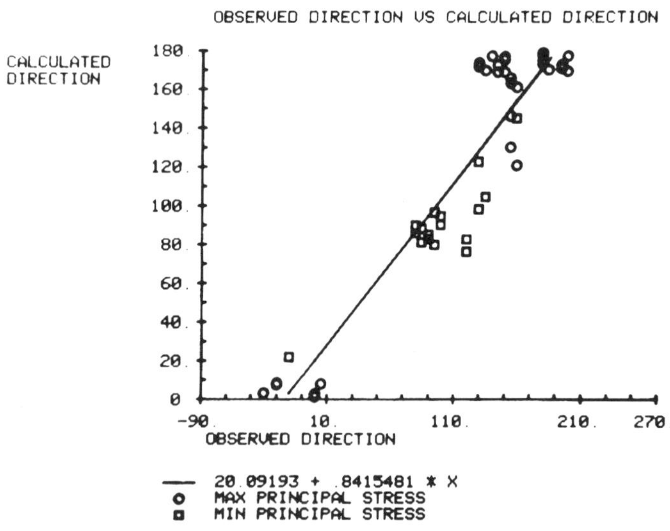

Figure 12.13 Principal stress directions as predicted from GIFTS model versus actual trabecular orientation from stereological measurements. Result confirms that trabeculae are aligned with principal stress directions

correlation ($R^2 = 0.865$, $P < 0.001$) confirms that the trabeculae are aligned with the principal stress directions as predicted by the finite element model.

12.4.4 Wolff's law hypothesis

The correlation between effective stress and trabecular density suggests that trabecular bone is most dense in regions of maximum principal stress difference, i.e. maximum shear. The correlation between trabecular orientation and principal stress direction suggests that the trabeculae are aligned with the principal stress directions, i.e. at 45° to the plane of maximum shear. These two results taken together suggest the dual hypothesis that: (1) individual trabeculae tend to align themselves so as to minimize bending deformations in individual trabeculae; and (2) trabecular density tends to increase in regions of high principal stress difference, i.e. maximum shear.

This hypothesis is illustrated graphically in Figure 12.14 showing that, under optimal alignment conditions, the trabeculae are aligned to minimize bending deformations in individual trabeculae. The non-optimal alignment conditions

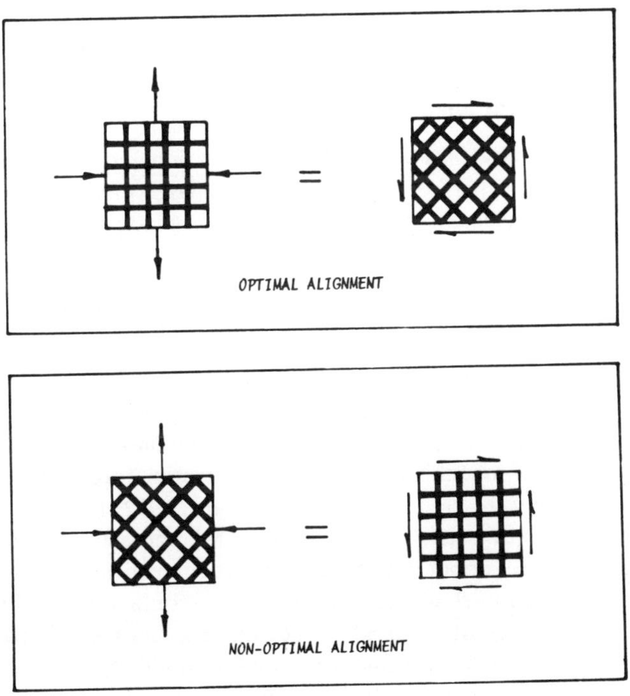

Figure 12.14 Wolff's law hypothesis: trabeculae are aligned optimally to minimize bending deformations in individual trabeculae

result in maximum bending of individual trabeculae. This suggests the working hypothesis that trabeculae exposed to stress fields resulting in non-optimal alignment would remodel toward a condition of optimal alignment if the principal stress differences are at an appropriate level. The latter requirement of course implies that stress levels must be sufficiently large to maintain a viable trabecular architecture (i.e. to prevent osteoporosis) but not so large so as to cause unreparable damage.

12.5 DISCUSSION

The utility of the working hypothesis of Wolff's law and the validity of the conclusions concerning the controlling mechanical parameters governing the structural behaviour of synovial joints both depend on the accuracy of the finite element models. These in turn are dependent on the quality of the input data and on the validity of a number of simplifying assumptions. In the following discussion, these assumptions are evaluated in order to assess the reliability of the analytical predictions. The biological implications of the studies of subchondral stiffening and the structural role of the cortical shell are discussed. Finally, the limited nature of the resulting working hypothesis on the mechanical factors controlling trabecular architecture is emphasized.

12.5.1 Finite element model validity

The finite element models used in this investigation required input data on geometry, material properties, applied loads and patello-femoral contact areas. A plane-strain finite element model was then created based on these input data.

12.5.1.1 Dependence on input data

Literature values for joint forces, contact areas, and material properties were used when available. Unfortunately, little data exists on patello-femoral forces and contact areas. Most previous studies predict resultant forces in the range of zero to four times body weight, depending primarily on the flexion angle of the knee. For the patellar model, literature values for tendon forces corresponding to a 55° flexion angle were used (Reilly and Martens, 1972; Perry *et al.*, 1975).

Since so little data exists for patello-femoral contact area, loading was assumed to occur over about one-third of the articular surface. Previous studies of patello-femoral contact suggest that maximum contact areas of this magnitude occur in the range of 40 to 60° of flexion and that the mid-patellar ridge remains in contact until flexion angles of about 90° (Wiberg, 1941; Goyman and Muller, 1974; Burke *et al.*, 1979). It should be noted that the mid-sagittal section used to generate the finite element mesh corresponds to an anterior–posterior (AP) plane through the mid-patellar ridge.

Reported literature values for the mechanical properties of articular cartilage and trabecular bone also vary over wide ranges. Reported values for the initial elastic modulus of normal human articular cartilage range from about 1 to 15 MPa (Hayes and Bodine, 1978). A value of 10 MPa was used for the finite element model. Accurate data are available on the elastic modulus and strength of trabecular bone spanning a wide range of apparent densities (Townsend *et al.*, 1975; Carter and Hayes, 1976; 1977). These were used in the patellar models to investigate the relative importance of density variations and anisotropy in modelling trabecular strectures.

As is apparent from the effective stress contour plots of Figure 12.7, it is critically important to include trabecular density variations in any attempt to predict stresses in these systems. Wide differences are observed between models based on the assumption of uniform density and those where density variations are included. Because of the cubic dependence of trabecular modulus on density, these density variations even more strongly influence modulus variations throughout the structure. The result on stress distributions is not unexpected, in that high-modulus regions in the anterior patella and in the subchondral bone tend to concentrate stress in models with non-uniform density.

The effect of assuming anisotropic properties for trabecular bone is far less significant, with only about 10 % differences in predicted stresses observed with the 3 : 1 anisotropy ratios reported by Townsend *et al.* (1975). The reason for this insensitivity to the assumption of anisotropy is that these 3 : 1 ratios are small in comparison with the variations of two orders of magnitude in modulus resulting from density variations. The conclusion to be gained from these parametric studies of density and anisotropy is that it is critically important to include density variations in finite element models but that by comparison, trabecular anisotropy may be neglected without major degradations in model accuracy.

From the wide variations reported in the literature on patellar loads and contact areas, it is also obvious that considerable uncertainty exists about the absolute accuracy of these input data. Mid-range values for the parameters were therefore assumed and held constant for all models in order to focus attention on modulus variations. The range of values assumed for trabecular modulus is considered accurate since it is based on direct stereologic measurements of areal density and our experimentally determined relationships between density and modulus. Thus, while the absolute stress levels predicted by the models may be inaccurate, the relative importance of the variables investigated in the parametric studies is assumed to reflect accurately the structural behaviour of trabecular systems.

12.5.1.2 Plane-strain modelling

In the two-dimensional, plane-strain model of the patella used in this investigation, a number of important aspects of patellar geometry and loading were neglected. The patella is actually a complex three-dimensional structure with medial and lateral facets which articulate with the femoral condyles. While the patella tendon runs in a nearly vertical line, the oblique insertion of the quadriceps mechanism is thought to result in a slightly oblique line of action (in the frontal plane) for the quadriceps force.

For these finite element models, a mid-sagittal slice was used to generate a plane-strain model of uniform cross-section. In this case, the model was assumed to be relatively long (in the medio-lateral direction) in comparison with its uniform, cross-sectional dimensions and to be loaded with forces which are uniform in the medio-lateral direction. The model thus did not reflect the existence of the patellar facets or the fact that the articular surface forces are perpendicular to the oblique angles of the medial and lateral facets (and thus directed obliquely toward the centre of the patella). Finally, the planar model did not reflect the oblique angle at which the quadriceps tendon inserts into the proximal pole of the patella. Presumably, this Q angle results in an unequal partitioning of the resultant patello-femoral force between the two facets.

However, the assumption of two-dimensional, plane-strain conditions resulted in a model of markedly reduced complexity and cost, and thus allowed an investigation of more of the mechanical parameters which influence the stress distributions in synovial structures. The models are less an accurate representation of actual joint stresses and more an efficient means for studying the relative importance of the various physical parameters. They thus allowed detailed explorations of variations in density, contact area, subchondral stiffness, and anisotropy. Such parametric studies would have been prohibitively expensive using three-dimensional modelling techniques. Obviously, however, to provide accurate predictions of actual stress levels, three-dimensional models must be used in future studies.

It is interesting to note that such strong correlations were noted between predicted stresses and the observed trabecular architecture despite these uncertainties in input data and the assumption of two-dimensional plane-strain conditions. The plane-strain assumption is illustrated graphically in Figure 12.15, which compares a vertical view of the planar model (with the constraint forces imposed by the plane strain condition) and a similar view of the actual patella. The major differences emphasized by this figure are the oblique articular contact forces that occur in the actual patella. The inclusion of these forces in a three-dimensional model would be expected to increase the compressive stress levels in the subchondral bone over the predictions of the planar models used here. This increase would be reflected in increased effective stress levels in the subchondral bone. Referring to Figure 12.12, it is precisely this subchondral area that produces the major departures from the regression line of effective stress on areal density. The data points which fall significantly below the line at high areal density values are all from the subchondral region. Effective stresses would be

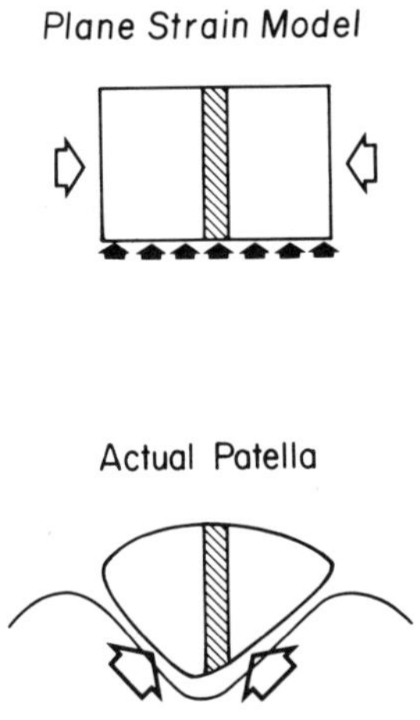

Figure 12.15 Plane-strain model compared with actual patella

raised in this region in a three-dimensional model, thus providing an improved correlation between stresses and the trabecular morphology.

12.5.2 Significance

The results of these stress analyses may be compared with previous studies of the patella and then used to critically evaluate their significance regarding the aetiologies of osteoarthritis and chondromalacia and the structural importance of the cortical shell. The temporal limitations on the working hypothesis for Wolff's law must also be examined and the hypothesis compared with existing hypotheses in the literature.

12.5.2.1 Comparison with previous stress analysis

Although many studies have been conducted on the resultant force systems acting on the patella, analyses of the internal stresses existing in trabecular bone have not been numerous. Simplified photoelastic models have been used to investigate trabecular stresses in the frontal plane (Silva and Bratt, 1970), mid-sagittal plane (Haasters, 1974) and horizontal plane (Maquet, 1976). The last two studies noted a remarkable qualitative correspondence between the photoelastic stress trajectories and the trabecular orientation, but quantitative correlations were not established. It should also be noted that it is impossible to produce photoelastic models which reflect the non-uniform densities which actually exist in trabecular bone. Thus, a major feature of the structural behaviour of trabecular systems is not in evidence in photoelastic studies.

Recently, Minns *et al.* (1979) presented a simple three-dimensional beam model for the patella to investigate the pathogenesis of chondromalacia and to determine the effect on patellar stresses of patellar tendon length and Q angle. Uniform materials were assumed throughout the model and no attempt was made to model separately articular cartilage or trabecular bone, and thus the results are of limited applicability in modelling trabecular stresses.

It should also be emphasized that in the present study, no attempt was made to model stresses in individual trabeculae. Instead, trabecular bone was assumed to be macroscopically homogeneous with modulus values equivalent to the bulk behaviour of specimens of trabecular bone. This is a well-established procedure used by many workers in the present volume. Pugh *et al.*, (1974), by contrast, modelled trabecular bone as a network of thin plates and used a finite element analysis to calculate individual trabecular stresses and displacements. The Pugh model has been criticized as unrepresentative of trabecular architecture and adjustments in trabecular spacing and width were necessary to match macroscopic modulus measurements. However, his study remains one of the few attempts to model trabecular stresses, and future studies must attempt to couple macroscopic, homogeneous models with microstructural models.

12.5.2.2 Aetiology of osteoarthritis

These finite element models of the patella also provide an opportunity to examine existing hypotheses on mechanical factors involved in the development of osteoarthritis. Two major hypotheses have been advanced, with most investigators suggesting that the earliest lesions occur in articular cartilage and that subsequent trabecular changes are secondary to abnormal load transmission through cartilage. An alternative hypothesis advanced by Radin and his co-workers is that the initiating factor is a stiffening of the subchondral trabecular bone which in turn results in increased stresses and accelerated degeneration in the articular cartilage. Several studies by this group tended to support an association between cartilage destruction and subchondral stiffness, but did not determine the exact sequence of the changes or establish the causative role of trabecular stiffening (Radin *et al.*, 1972; Raux *et al.*, 1975; Townsend *et al.*, 1975).

The finite element predictions of the present investigation do not support the hypothesis that subchondral stiffening is a causative factor in the pathogenesis of osteoarthritis. Instead, articular cartilage stresses were found to be relatively insensitive to variations in trabecular stiffness even over a much wider range than can be expected physiologically. The density variations used in the models spanned two orders of magnitude in both directions and yet the stresses in cartilage changed only by a few percent. The reason for this insensitivity is the large modulus difference between bone and cartilage that exists initially, prior to initiating the degenerative process.

Thus, it appears unlikely that trabecular stiffening is a significant causative factor which increases cartilage stresses. Instead, it seems more plausible that the initial lesion occurs in articular cartilage and that the associated changes in trabecular density are secondary to softening and fibrillation of the articular cartilage.

It should also be noted that no attempt was made in these models to investigate the changes in the stiffness gradient that these workers have implicated as significant in more recent discussions of the pathogenesis of chondromalacia (Abernathy, *et al.*, 1978). Instead, a generalized stiffening was assumed and this was shown to have little effect on cartilage stresses. According to the stiffness-gradient view, it is sharp changes in subchondral stiffness that lead to locally increased cartilage shear stresses. It is doubtful that the relatively gradual stiffness gradients noted by these workers in their studies of the patella would have more significant influences on cartilage stresses than do the changes in the current models. It is possible, however, that highly localized stiffness variations such as from a disruption of the subchondral plate could have a strong influence on cartilage stresses, and these should be investigated in future models.

12.5.2.3 Cortical shell

The finite element models used to investigate the structural significance of the cortical shell suggest that the presence of a continuous shell has little influence on trabecular stresses in the interior of the patella. In addition, however, the results suggest that the shell is a highly efficient structural member which, like engineering shells, supports loads primarily by in-plane membrane forces and not by bending moments. This situation is a consequence of the oblique insertion of tendon and ligaments in most locations throughout the skeleton.

The relative insensitivity of the interior stresses to the presence of the shell may also be a consequence of the modulus value assumed for the shell. A value of 1000 MPa was used although no data are available from measurements of the cortical shell. If cortical bone values had been used, a much stronger influence would be expected. This uncertainty in the appropriate value to assume for the shell modulus has been noted by several workers and points out the need for experimental data on this region.

12.5.2.4 Wolff's law

The results of these finite element modelling efforts are consistent with the long-standing hypothesis that trabecular architecture is related to the imposed stress fields. In the mid-sagittal plane used as the basis for these studies, the primary features of the trabecular architecture are a markedly oriented anterior trabecular group, a central region of reduced density, and a posterior, radially oriented region, presumably associated with the patello-femoral contact area. The anterior trabecular group obviously serves to transmit the large forces generated in the quadriceps mechanism to the patellar tendon. The posterior group is related to the large contact stresses developed at the articular surface and is characterized by near isotropy and a marked increase in density. Interior regions, which are presumably lightly loaded, exhibit relatively low trabecular density and a modest radial orientation.

The stress predictions of the planar finite element models developed in this investigation exhibit a remarkable correspondence to these architectural features. The maximum principal stresses are tensile and correspond closely in direction to the orientation of the anterior trabeculae. Minimum principal stresses are compressive and fan out radially from the articular cartilage contact area toward the anterior group.

These results on orientation thus confirm in quantitative terms the long-standing hypothesis that trabeculae are aligned with the principal stress directions (Roux, 1895; Koch, 1917; Murray, 1936; Kummer, 1972; Toridis, 1968; Rybicki *et al.*, 1972; Simon *et al.*, 1974). This is usually referred to as the 'trajectorial theory' of bone architecture, primarily with reference to the proximal femur. In the present study, confirmation for the theory is also found for the trabecular architecture of the patella. Moreover, for the first time, these results are expressed quantitatively in terms of highly significant correlations between predicted stresses and stereologically determined architectural parameters.

The mechanical factors controlling the density, spacing, and average width of trabeculae appear unrelated to either tensile or compressive stresses alone. Instead, areal density is most strongly related to a combined stress parameter, the effective stress, which in turn is related to the principal stress differences. Since the principal stress difference is also proportional to the maximum shear stress, the results of this investigation are compatible with the hypothesis that trabecular density is related to maximum shear stress. The results thus confirm Koch's (1917) hypothesis, derived from the proximal femur, that the thickness and spacing of trabeculae vary with the shear stress magnitude. Based on a strain gauge study of the sheep calcaneus, similar conclusions were reached by Lanyon *et al.*, (1976).

Several limitations in the modelling techniques and the experimental design should be emphasized in any attempt to assess the significance of the above correlations. First, the finite element models were two-dimensional and based on a mid-sagittal slice from a single human patella. The patella actually exhibits a complex, three-dimensional external geometry and internal architecture with non-symmetric loading on medial and lateral facets. Second, a single loading configuration based on an expected maximum load was assumed. In actuality, the patella is subjected to a statistical distribution of loading patterns which must also exert some influence on trabecular architecture. The effect of these assumptions is not known, although the three-dimensional loading would tend to increase the effective stresses in the subchondral region and thus further improve the correlations shown in Figure 12.12.

The most important limitation of the current study is that it represents a comparison of stress and morphology at one point in time in a single human specimen. The comparisons are thus static, in a temporal sense, and provide no opportunity to test the hypothesis that trabeculae will re-align to minimize bending individual trabeculae. Obviously, to examine this hypothesis in a dynamic, *in vivo* environment, it would be desirable to design experiments whereby loads could be altered in a living skeletal structure and the resulting alterations in trabecular architecture observed.

Despite the above limitations, the current study provides powerful support for the trajectorial theory of bone architecture. In the patella, as in the proximal femur, the trabecular architecture is strongly related to the imposed stresses. Since the architecture of trabecular bone plays an important role in osteoporosis, fracture healing, growth deformities, and total joint replacement, the stress–morphology relations suggested here may be of significance in both understanding and treating the above clinical conditions. In addition, our results suggest a hypothesis for the microstructural remodelling mechanisms which may be invoked to alter trabecular architecture in response to changes in the imposed stress fields. With confirmation of this hypothesis in an *in vivo* environment, it may well be possible to exert control on trabecular architecture through the appropriate alterations in the applied stresses.

12.5.3 Application to total joint replacement design

Emerging problem areas with total joint replacement of the hip and knee (such as loosening of tibial and acetabular components) involve interfacing prosthetic components to trabecular bone. The large modulus differences between the relatively rigid prosthetic component and the low-modulus trabecular bone are generally not appreciated, and these differences may span several orders of magnitude. The result is that, even with careful cement techniques, the prosthesis and PMMA tend to act as a rigid, composite structure floating in a very low-modulus region of trabecular bone. This situation can lead to large stress concentrations at the cement–bone interface and the stress concentrations may be related to the loss of component fixation.

The problem may be exacerbated by the surgical techniques used with currently available prosthesis. For instance, the degenerative process in the proximal tibia results in subchondral regions of markedly increased density as well as in a marked reduction in the stiffness and strength of trabecular bone in a region deep to the subchondral bone. However, current tibial component insertion techniques require the removal of this sclerotic subchondral bone. Thus, the tibial component is actually inserted in a region of trabecular bone of markedly inferior mechanical properties. Similar effects probably occur with acetabular components in total hip replacement and certainly are active in patellar resurfacing procedures.

The finite element modelling and stereological analysis techniques developed in this investigation may be applied to the optimized design of total joint replacements. Using such an approach, attention is focused not only on the stress distributions in the prosthetic components and cement, but also on the stress field created in trabecular bone. It may be possible to 'tune' the prosthesis through appropriate selection of geometry and material properties, to the particular region of trabecular bone into which the prosthesis is inserted. The approach thus recognizes the interactive nature of bone-implant systems and raises the possibility that the remodelling potential of trabecular bone may be used to achieve more secure fixation.

The criteria for this optimization process may be based on the maximum shear stress criterion which is our hypothesized mechanism for Wolff's law in trabecular bone. In this view, the insertion of a prosthetic component alters stress distributions in trabecular bone. These altered stress distributions may be examined on the basis of finite element analysis and compared with the stress distributions and morphology in normal and degenerative synovial joints. If the stress distributions imposed by the prosthesis exceed either the maximum or minimum limits required to maintain a viable trabecular architecture, or if they alter the principal stress directions so much that they exceed the ability of trabeculae to alter direction, then loosening and failure is to be expected. On the other hand, if prosthetic components and prosthetic/bone interfaces are designed appropriately, then the stress distributions created in trabecular bone will be appropriate for maintaining a viable trabecular architecture and for assisting in the fixation process.

These techniques are the basis for a comprehensive design and development program currently under way in our laboratory. Particular emphasis is placed on those components most involved in clinical loosening, such as the proximal tibia, the patella, the acetabulum, and the ankle. The finite element analysis techniques have been applied to the problem of patellar surface replacement (Hayes and Levine, 1980). The analysis results were consistent with the permanent deformations noted in patellar surface components removed at revision surgery at the Hospital for Special Surgery. Alternative methods for patellar resurfacing were also examined, in particular for their potential to provide more appropriate stress distributions in the trabecular architecture of the patella. It was found that with removal of the central fixation peg, stress distributions could be created which were much more in line with the trabecular architecture. Results with this model indicated a more physiologic loading pattern in trabecular bone and marked reductions in the stress concentration in PMMA. Analyses with these models are continuing in an attempt to quantitatively relate the analytical predictions of the currently available design to the permanent deformations observed with implanted patellar resurfacing components.

12.5.4 Quantitative formulation of Wolff's law

The long-term objective of our future work is to use macrostructural finite element predictions and quantitative stereologic measurements to provide a synthesis of the mechanical factors which control the remodelling of trabecular bone. For this purpose, it is planned to use the working hypothesis that bone forms in regions of maximum shear and is aligned so as to minimize bending of individual trabeculae. Microstructural models and local histomorphometric techniques will be used to suggest likely mechanisms for the control of remodelling activity. In this respect, it is hoped particular to explore piezoelectric effects in local adaptive remodelling procedures. It is thus hoped to provide a synthesis of the many conflicting views on the mechanical factors controlling trabecular architecture. Such a synthesis would provide a contribution to the understanding of Wolff's law, a long-standing source of controversy in the literature on musculoskeletal physiology. An ability to specify the stress fields necessary to control the architecture of trabecular bone could also be expected to have valuable clinical applications in prosthetic joint replacements, fracture healing, growth deformities, and in the treatment of osteoporosis.

ACKNOWLEDGEMENTS

It is a pleasure to acknowledge those who have supported and contributed in various ways to the work presented here. These studies were supported by NIH AM18376 at Stanford University, by AM20135 at the University of Pennsylvania, by AM26740 at Beth Israel Hospital and Harvard Medical School, and by Research Career Development Award AM00749. Additional support was provided by BRSG grants RR-05415 at the University of Pennsylvania Medical School Computer Facility and RR-01032 to Beth Israel Hospital. Data organization and analysis were performed on the PROPHET system, a national resource sponsored by the Chemical/Biological Information Handling Program, National Institutes of Health.

Many students and colleagues contributed to this work. Brian Snyder, in particular, developed and implemented many of the stereological techniques as part of his Senior Thesis at the University of Pennsylvania. The elucidation of these strees–morphology relationships would have been impossible without his important contributions. Dennis Boyle, Barry Levine, and Seshadri Ramaswamy worked on various aspects of the finite element analyses. It has been a particular pleasure to work with David J. Schurman, M. D., David Nunamaker, V. M. D., and Paul J. Lotke, M. D.. Finally, we wish to thank, Carl T. Brighton, M. D., and Augustus A. White, M. D., for their counsel, their scientific contributions to this work, and their development of supportive, multidisciplinary research environments.

REFERENCES

Abernathy, P. J., Townsend, P. R., Rose, R. M., and Radin, E. L. (1978) Is chondromalacia patellae a separate clinical entity?, *J. Bone and Joint Surg.* **60**(B), 205–210.

Akeson, W. H., Woo, S. L. -Y., Rutherford, L., Coutts, R. D., Gonsalves, M., and Amiel, D. (1976) The effects of rigidity of internal fixation plate, *Acta Orthop Scand.*, **47**, 241–249.

Albright, J. A., and Brand, R. A. (1979) *The Scientific Basis of Orthopaedics*, Appleton–Century–Crofts, New York.

Albright, J., Martin, R., and Flohr, R. (1978) Automatic image analysis of the bone biopsy: variations in rib architecture, *Microscope*, **26**, 15–34.

Bartel, D. L., Santavicca, E. A., and Burstein, A. H. (1980) The effects of pegs and trays on stresses associated with loosening of knee prostheses, *Trans. O. R. S.* **5,** 165.

Bassett, C. A. L. (1968) Biologic significance of piezoelectricity, *Calc. Tiss. Res.* **1,** 252.

Bassett, C. A. L. (1971) Biophysical principles affecting bone structure, Chapter 1 in *The Biochemistry and Physiology of Bone* (2nd edn), Vol. 3, (ed. G. H. Bourne), Academic Press, New York.

Bassett, C. A. L., and Becker, R. O. (1962) Generation of electric potentials by bone in response to mechanical stress, *Science*, **137,** 1063.

Becker, R. O., Spadaro, J. A., and Marino, A. A. (1977) Clinical experience with low intensity direct current stimulation of bone growth, *Clin. Orthop.*, **127,** 75.

Black, J., and Korostoff, E. (1974) Strain-related potentials in living bone, *Ann. N. Y. Acad. Sci.*, **238,** 95.

Booth, F. W., and Gould, E. W. (1975) Effects of training and disuse on connective tissue, In: *Exercise and Sports Science Reviews* (ed. J. F. Keogh), New York, Academic Press, pp. 83–112.

Brighton, C. T., Black, J., and Pollack, S. R. (1979) *Electrical Properties of Bone and Cartilage*, Grune and Stratton, N. Y.

Brighton, C. T., Friedenberg, Z. B., Mitchell, E. I., and Booth, R. E. (1977) Treatment of nonunion with constant direct current, *Clin. Orthop.*, **124,** 106.

Brown, T. D., Way, M. E. and Ferguson, A. B. (1980a) Stress distribution changes in adult femoral heads by avascular necrosis, In: *Finite Elements in Biomechanics* (ed. B. R. Simon), pp. 147–167.

Brown, T. D., Way, M. E., Fu, F. H., and Ferguson, A. B. (1980b) Some growth-related changes in the distribution of stress in the proximal juvenile femur, In: *Finite elements in Biomechanics*. (ed. B. R. Simon), pp. 129–145.

Burke, D. L., Ahmed, A. M., and Miller, J. (1979) The effects of patellar tendon transfer on patello-femoral load-bearing characteristics, *Trans. O. R. S.*, **4,** 153.

Carey, E. J. (1929) Studies in the dynamics of histogensis, *Radiology*, **13,** 127.

Carter, D. R., and Hayes, W. C. (1976a) Bone compressive strength: the influence of density and strain rate, *Science*, **194,** 1174–1176.

Carter, D. R., and Hayes, W. C. (1976b) Fatigue life of compact bone. I. Effects of stress amplitude, temperature, and density, *J. Biomechanics*, **9,** 27–34.

Carter, D. R., and Hayes, W. C. (1977) The compressive behavior of bone as a two-phase porous material, *J. Bone and Joint Surg.*, **49**(A), 954–962.

Carter, D. R., Hayes, W. C., and Schurman, D. J. (1976) Fatigue life of compact bone II. Effects of Microstructure and density, *J. Biomechanics*, **9,** 211–218.

Carter, D. R., Vasu, R., Spengler, D. M., and Dueland, R. T. (1980) Stress fields in the unplated and plated canine femur calculated from *in-vivo* strain measurements, *J. Biomechanics* (submitted for publication).

Chamay, A., and Tschantz, P. (1972) Mechanical influences in bone remodeling. Experimental research on Wolff's law, *J. Biomechanics*, **5**, 173–180.

Chavapil, M., Bartos, D., and Bartos, F. (1973) Effect of long-term physical stress on collagen growth in the lung, heart and femur of young and adult rats, *Gerontologia*, **19**, 263–270.

Churches, A. E., Howlett, C. R., and Waldron, K. J. (1979) The response of living bone to controlled time-varying loading: Method and preliminary results, *J. Biomechanics*, **12**, 35–45.

Cochran, G. V. B., Pawluk, R. J., and Bassett, C. A. L. (1968) Electromechanical characteristics of bone under physiologic moisture conditions, *Clin. Orthop.*, **58**, 249.

Cook, S. D., Klawitter, J. J., Weinstein, A. M., and Lavernia, C. J. (1980) The design and evaluation of dental implants with finite element analysis, In: *Finite Elements in Biomechanics* (ed. B. R. Simon), Univ. Arizona Press, Tucson., pp. 169–78.

Cowin, S. C., and Hegedus, D. H. (1976) Bone remodeling. I: Theory of adaptive elasticity, *Journal of Elasticity*, **313**, 326.

Cowin, S. C., and Van Buskirk, W. C. (1978) Internal bone remodeling induced by a medullary pin, *J. Biomechanics*, **11**, 269–275.

Currey, J. D. (1968) The adaptation of bones to stress, *J. Theoret. Biol.*, **20**, 91–106.

Dalen, N., and Olsson, K. E. (1974) Bone mineral content and physical activity, *Acta Orthop. Scand.*, **45**, 170–174.

Daniels, A. V., McKenna, G. B., Bradley, G. W., and Dunn, H. K. (1978) Effects of plate stiffness on bone remodeling, *Trans. 4th Soc. Biomaterials*, **2**, 52–53.

Da Silva, O. L., and Bratt, J. F. (1970) Stress trajectories in the patella, *Acta. Orthop. Scand.*, **41**, 608–618.

Dehoff and Rhines (1968) *Quantitative Microscopy*, McGraw-Hill, New York.

Dixon, A. T. (1910) Architecture of the cancellous tissue forming the upper end of the femur, *J. Anat. Lond.* (3) **44**, 223–30.

Eriksson, C. (1976) Electrical properties of bone, In: *The Biochemistry and Physiology of Bone*, Vol. 4, Academic Press, New York.

Friedenberg, Z. B., Zensky, L. M., Pollis, R. P., and Brighton, C. T. (1974) The response of non-traumatized bone to direct current, *J. Bone and Joint Surg.*, **56**(A), 1023

Frost, H. M. (1964a) Dynamics of bone remodeling, In: Frost, H. M. (ed.) *Bone Biodynamics*, Little, Brown and Co., Boston, MA., pp. 315–333.

Frost, H. M. (1964b) *Mathematical Elements of Lamellar Bone Remodeling*, Thomas, Springfield, Ill.

Frost, H. M. (1973) Bone modeling and skeletal modeling errors, *Orthopaedic Lectures*, Vol. IV. Charles C. Thomas, Springfield, Ill.

Fukada, E. and Yasuda, (1957) On the piezoelectric effect of bone, *J. Phys. Soc. Japan*, **12**, 1158.

Galileo, G. (1638) *Dialogues concerning two new sciences*, (Grew and Salvio's translation, New York, 1914).

Gjelsvik, A. (1973) Bone remodeling and piezoelectricity – I, *J. Biomechanics*, **6**, 69–77.

Glucksmann, A. (1938) Studies on bone mechanics *in-vitro*, *The Anatomical Record*, **72**, (1), 97–113.

Goymann, V., and Muller, H. G. (1974) New calculation of the biomechanics of the patello-femoral joint and its clinical significance, In: *The Knee Joint*, O. S. Ingwersen, *et al.*, (eds.), American Elsevier, New York, pp. 16–21.

Hall, B. K. (1968) *In-vitro* studies on the mechanical evocation of adventitious cartilage in the chick, *J. Exp. Zoology*, **168**, 283–306.

Hassler, C. R., Rybicki, E. F., Cummings, K. D., and Clark, L. C. (1977) Quantification of compressive stress and its effects on bone remodeling, *Bull Hosp. Joint Dis.*, **38**, 90–93.

Haasters, J. (1974) Functional analysis of the spongiosa structure of the human patella, In: *The Knee Joint*, O. S. Ingwersen, *et al.*, (eds.), American Elsevier, New York, pp. 25–29.
Hayes, W. C., and Levine, B. M. (1980) Finite element analysis of patellar resurfacing procedures, *Trans. O. R. S.*, **5,** 163.
Hayes, W. C. and Bodine, A. J. (1978) Flow-independent viscoelastic properties of articular cartilage matrix, *J. Biomechanics*, **11,** 407–419.
Hayes, W. C., Snyder, B., and Levine, B. M. (1980) Stress–morphology relationships in trabecular bone of the patella, In: *Finite Elements in Biomechanics* (ed. B. R. Simon), pp. 107–127.
Hayes, W. C., Swenson, L. M., and Schurman, D. J. (1978) An axisymmetric finite element model of the lateral tibial plateau, *J. Biomechanics*, **11,** 21–23.
Hert, J., Liskova, M., and Landrgot, B. (1969) Influence of long-term continuous bending on the bone, *Folia Morph.*, **17,** 389–399.
Jansen, M. (1920) *On Bone Formation*, University Press, Manchester.
Jones, H. H., Priest, J. D., Hayes, W. C., Tichenor, C. C., and Nagel, D. A. (1977) Humeral hypertropy in response to exercise, *J. Bone Joint Surg.*, **59**(A), 204–208.
Jowsey, J. (1977) *Metabolic Diseases of Bone*, W. B. Saunders Co., Phla.
Justus, R., and Luft, J. (1970) A mechanochemical hypothesis for bone remodeling induced by mechanical stress, *Calc. Tiss. Res.*, **5,** 222–235.
Kaplan, F. S., Bodine, A. J., Hayes, W. C., and Brighton, C. T. (1979) Complex shear modulus of proteoglycan depleted articular cartilage matric, *Trans. Orthop. Res. Soc.*, **4,** 140.
Keith, A. (1918) Hunterian lecture on Wolff's Law of bone transformation, *The Lancet*, 16, Feb. 250–252.
Kiiskinen, A. (1977) Physical training and connective tissues in young mice physical properties of achilles tendon and long bones, *Growth*, **41,** 123–137.
Kiiskinen, A., and Keikkinen, E. (1978) Physical training and connective tissues in young mice: Biochemistry of long bones, *J. Appl. Physiol: Respirat. Environ. Exercise Physiol.*, **44,** 50–54.
King, D. W., and Pengelly, R. G. (1973) Effect of running on the density of rat tibias, Abstract published In: Medicine and Science in Sports, **5,** 68–69.
Koch, J. C. (1917) The Laws of bone architecture, *Am. J. Anat.*, **21,** 177–298.
Korostoff, E. (1979) A comprehensive model for characterizing stress generated potentials, In: *Electrical Properties of Bone and Cartilage*, (eds. Brighton, C. T., Black, J. and Pollack, S. R.), Grune and Stratton, New York.
Kummer, B. (1966) Photoelastic studies on the functional structure of bone, *Folia Biotheoretica*, **6,** 31–40.
Kummer, B. (1972) Biomechanics of bone mechanical properties, functional structure, functional adaptation, In: *Biomechanics, Its Foundations and Objectives*, Fung, Y. C., Perrone, N. and Anliker, N. (eds.) Prentice-Hall, Inc. Englewood Cliffs, N. H.
Lanyon, L. E., (1974) Experimental support for the trajectorial theory of bone structure, *J. Bone and Joint Surgery*, **56**(B), 160–166.
Lanyon, L. E., Goodship, A. E., and Baggot, D. G. (1976) The significance of bone strain *in-vivo*, *Acta Orthopaedica Belgica*, Tome 42, Suppl. 1. Quatrieme Symposium de Biomechanique osseuse, CIBO, 21 mai 1976.
Lanyon, L. E., and Hartman, W. (1977) Strain related electrical potentials recorded *in-vitro* and *in-vivo*, *Calc. of Tissue Res.*, **22,** 315.
Lavernia, C. J., Cook, S. D., Klawitter, J. J., Weinstein, A. M., and Das, S. C. (1980) The effect of implant elastic modulus on the stress distribution surrounding dental implants, In: *Finite Elements in Biomechanics*. (ed. B. R. Simon) Univ. Arizona Press, Tuscon, pp. 179–192.

Lewis, J. L. (1977) Analytical methods in prosthesis design, *Proc. Workshop on Internal Joint Replacement*, Northwestern Univ. Chicago Press, pp. 127–134.

Lieb, F. J., and Perry, J. (1968) Quadriceps function, *J. Bone Joint Surg.*, **50**(A), 1534–1548.

Liskova, M., and Hert, J. (1971) Reaction of bone to mechanical stimuli. Part 2. Periosteal and endosteal reaction of tibial diaphysis in rabbit to intermittent loading, *Folia Morphologica*, **29,** 301–317.

Lloyd, E., Rowland, R. E., Hodges, D., and Marshall, J. H. (1968) Surface to volume ratios of bone determined by computer analysis of microradiographs, *Nature*, **218,** 365–366.

Maquet, P. G. J. (1976) *Biomechanics of the Knee*, Springer-Verlag, Berlin.

Marino, A. A., and Becker, R. O. (1970) Piezoelectric effect and growth control in bone, *Nature* **228,** 473.

Martin, R. B. (1972) The effects of geometric feedback in the development of osteoporosis, *J. Biomechanics* **5,** 447–455.

Martin, R. B. (1979) Analysis of bone and other piezoelectric textures, In: *Electrical Properties of Bone and Cartilage*, Brighton, C. T., Black, J. and Pollack, S. R. (eds.), Grune and Stratton, New York.

McMaster, J. H., and Weinert, C. R. (1970) Effect of mechanical forces on growing cartilage, *Clinical Orthopaedics*, **72,** 308–314.

Merz, W. A. (1967) Die Streckenmessung an gerichteten Strukturen im Mikroskop und ihre Anwendung zur Bes immung von Oberflachen-volumen-rationem im Knochengewebe, *J. Mikro kopie*, **22,** 132–142.

Merz, W. A., and Schenk, R. K. (1970a), Quantitative structural analysis of human cancellous bone, *Acta Anat.*, **75,** 54–66.

Merz, W. A., and Schenk, R. K. (1970b), A quantitative histological study on bone formation in human cancellous bone, *Acta Anat.*, **76,** 1–15.

Meyer, H. (1867) Die architektur der spongiosa, *Arch. Anat. Phys.* **47,** 615–628.

Milch, H. (1940) Photo-elastic studies of bone forms, *J. Bone and Joint Surg.*, **22**, 621–626.

Minns, R. J., Birnie, A. J. M., and Abernathy, P. J. (1979) A stress analysis of the patella and how it relates to patella articular cartilage lesions, *J. Biomech.*, **12,** 699–711.

Morey, E. R., and Baylink, D. J. (1978) Inhibition of bone formation during space flight, *Science*, **201,** 1138.

Morrison, J. B. (1968) Bioengineering analysis of force actions transmitted by the knee joint, *Bio. Med. Eng.*, **3,** 164.

Mow, V. C., and Lai, W. M. (1979) Mechanics of animal joints, *Ann. Rev. Fluid Mech.*, **11,** 247–288.

Moyen, B. J-L., Lahey, R. J., Weinberg, E. H., and Harris, W. H. (1978) Effects on intact femora of dogs of the application and removal of metal plates, *J. Bone and Joint Surg.* **60**(A), 940–947.

Murray, P. D. F. (1936) *Bones. A Study of the Development and Structure of the Vertebrate Skeleton*, Cambridge University Press.

Nargurka, M. L. and Hayes, W. C. (1980) Technical note: An interactive graphics package for calculating cross-sectional properties of complex shapes, *J. Biomech.* (in press).

Nilsson, B. E., and Westlin, N. E. (1971) Bone density in athletes, *Clin. Orthop. and Rel. Res.*, **77,** 179–182.

Outerbridge, R. E. (1961) The etiology of chondromalacia, *J. Bone and Joint Surg.*, **43(B)**, 752–757.

Oxnard, C. E. (1972) Tensile forces in skeletal structures, *J. Morph.*, **134,** 425–436.

Pauwels, F. (1948) Bedeutung und Kausale Erklarung der Spongiosaarchitektur in neuer Auffassung, *Arztl. Wschr.*, **3,** 379.
Perry, J., Antonelli, D., and Ford, W. (1975). Analysis of Knee-joint forces during flexed knee stance, *J. Bone and Joint Surg.*, **57**(A), 961–967.
Pugh, J. W., *et al.*, (1974) Quantitative studies of human subchondral cancellous bone, *J. Bone and Joint Surg.*, **56**(A), No. 2, 313–321.
Radin, E. L., Pugh, J. W., Steinberg, R. S., Parker, H. G., and Rose, R. M. (1972), Trabecular microfractures in response to stress: the possible mechanism of Wolff's Law, International Congress series no. 291 *Orthopaedic Surgery and Traumatology*, 59–65.
Radin, E. L., *et al.*, (1973), Response of joints to impact loading. III, *J. Biomechanics*, **6,** 51–57.
Rambaut, P. C., and Johnson, R. S. (1979) Prolonged weightlessness and calcium loss in man, *Acta Astronautica*, **6,** 1113–1122.
Raux, P., Townsend, P. R., Piegel, R., Rose, R. M., and Radin, E. L. (1975) Trabecular architecture of the human patella, *J. Biomechanics*, **8,** 1–7.
Reilly, D. T., and Martens, M. (1972) Experimental analysis of the quadriceps muscle force and patello-femoral joint reaction force for various activities. *Acta Orthop. Scand.*, **43,** 126–137.
Reinish, G. B., and Nowick, A. S. (1975) Peizoelectric properties of bone as a function of moisture content, *Nature*, **253,** 626.
Robb, R. A., and Jowsey, J. (1978) Quantitative measurement of fractional bone volume using digital scanning videodensitometry, *Calc. Tissue Res.*, **25,** 265–271.
Rodan, G. A., Mensi, T., and Harvey, A. (1975). A quantitative method for the application of compressive forces to bone in tissue culture, *Calc. Tissue Res.*, **18,** 125–131.
Rosenfeld, R., Rosenfeldova, A., Steiglova, G., and Kvapilova, I. (1973) Effect of exercise of acceleration on calcium metabolism in rats, *Physiol. Bohemoslov.*, **22,** 195–199.
Roux, W. (1895) *Ges Abhandlungen uber Entwicklungsmechanik der Organismen*, Bd. I., Funktionelle Anpassung (Leipzig).
Rybicki, E. F., Simonen, F. A., and Weis, E. B. (1972) On the mathematical analysis of stress in the human femur, *J. Biomechanics*, **5,** 203–215.
Saha, S., and Lakes, R. S. (1979) A new non-invasive device for monitoring the piezoelectric character of bone, In: *Electral Properties of Bone and Cartilage*, Brighton, C. T., Black, J. and Pollack, S. R. (eds.), Grune and Stratton, New York.
Salmons, S. (1969) Report on the 8th Int'l Conf. in Med. Biol. Eng., *Biomed Enging.*, **4,** 467–474.
Saltykov, S. A. (1958) *Stereometric Metallography* (2nd edn), Moscow; Metallurgizdat.
Saville, P. D., and Whyte, M. P. (1969) Muscle and bone hypertrophy – Positive effect of running exercise in the rat, *Clin. Orthop. and Rel. Res.*, **65,** 81–88.
Scott, J. H. (1957) The mechanical basis of bone formation, *J. Bone and Joint Surg.*, **39**(B), 134–144.
Shamos, M. H., Levine, L. S., and Shamos, M. F. (1963) Piezoelectric effect in bone, *Nature*, **197**, 91.
Simon, B. R., Stanley, G. M., Kamel, H. A., and Peltier, L. F. (1974) Relation of subchondral bone and articular cartilage stresses to joint function, *Proc. 5th Int'l Conf. Exp. Stress Anal.*, Paper 33, pp. 3.45–3.51.
Simon, B. R. (1980) *Finite Elements in Biomechanics*, Univ. Arizona Press, Tuscon.
Singh, M., *et al.*, (1970) Changes in trabecular pattern of the upper end of the femur as an index of osteoporosis. *J. Bone and Joint Surg.*, **52,** 457–467.
Snyder, B., and Hayes, W. C. (1980) Development of a computerized system for the

morphological analysis of trabecular bone, *Technical Report OBL 80–2*, Orthopaedic Biomechanics Laboratory, Beth Israel Hospital.

Spadaro, J. A. (1977) Electrically stimulated bone growth in animals and man, *Clin. Orthop.*, **122,** 325.

Takechi, H. (1977) Trabecular architecture of the knee joint, *Acta. Orthop. Scand.*, **48,** 637–681.

Thompson, Sir D'Arcy (1917) *On Growth and Form* (abridged edn), Cambridge University Press, 1961.

Tipton, C. M., Matthes, R. D., and Maynard, J. A. (1972) Influence of chronic exercise in rat bones. Abstract published in *Medicine and Science in Sports*, **4,** 55.

Tobin, W. J. (1968) Further evidence and confirmation of Wolff's Law of bone transformation. Part 1 – An atlas of the comparative anatomy of the upper end of the femur, *Clin. Ortho.*, **56,** 83–103.

Todd, R., Freeman, M. A. R., and Pirie, C. (1972). Isolated trabecular fatigue fractures in the femoral head, *J. Bone and Joint Surg.*, **54**(B), 723–728.

Tonino, A. S., Davidson, C. L., Klopper, P. J., and Linclau, L. A. (1976) Protection from stress in bone and its effects, *J. Bone and Joint Surg.*, **58**(B), 107–113.

Toridis, T. G. (1969) Stress analysis of the femur, *J. Biomechanics*, **2,** 163–174.

Townsend, P. R., Raux, P., Rose, R. M., Meigel, R. E., and Radin, E. L. (1975) The distribution and anisotropy of the stiffness of cancellous bone in the human patella, *J. Biomechanics*, **8,** 363–367.

Trueta, J. (1962) A theory of bone formation, *Acta Orthop. Scand.*, **32,** 190.

Trueta, J. (1968) *Studies of the Development and Decay of the Human Frame*, Heinemann, London.

Underwood, E. E. (1970) *Quantitative Stereology*, Addison-Wesley, Reading, Mass.

Urovitz, E. P. M., Fornasier, V. L., Risen, M. I., and McNab, I. (1977) Ethiological factors in the pathogenesis of femoral fatigue fractures, *Clin. Orthop. Rel. Res.*, **127,** 275–280.

Wakamatsu, E., and Sisson (1969) The cancellous bone of the illiac creast, *Calc. Tiss. Res.*, **4,** 147–161.

Walonsley, B., Hodgson, J. A., and Burke, R. E. (1978) Forces produced by the medial gastrocnemius and soleus muscles during locomotion in freely moving cats, *J. Neurophysiology*, **41,** 1203–1216.

Weibel, E. R., and Elias, H. (1967) *Quantitative Methods in Morphology*, Springer-Verlag, N.Y.

Weibel, E. R. (1973) Stereological techniques for election microscopic morphometry, In: *Principles and Techniques of Electron Microscopy* (ed. M. A. Hayat) Van Nostrand Reinhold Co., N.Y.

Whitehouse, W. J. (1974) The quantitative morphology of anisotropic trabecular bone, *J. Microscopy*, **101,** 153–168.

Whitehouse, W. J. (1975) Scanning electron micrographs of cancellous bone from the human sternum, *J. Path.*, **116,** 213–223.

Whitehouse, W. J. (1977) Cancellous bone in the anterior part of the illiac crest, *Calc. Tissue Res.*, **23,** 67–76.

Whitehouse, W. J., and Dyson, E. D. (1974) Scanning electron microscope studies of trabecular bone in the proximal end of the human femur, *J. Anat.*, **118,** 417–444.

Wiberg, G. (1941) Roentgenographic and anatomic studies on the femoral patellar joint. *Acta Orthopaedica*, **12,** 319–404.

Williams, J. L., and Lewis, J. L. (1979) An anisotropic model of cancellous bone, *A.S.M.E. 1979 Biomechanics Symp.* AMD **32,** 181–182.

Wolff, J. (1870a) Die innere architekture der knochen, *Arch. Anat. Phys.*, **50.**

Wolff, J. (1870b) Ueber die innere architectur der knochen und'ihre Bedeutung fur die Frage von Knochenwachstum, *Uirchows Arch.*, **50,** 389–450.
Wolff, J. (1892) *Das Gesetz der Transformation der Knochen*, A. Hirchwild, Berlin.
Woo, S. L-Y., Kuei, S. C., Dillon, W. A., White, F. C., and Akeson, W. H. (1979) The effects of stresses on long bone remodelling: A study of Wolff's Law, *Trans. 25th Ortho. Res. Soc.*, **4,** 93.
Woo, S. L-Y., Kuei, S. C., Dillon, W. A., Amiel, D., White, F. C., and Akeson, W. H. (1980) The effect of prolonged physical training on the properties of long bone – a study of Wolff's Law, *J. Bone and Joint Surg.* (submitted for publication).
Wyman, M. (1849) Animal mechanics: on the cancellated structure of some of the bones of the human body, Boston Soc. of Nat. Hist.
Yagodovsky, V. S., Trifranidi, L. A., and Goroklova, G. P. (1976) Space flight effects on skeletal bones of rats, *Aviat. Space Environ. Med.*, **47,** 734.

Finite Elements in Biomechanics
Edited by R. H. Gallagher, B. R. Simon, P. C. Johnson, and J. F. Gross

Chapter 13

Some Femoral Head Load Transmission Changes Occurring with Infarction and with Normal Growth

Thomas D. Brown, Margaret E. Way, Freddie H. Fu, and Albert B. Ferguson, Jr.

13.1 INTRODUCTION

Hip joint destruction secondary to aseptic necrosis of the femoral head is currently a major cause of disability in both the juvenile and adult populations. Regardless of whether the infarction arises spontaneously (Legg–Perthes disease in children; idiopathic avascular necrosis in the adult) or is due to external insult (trauma, synovitis, alcoholism, steroids, radiation, etc.), the usual result is gradual collapse of the weakened infarct region. Orthopaedic management of this condition has generally aimed to retard or arrest the collapse process by minimizing load transmission through the affected head segment, which is unfortunately most often located superio-centrally in the major weight-bearing tract. To this end, treatments employed have included months of complete bed rest, abduction bracing, and various femoral or pelvic osteotomies (Curtis, 1979). In this context, detailed information concerning stress transmission anomalies associated with infarcted femoral head segments would be highly desirable. This chapter consists of two closely related parts. In Part I, a two-dimensional finite element model was used to assess the salient alterations in stress distribution occurring within and around a weakened central necrotic segment of cancellous bone in the adult femoral head. In Part II a related formulation was employed to explore some stress distribution phenomena peculiar to various stages of femoral head ossification encountered during the normal growth process.

13.2 PART I: STRESS CHANGES WITH OSTEONECROSIS IN THE ADULT PROXIMAL FEMUR

Alterations in load transmission in the osteonecrotic adult femur result from rheological deficits in the avascular region which occur because of the accumulation of unrepaired trabecular microfractures, as well as the creeping substitution accompanying revascularization and fibrous invasion (Catto, 1976).

A series of computations was performed to delineate the spread of load transmission changes as stiffness deficits accumulate during the pathogenesis of a typical wedge-shaped infarct. The effects of lesion size and shape were then explored, for a fixed stiffness deficit, by serial parametric variations of the infarct width, depth, and apex-to-base aspect ratio. Of particular interest is the degree of stress-shielding afforded by the relatively stiff viable bone surrounding the infarct. The finite element data also reflect the tendency toward gross congruity changes (especially head flattening) predisposing to secondary osteoarthritis, and they identify regions of stress elevation susceptible to fracture outside the infarct.

13.2.1 Method

Two-dimensional (plane-strain) stress fields were computed for a coronal midsection of the adult proximal femur. The finite element zoning employed 120 curved quadratic isoparametric elements of the 'serendipity' type (810 global degrees of freedom). The loading conditions applied were characteristic of single-legged stance, and consist of normal compressive stress at the articular surface, abductor muscle (gluteus medius) traction at the greater trochanter, and zero displacement across a horizontal section just distal to the lesser trochanter. Allocation of articular node loading forces was performed on a virtual work basis, assuming the articular contact stress to be uniformly distributed. The global element structure and boundary condition arrangement for the study section considered are shown in Figure 13.1.

Regions of the proximal femur occupied by cancellous bone were modelled as a non-homogeneous, anisotropic continuum. Material property data for viable cancellous bone were input from contour mappings (see Figure 13.2) of spatial and directional modulus variations obtained from previous laboratory studies of large numbers of side-by-side cubic test specimens (Brown and Ferguson, 1979).

To explore the rheological deficits associated with necrosis of cancellous bone, quasi-static compression tests were performed on several hundred 5-mm cubes taken from involved regions of nine freshly resected necrotic human femoral heads (Brown *et al.*, 1979). The tests on specimens with well-established necrosis showed that the average strength and stiffness deficits (relative to healthy bone from similar regions) were 52% and 72%, respectively. To study the effects of necrosis within specific femoral head segments, modulus values for all nodes within the infarct were reduced to a specified fraction, C_N, of the value locally characteristic of viable cancellous bone. Material property values for cartilage and cortical bone were taken from the literature.

For a standard infarct geometry (shaded region in Figure 13.3b), nine computations were performed for C_N values ranging from 1.0 (normal case) to 0.1 (very severe deficit). Five computations were undertaken to study the role of

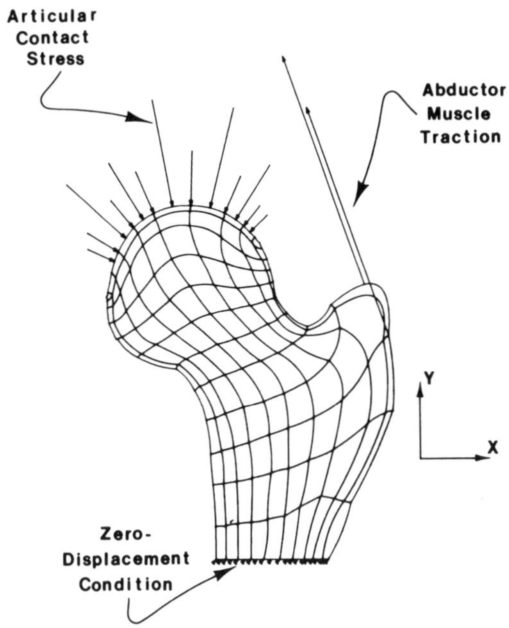

Figure 13.1 Finite element zoning used to study stress distributions in the proximal femur. This study section (a coronal midsection from an eighteen-year-old) is subjected to normal compressive stress at the articular surface and to abductor muscle traction at the greater trochanter (arrows). Distally, a zero displacement condition is assumed across a transverse section of the shaft

base widening for an infarct of fixed depth and five computations were performed to study the role of increasing depth for an infarct of fixed base width. Finally six cases were considered in which the base width and depth were both held constant, but in which the apex of the lesion had finite width. In all cases, the infarct was assumed to be symmetrical about an axis coincident with the line of action of the resultant joint load vector. A single C_N value of 0.3, which is typical of clinically symptomatic aseptic necrosis, was used in all three series of geometry-change computations.

13.2.2 Results

Comparison of the principal stress field for an involved proximal femur (Figure 13.3) with that for the normal structure (Figure 13.8, eighteen-year-old) shows that the salient load transfer mechanisms remain operative. The compressive stresses of weight bearing are progressively concentrated as they pass distally, along the primary trabeculation system, from the articular surface to the medial

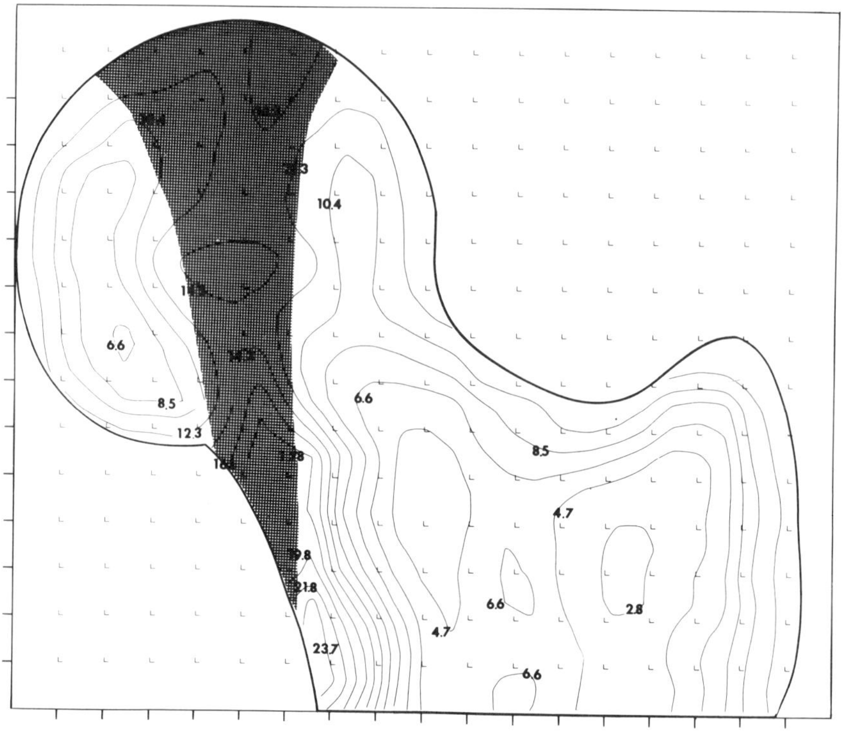

Figure 13.2 Spatial distribution of elastic modulus values in the study section. Values for the isopleths are in 10^4 lbs/in^2. The contour plot was generated from quasi-static Instron compression tests of approximately 150, 5 – mm cubic test samples, each loaded in the superior-inferior direction. The shaded area indicates the extent of the primary 'weight-bearing' trabeculation system

cortex. During this passage, the stresses also change from essentially biaxial to essentially uniaxial. Tensile forces in the abductor muscles induce a bending moment, which is resisted by large longitudinal stresses (compressive medially, tensile laterally) in the femoral cortices. The muscle pull also causes substantial compressive stress at the neck–trochanter junction. The major changes associated with necrosis were seen to be:

(a) general reductions in stress levels within the infarct region (especially in its central core), coupled with stress elevations in the surrounding viable bone;
(b) increased (elastic) head flattening, associated with increased strain in the relatively compliant infarct; and
(c) substantially increased compressive hoop stresses in the shell-like viable subchondral bone layer overlying the infarct.

The load transmission aberrations associated with progressively increasing infarct stiffness deficits are summarized in Figure 13.4. Along the axis of the joint

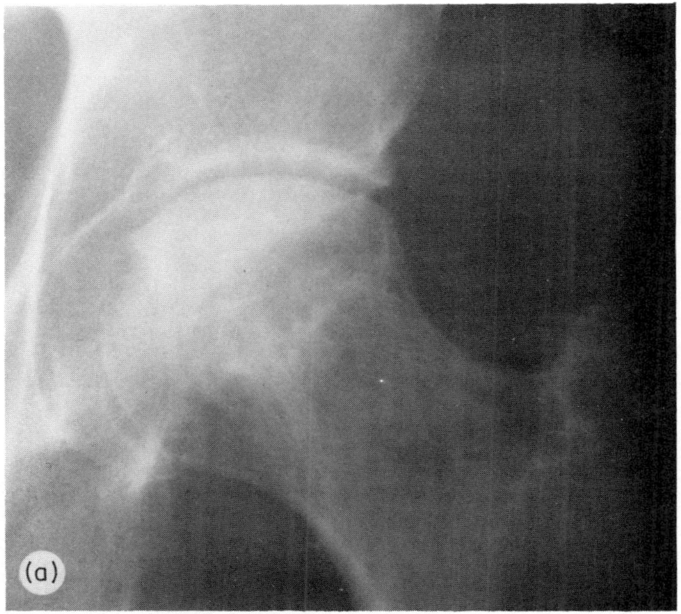

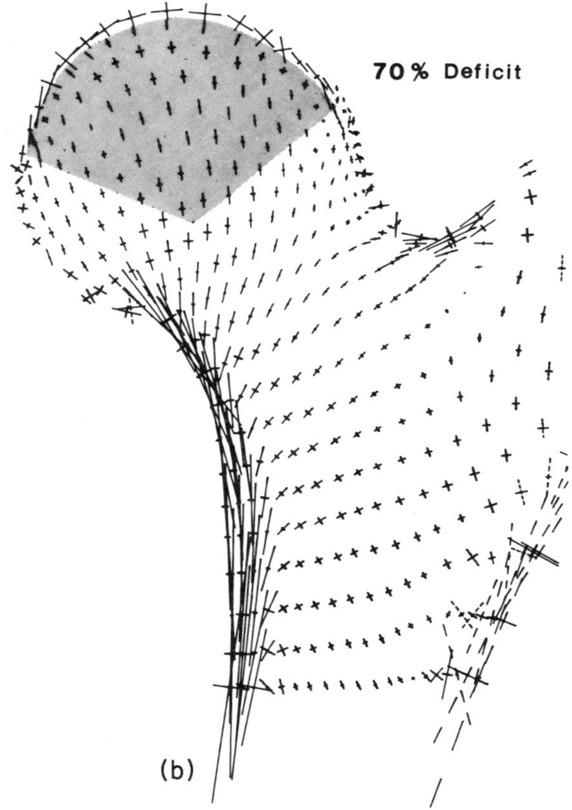

Figure 13.3 (a) Roentgenogram of an adult hip afflicted with aseptic necrosis. Note the wedge-shaped radiodense infarct region in the femoral head. (b) Vectorial representation of the principal stress field in a femoral head with a (shaded) wedge-shaped infarct with a clinically typical stiffness deficit of 70%

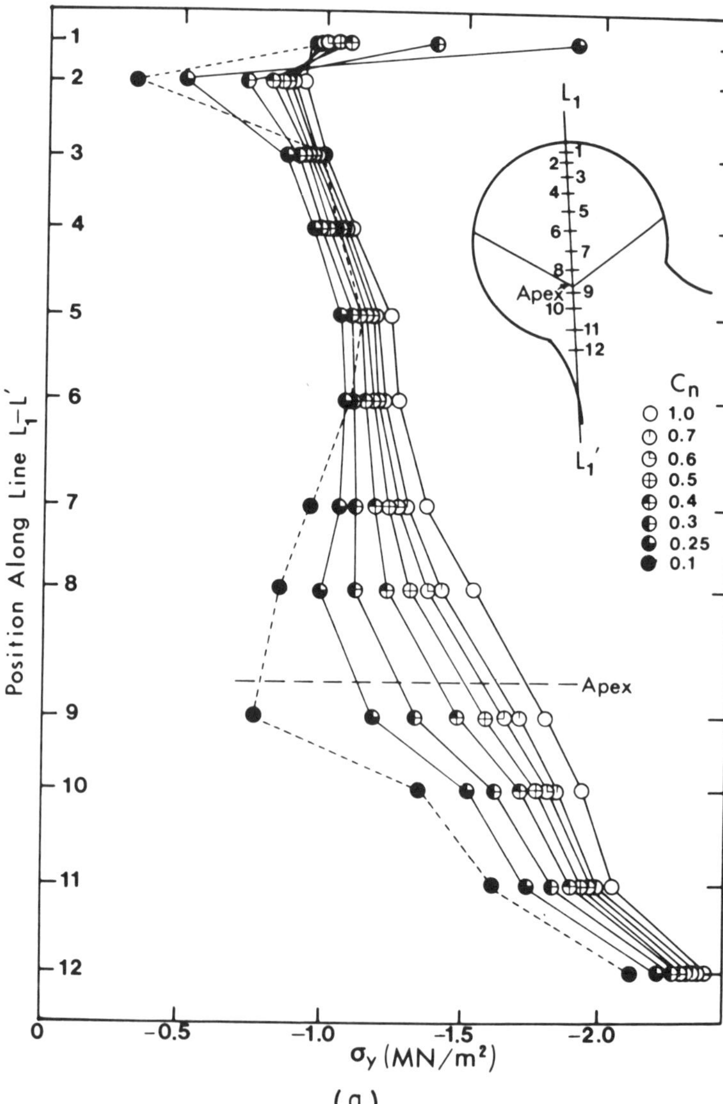

Figure 13.4(a) The effects of accumulating infarct stiffness deficit on longitudinal stress variation along a line (L_1–L_1') coincident with the joint load resultant

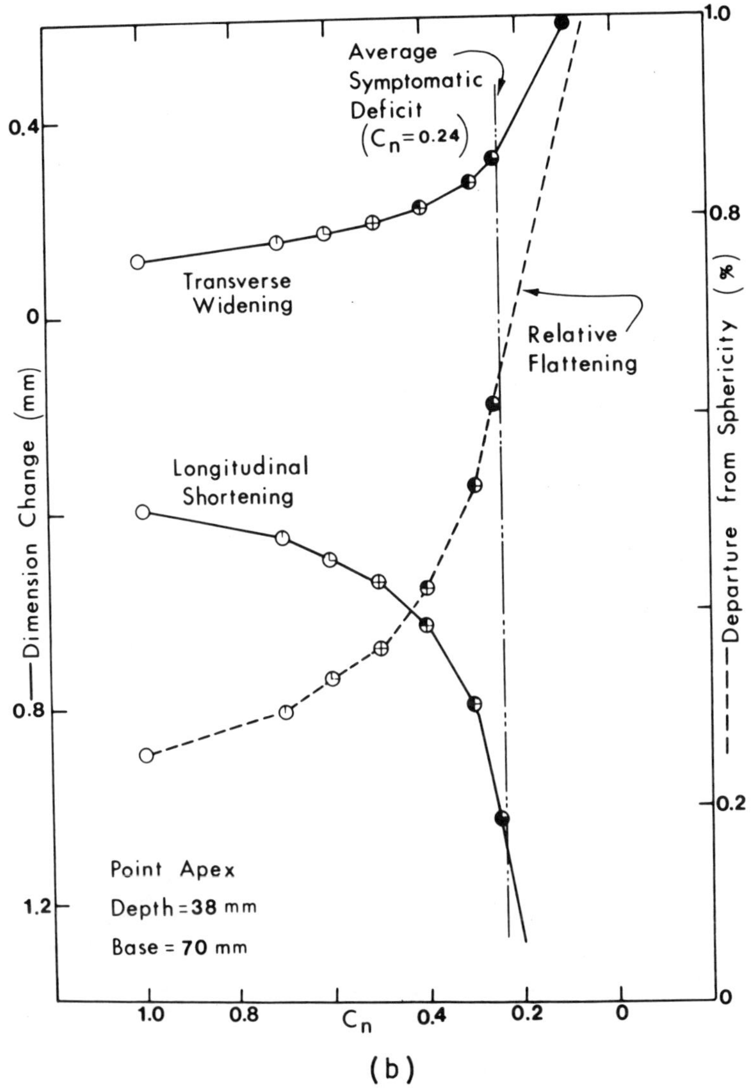

Figure 13.4(b) The effects of accumulating infarct stiffness deficit on gross (elastic) shape change in the femoral head

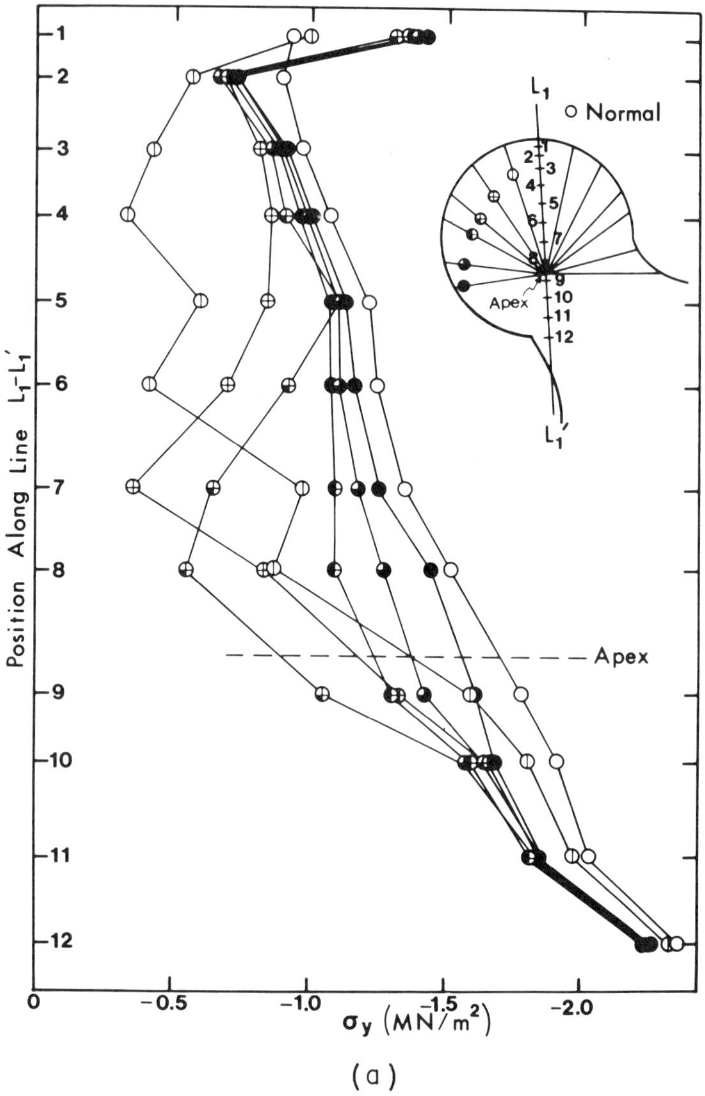

Figure 13.5(a) The role of variations in the infarct base width, as reflected in longitudinal stress variations along a line L_1–L_1' coincident with the joint load resultant ($C_N = 0.3$)

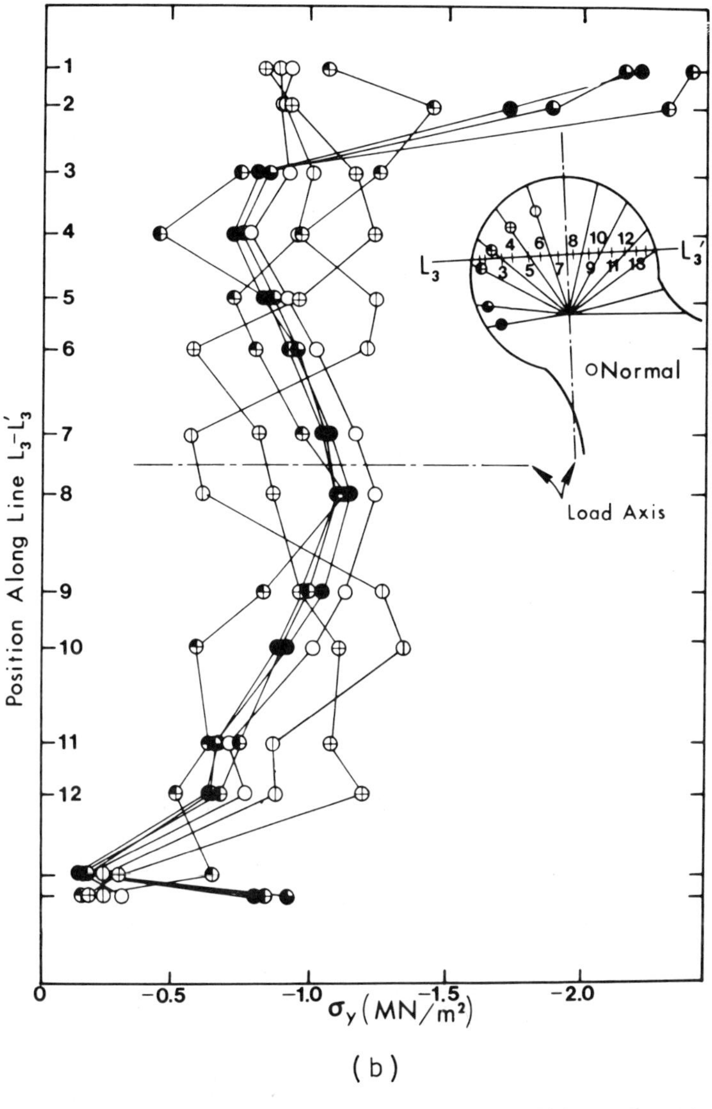

Figure 13.5(b) The role of variations in the infarct base width, as reflected in longitudinal stress variations along a line L_3–L_3 running transversely across the superio-central portion of the femoral head. Stiffness deficit levels in all in infarcts

load resultant (Figure 13.4a), longitudinal stresses in the healthy femoral head (open symbols) progressively increase, from about 1MN/m^2 under the articular surface, to about 2.5MN/m^2 just above the medial cortex. Progressively increasing stiffness deficits (i.e. reductions in C_N) in the infarct cause progressive reductions in the rate of distal stress concentration, their effect being most pronounced (for this particular lesion geometry) at about the level of the infarct apex. The magnitude of transverse subchondral compression increases rapidly as the infarct stiffness shrinks toward zero, leading toward elastic instabilities (for the particular element zoning considered) in the peripheral subchondral regions for C_N values below 0.2.

Examination of longitudinal stresses along lines parallel to the medial infarct margin reveals that the stiffness deficit–related stress relief experienced by the infarct's core is rapidly diminished toward the infarct periphery.

The tendency toward hip congruity changes, as reflected in the (elastic) departure from head sphericity, is seen to accelerate very strongly as stiffness deficits approach clinically symptomatic levels (Figure 13.4b). The head flattening is caused by both transverse widening and longitudinal shortening, the latter being responsible for about two-thirds of the overall flattening effect.

The role of changes in the infarct base width can be appreciated from Figure 13.5. Along the loading axis (Figure 13.5a), the point of maximal stress relief shifts distally as the infarct base widens. The maximum relative stress relief is rather insensitive to the infarct base width, unless the infarct is so wide that the stress-relief peak occurs in the viable bone below the lesion apex. A comparison of longitudinal stress profiles across a transverse section (Figure 13.5b) shows that the preferential load bearing and stress concentration occurring along the principal trabecular corridor is altered by the necrosis. As the infarct base width decreases, the central stress peak is progressively blunted and then dimpled, the intensity of the peripheral load transfer increasing with the narrowness of the infarct. Across head sections, similar trends are observed, but their effect is much smaller, producing only minor perturbations (except near the load axis) on the highly protuberant stress profile characteristic of the normal femoral head. The relative head flattening increases roughly linearly with base width from about 0.34 % for the narrowest infarct to about 0.62 % for an infarct extending across the entire head width. Again, longitudinal shortening effects dominate.

The role of changes in the depth of an infarct of fixed base width is shown in Figure 13.6. Along the load axis, the point of maximum stress relief moves distally, and the relative magnitude of stress relief increases quite markedly, as the depth of the infarct is increased. Across a central transverse head section, peripheral load shifts are again apparent; in general, those infarcts whose apices lie above the section level considered exhibit stress-profile blunting, while those whose apices lie below the section level exhibit stress-profile dimpling. The degree of head flattening again increases linearly with the infarct geometry change, reaching a maximum of 0.55 % for a full head-depth lesion.

The final parametric series (Figure 13.7) involved a study of the effects of changes in the width of the infarct apex. Along the load axis, it is apparent that appreciable stress relief can be obtained only if the apex is narrow, although a small degree of stress relief persists even for the widest lesions. Across the central transverse section, a rather narrow range (about 0.4 MN/m^2) of stress variation occurred near the load axis, although peripheral load transfers at the (relatively distant) necrotic/viable interfaces varied considerably from case to case. Variability across the entire lower transverse section was much more pronounced, owing to uniform distribution of interface stress transfer points along almost the entire length of the section. Head flattening was found to become progressively insensitive to apex widths greater than about 30 mm, reaching its plateau at about 0.62 ‰.

13.2.3 Discussion

This part of the study explores the salient load transmission changes occurring within and around a necrotic segment of cancellous bone in the human femoral head. Assessment of the implications of the findings must include an appreciation of the limitations of the mathematical model employed. In addition to the usual restrictions inherent in the finite element approach for complex biological structures, the following points are noteworthy:

1. The choice of a two-dimensional plane-strain formulation precludes assessment of that portion of load transfer which occurs across portions of the necrotic/viable interface which are not normal to the study section.
2. The model assumes that no frank joint destruction has yet occurred. Even modest collapse of the infarct region, however, will appreciably alter the joint congruity; contact stress will thus tend to be shifted progressively toward the joint periphery.
3. The infarcts themselves have been modelled as smoothly bounded regions of uniform material property deficit. In view of the irregular boundaries and the high variability of stiffness deficit in actual infarcts, one should recognize the likely presence of many small foci of high stress gradients (quite apart from those associated with the trabeculum-to-marrow modulus discontinuities not addressed in a continuum model) occurring inside the lesion and around its borders.

The computational results nevertheless indicate a clear tendency for stress transfer from necrotic to neighboring viable regions. In general, infarcted bone in the superior weight-bearing region must continue to carry near-physiological stresses even though load is progressively transferred out through the distal lesion periphery. Hence, vulnerability to mechanical overload would appear to be more pronounced superiorly than inferiorly. Unfortunately, aseptic necrosis of the femoral head is generally not diagnosed until the pathology is well

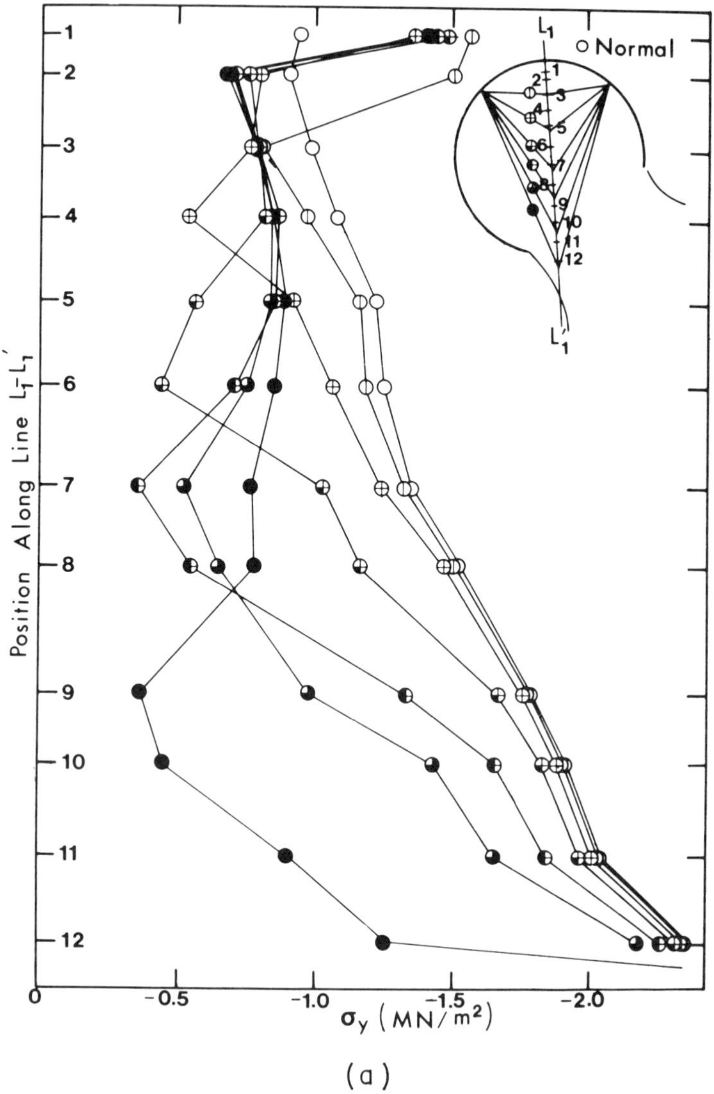

Figure 13.6(a) The effects of variations in the infarct depth upon longitudinal stress variations along the line of the joint load resultant ($C_N = 0.3$)

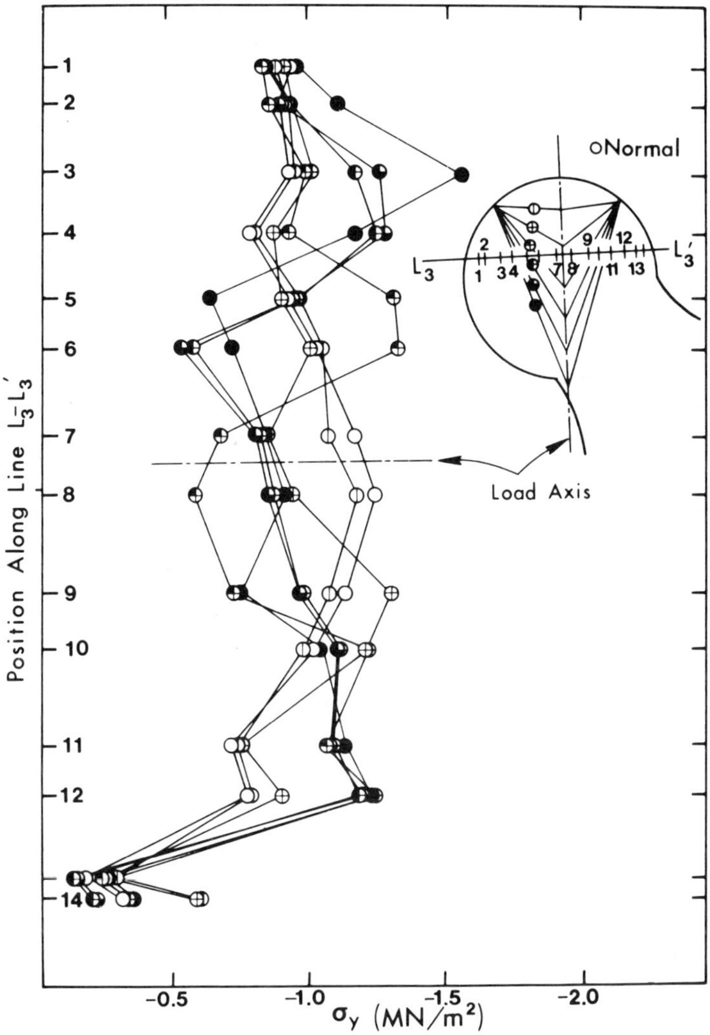

(b)

Figure 13.6(b) The effects of variations in the infarct depth upon longitudinal stress variations along a transverse superio-central section ($C_N = 0.3$)

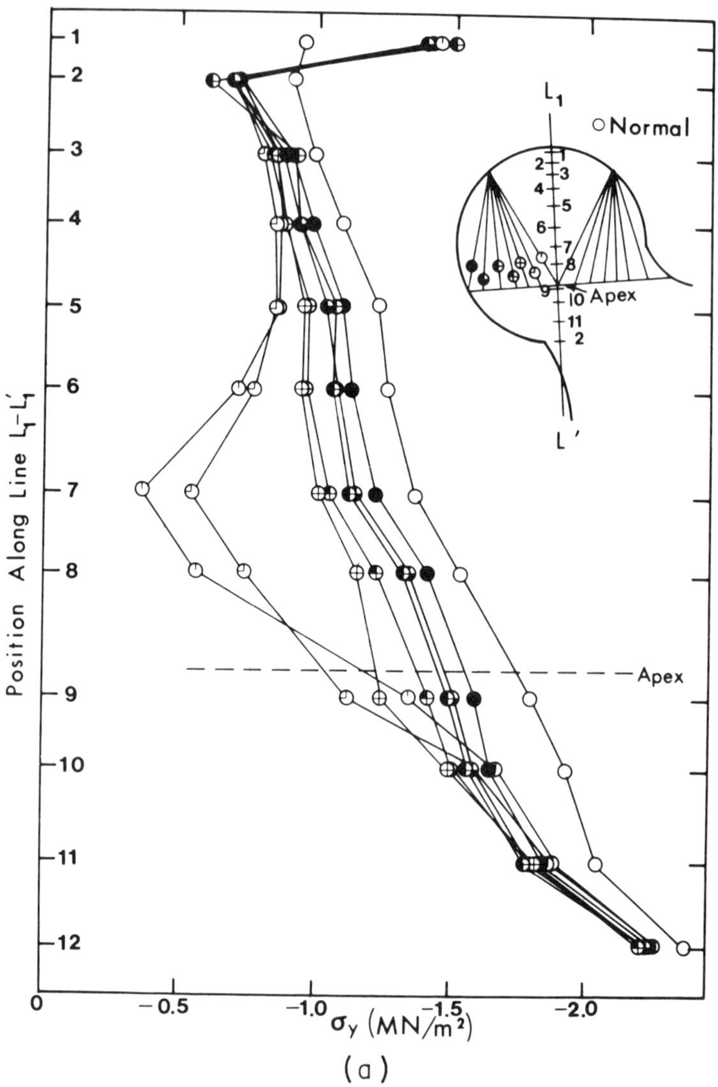

Figure 13.7(a) The effects of changes in the infarct apex width upon longitudinal stress variations along the line of the joint load resultant ($C_N = 0.3$)

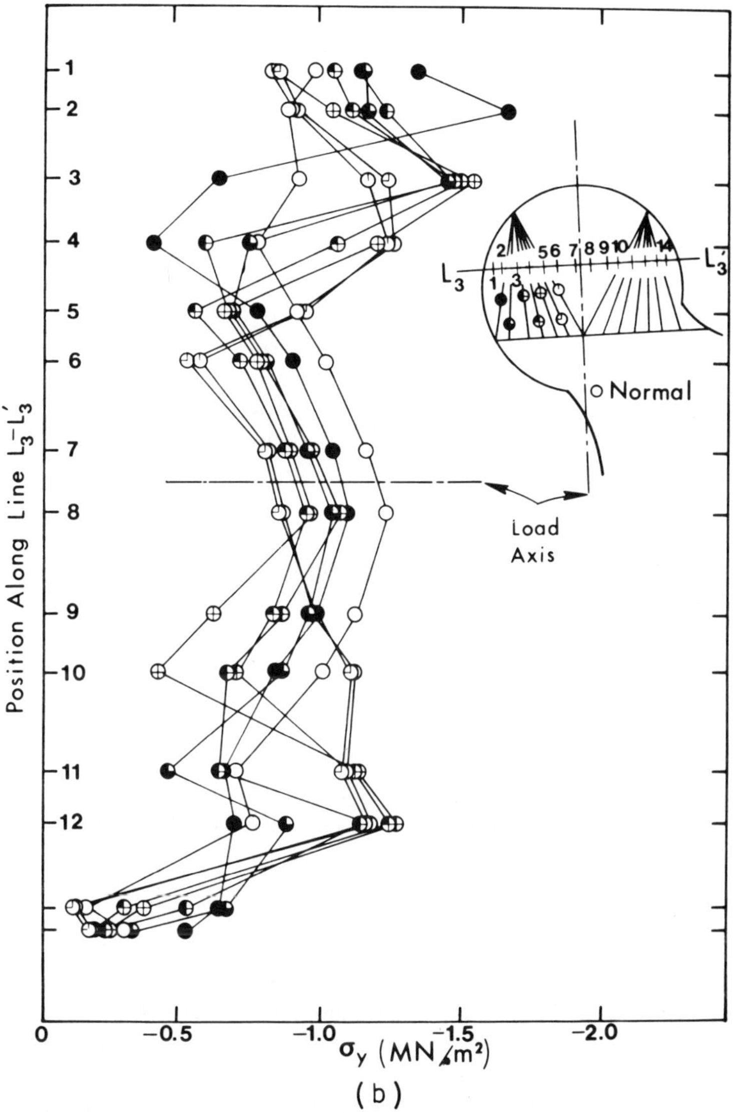

Figure 13.7(b) The effects of changes in the infarct apex width upon longitudinal stress variations along a transverse superio-central section ($C_N = 0.3$)

advanced, and the lesion morphology which does present is a *fait accompli* (Ferguson, 1975). Even so, the computational data indicate that the surgeon planning an osteotomy to provide stress relief for necrotic regions should place his greatest emphasis on repositioning all or most of the superior aspect of the lesion to a new location outside the weight-bearing tract.

13.3 Part II: STRESS CHANGES WITH GROWTH IN THE NORMAL JUVENILE PROXIMAL FEMUR

Mechanical stress is widely recognized as a major stimulus in the development and growth of articular joints. Biomechanical anomalies are implicated in a broad variety of juvenile hip disorders, including not only Legg–Perthes disease, but coxa vara and congenital hip dislocation as well.

Biomechanical changes at the hip during growth do not result simply from increases in stature and body weight, or from the concommitant muscle force changes; rather, profound morphological changes are at work as well. The femoral neck undergoes large relative lengthening, and the neck-shaft angle progressively decreases from nearly 180° in infancy to about 125° at maturity. Ossification of the chondroepiphysis of the femoral head begains between the third and seventh month. This ossific nucleus, initially small and ovoid, gradually assumes a hemispherical contour as it increases in size. The superior surface of the neck acquires a pronounced meridional concavity. Enlarging ossification centres appear at the greater trochanter between the ages of three and four years. Finally, the epiphyseal plate, which is wide in infancy, is gradually obliterated at maturity.

Although finite element analysis has been extensively applied in recent years to the study of stresses in the adult hip, these major structural differences between child and adult suggest that much of that information is inapplicable to the juvenile hip. Here we have attempted to quantify the salient load transmission changes accompanying the complex morphological alterations of growth.

13.3.1 Method

The 120-element plane-strain model described in Part I was considered representative of the geometrical and rheological characteristics and the applied external loading and support conditions of the juvenile proximal femur. Input data were prepared based upon serial roentgenograms of the male hip taken at ages 1, 2, 3, 4, 5, 7, 9, 13, and 18 years. Laboratory compression tests of fresh juvenile autopsy material allowed rheological characterization of the endochondral ossification centres, the growth plates, and the chondroepiphyses (see Table 13.1). Modulus data for articular cartilage, cortical and cancellous bone were taken from previous studies and from the literature (Brown and Ferguson, 1979; Hayes and Mockros, 1971; Reilly and Burstein, 1974). The

Table 13.1 Material property values for various tissue types included in the finite element model of the juvenile proximal femur (all modulus values in MN/m^2)

Elastic modulus for age	Articular cartilage (A.C.)	Epiphyscal plate	Chondro-epiphysis (C.E.)	Subchondral bone	Cortical bone	Capital ossifi-cation centre	Age attenu-ation factor	Capital ossifi-cation fringe	A.C.–C.E. boundary
1	17.25	134.55	108.12	345.00	11 500	69.00	0.07	48.30	55.20
2	17.25	134.55	108.12	345.00	11 500	106.63	0.11	48.30	55.20
3	17.25	134.55	108.12	345.00	11 500	376.44	0.39	48.30	55.20
4	17.25	134.55	108.12	345.00	11 500	541.11	0.56	48.30	55.20
5	17.25	134.55	108.12	345.00	11 500	641.03	0.66	48.30	55.20
7	17.25	134.55	108.12	379.50	11 500	765.94	0.79	172.50	69.00
9	17.25	134.55	108.12	414.00	11 500	815.90	0.84	276.00	86.25
13	17.25	134.55	108.12	690.00	11 500	915.83	0.95	431.25	110.40
18	17.25	N/A	N/A	690.00	11 500	966.00	1.00	690.00	N/A
Poisson ratio	0.45	0.40	0.34	0.30	0.30	0.24	N/A	0.24	0.40

elemental shape functions were used to interpolate the elemental elasticity matrices from discrete input values for the nodes.

Since the chondroepiphysis of the juvenile hip is highly compliant compared with bone, it was assumed (following Greenwald and O'Connor, 1971) that the joint stress distributes itself uniformly over the contact surface. The actual joint surface contours and distributions of muscle fibres at the abductor attachment, not visible radiographically, were estimated from experience on the basis of typical appearance at surgery.

13.3.2 Results

Age-related variations in the overall load transmission patterns can be appreciated from Figure 13.8, which shows the calculated principal stress fields for ages 1, 3, 7, and 18. The following major qualitative features in the stress distribution can be seen for all of these ages:

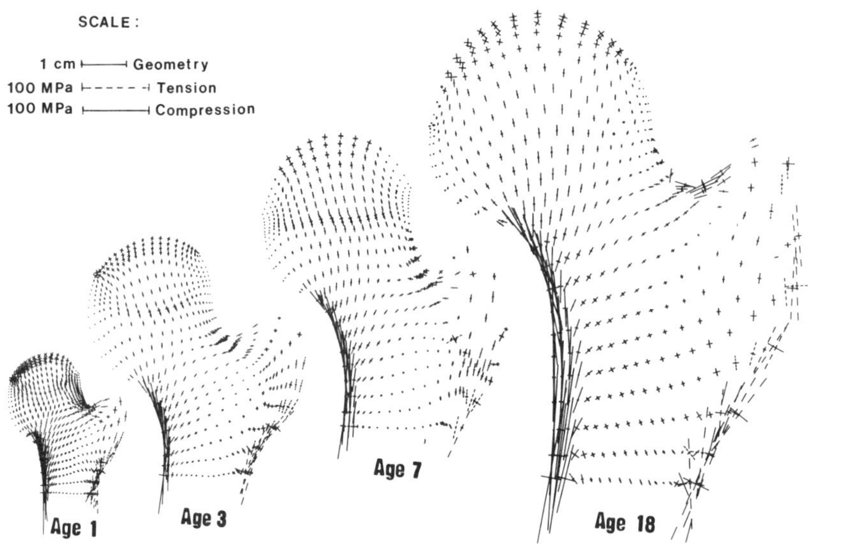

Figure 13.8 Principal stress fields for ages one, three, seven, and eighteen

(a) distal transmission and progressive concentration of the articular contact stress, through the primary trabeculation system of the femoral head, to the medial cortex;
(b) strong longitudinal compressive stress along the medial cortex, with very low stress in the medullary cavity;
(c) strong longitudinal tensile stress in the lateral cortex:

(d) moderate tensile stress in the greater trochanter, roughly aligned with abductor traction;
(e) moderate transverse compression in the neck, near the trochanter junction; and
(f) abrupt elevation in transverse compressive stress in the growth plate (disappears after closure).

Comparatively speaking, several age-related trends are apparent. The compressive weight-bearing stresses become progressively less uniformly distributed across the femoral head as the primary trabeculation system develops centrally. Coupled with the increasingly preferential load transmission through the centre of the head is an increasingly rapid transition to a nearly uniaxial stress state, from the essentially biaxial stress state existing near the articular surface. There is an abrupt, but relatively age-invariant, local elevation in transverse compressive stress occurring with the growth plate. This elevation disappears with epiphyseal closure. This effect is clearly associated with the discontinuity of gross compressibility values (i.e. a Poisson ratio of 0.24 for cancellous bone compared with 0.4 for the physis) in this region. Also with increasing age, there is a very marked reduction in the transverse compressive stress developed in the superio-lateral aspect of the metaphysis. In the one-year-old, the magnitude of these stresses exceeds that of the main compressive weight-bearing stresses in the head by as much as 50%. At maturity (eighteen-year-old), by contrast, the relative magnitude of the transverse stresses in the proximal metaphysis has shrunk to less than one-half that of the main weight-bearing stresses. This progressive relative reduction in superio-lateral metaphyseal compression results from a progressive medial shift of the resultant loading force, caused by neck lengthening and neck-shaft angle reduction. These morphological changes, in turn, are associated with preferential lengthening and ossification in the lateral aspect of the neck.

Detailed age-wise parametric variations of stress levels across several sections of interest are presented in Figures 13.9 through 13.12. Longitudinal stresses along the axis of the resultant joint force generally increased with age (Figure 13.9), and showed a strong tendency toward distal concentration, beginning after age seven. Part of this trend toward distal stress concentration is of course due to the fact that the load resultant axis itself shifts with age toward the heavily stressed medial cortex. Examination of longitudinal stress profiles across a transverse neck section (Figure 13.10) reveals no consistent age-related changes in the low lateral stress levels. There is, however, a clear trend toward medially increasing stress levels, culminating in abrupt concentrations near the medial cortex at maturity. The longitudinal stress profiles across a transverse section passing through the centre of the head (see Figure 13.11) show a strong tendency for preferential transmission of stress along the developing primary trabeculation tract. Shearing stress levels within the growth plate are low and uniformly

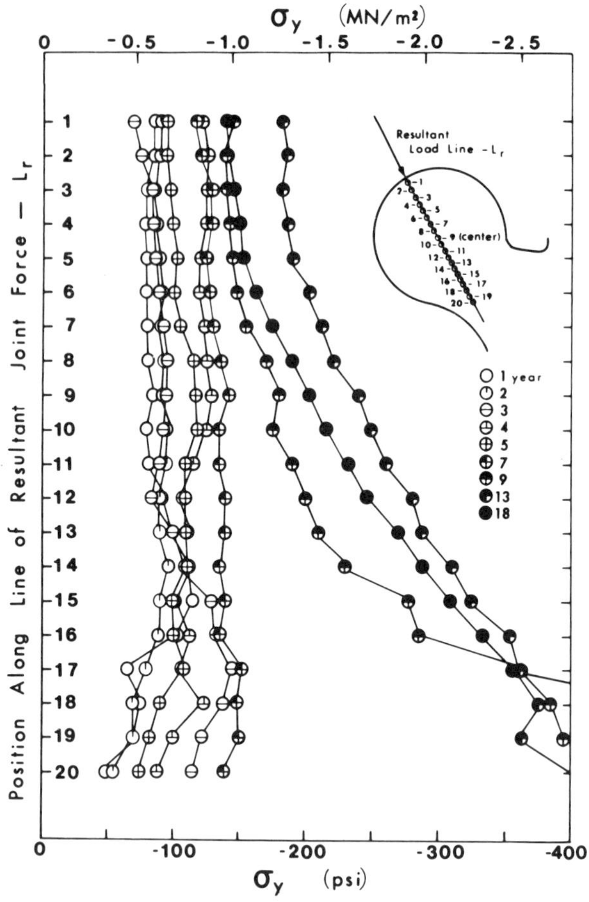

Figure 13.9 The effects of age upon longitudinal stress variations along a line coincident with the joint load resultant

distributed in the very young child (Figure 13.12), but with growth show a progressive tendency toward strong medial augmentation, approaching values of 0.5 MN/m^2 prior to closure.

In a previous study (Brown and Ferguson, 1978) it was shown that an appreciable elevation in tensile and shear stress occurred (with abduction or varus) in the cancellous bone surrounding the region of lateral epiphyseal artery branching. Since subclinical trauma is widely suspected as a possible aetiological factor in Legg – Perthes disease (Ferguson, 1978), the present data were examined for evidence of an age-related peak of stresses in this sensitive area, possibly coinciding with the usual age range (ages 3 to 7) for the onset of that

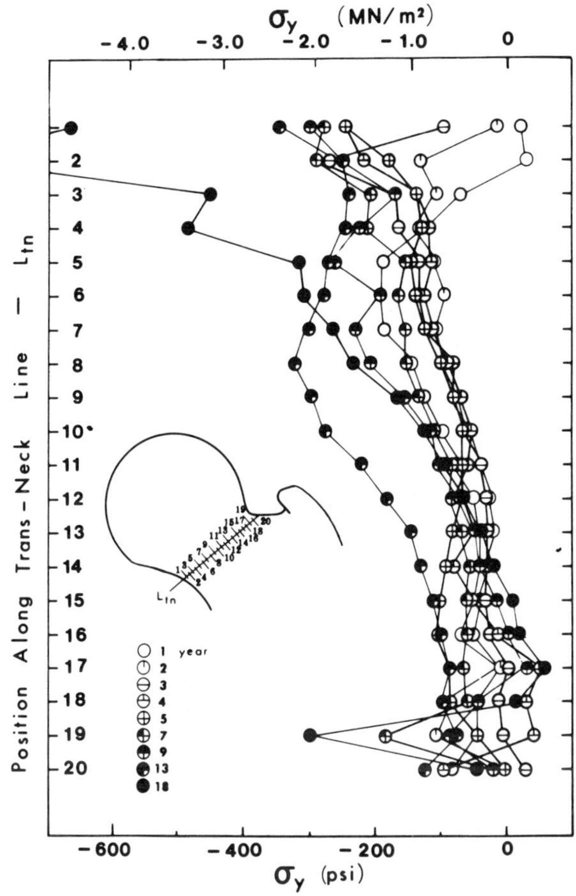

Figure 13.10 The effects of age upon the variation of longitudinal stress across a transverse neck section

condition. It can be seen from Figure 13.13, however, that at points both just inside (A) and just outside (B) the margin of the ossific nucleus, the levels at all stress components generally fall with increasing age, the rate of decline being greatest early in life.

13.4 DISCUSSION

The accuracy of the results in both Part I and Part II of this study is of course limited by the assumptions of problem formulation (continuum discretization, mesh resolution, two-dimensional plane strain, uniform contact stress

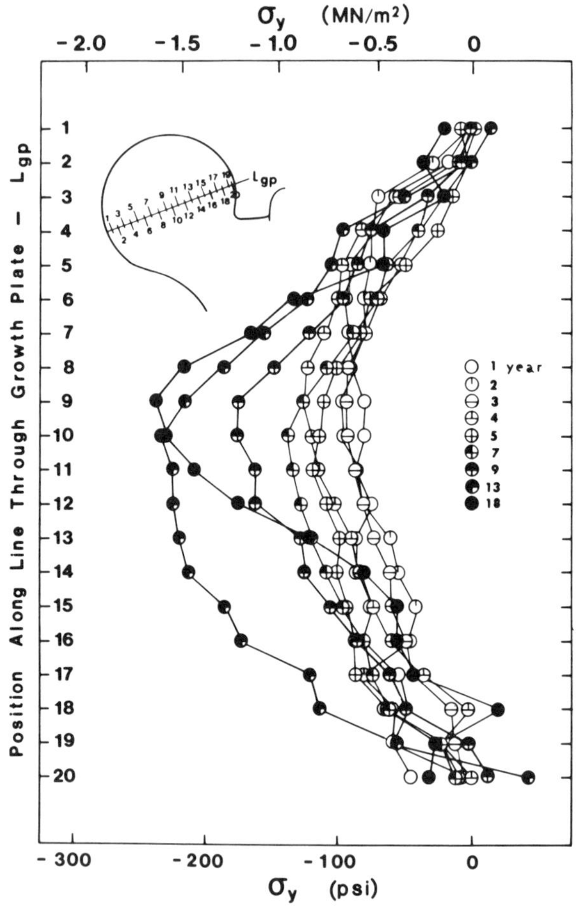

Figure 13.11 Longitudinal stress distributions across a transverse head section at the level of the growth plate, plotted as a function of age

distribution, zero-displacement distal condition, etc.) and by imprecision in the input data (material property distributions, joint configuration, X-ray positioning, etc.). Nevertheless, examination of Figure 13.14 reveals that the stress results calculated along the periosteal surfaces of the proximal cortices for the eighteen-year-old case (which also corresponds to the non-infarcted case for Part I) are in very reasonable agreement with values reported for the adult femur by other investigators using a variety of analytical and experimental techniques. Moreover, the principal stress patterns in Figure 13.8 correlate well with the bony architecture.

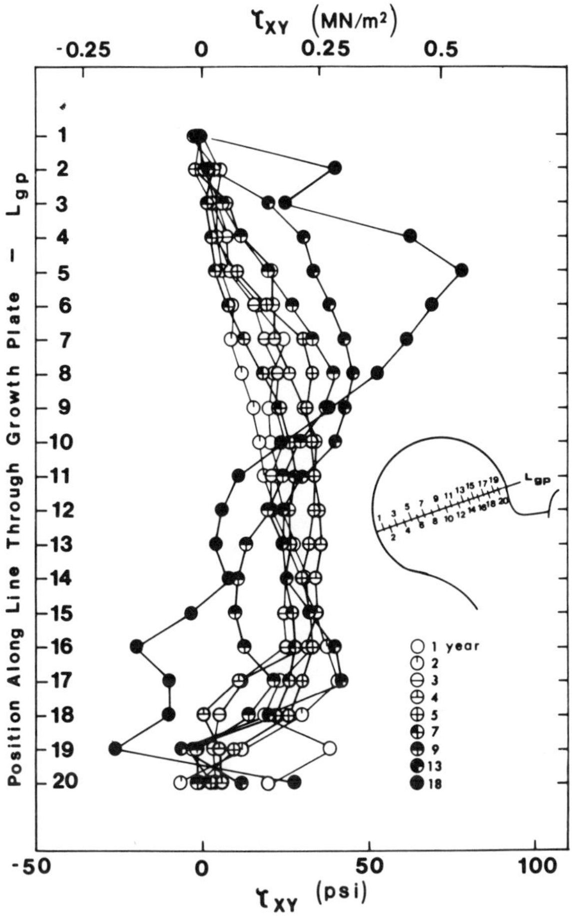

Figure 13.12 The effects of age upon the variation of shear stress in the growth plate

The juvenile hip data clearly indicate that the major features of the stress pattern (distal ward stress concentration in the primary trabeculation system, longitudinal stresses in cortices) persist from early ambulation through epiphyseal closure. However, the proximal femur does remodel with growth in such a manner as to reduce the relative importance of the transverse metaphyseal stresses which are present early in life. Indeed, changes in these stresses may be the cause/result of the normal decrease in the neck/shaft angle, with over compensation perhaps being one cause of coxa vara. The results also show that absolute levels of shear stress in the physis increase progressively, especially in the

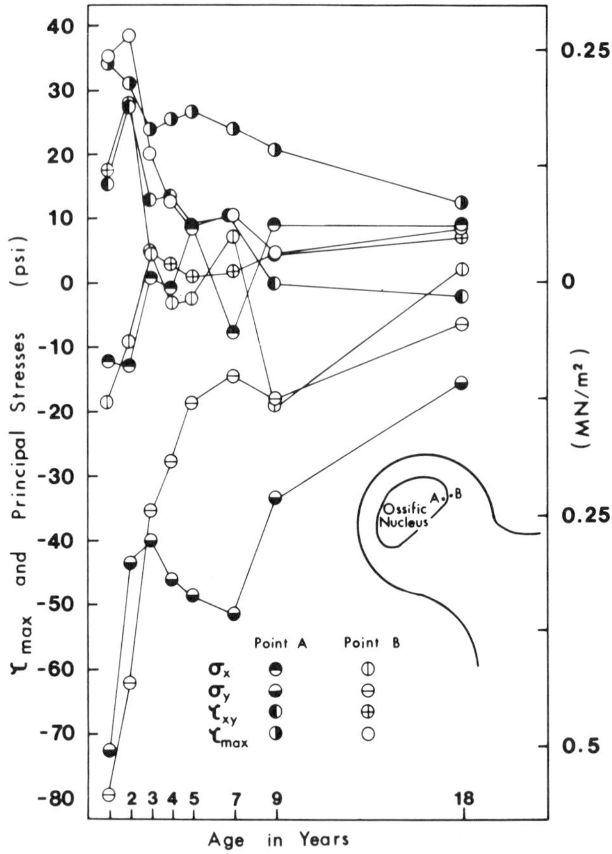

Figure 13.13 Age-related changes in stress components at points just inside (A) and just outside (B) the ossific nucleus, near the point of branching of the lateral epiphyseal artery

teenage years. While this finding does not explain the cause of epiphyseal slippage, it does indicate, other things being equal, that the age group which is clinically at risk is the most vulnerable to hormonally related material property deficits. Finally, there is an absence of any age-related elevations in stress near the lateral epiphyseal artery branch point. This finding does not rule out the subclinical trauma aetiology theory for Legg–Perthes disease, but it does imply that segmental ischemia is much more likely to be a result of the developing vascular patterns than of an age-related increase in the incidence of trabecular microfractures.

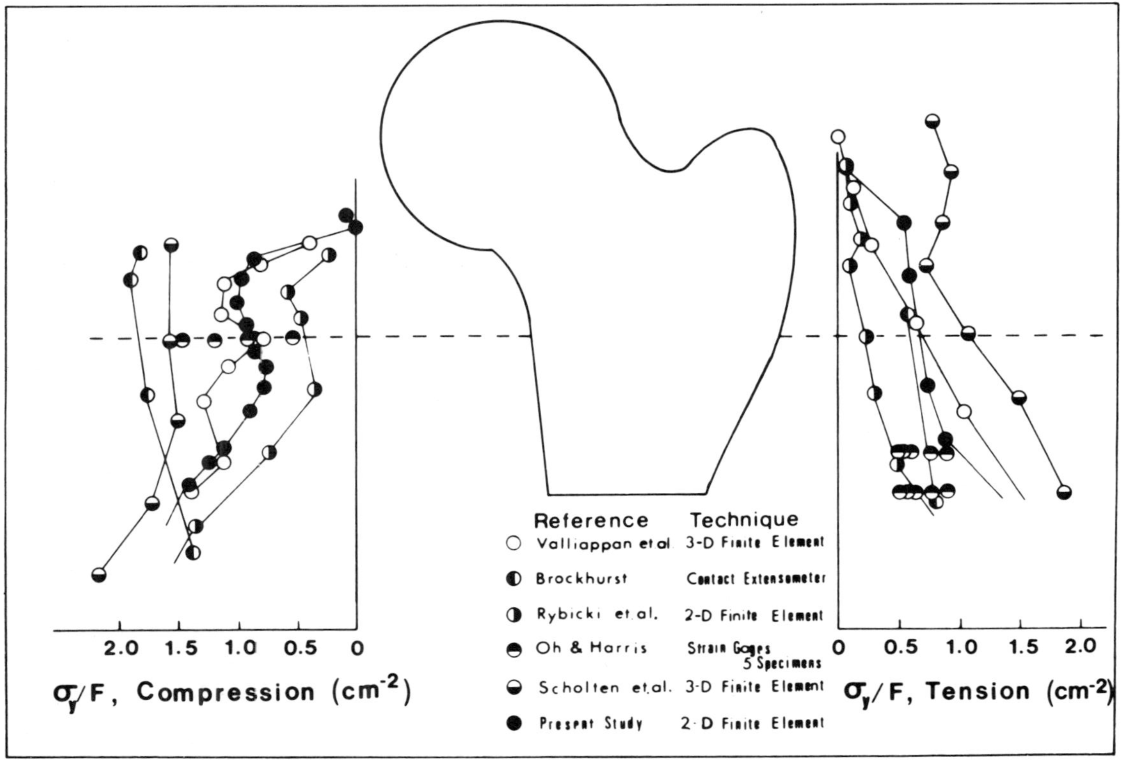

Figure 13.14 Comparison of eighteen-year-old longitudinal stress values (normalized to the joint load *F*) along the periosteal surfaces of the proximal femur, with adult values determined by several studies

REFERENCES

Andriacchi, T. P., Galante, J. O., Belytschko, T. B., and Hampton, S. (1976) A stress analysis of the femoral stem in total hip prostheses, *J. Bone Joint Surg.* **58**(A), 618–624.

Brockhurst, P. (1975) *Design of Total Hip Prosthesis–the Femoral Stem.* Centre for Biomedical Engineering Report, University of New South Wales.

Brown, T. D., and Ferguson, A. B., Jr. (1978) The development of a computational stress analysis of the femoral head – mapping tensile, compressive and shear stress for the varus and valgus positions, *J. Bone Joint Surg.* **60**(A), 619–629.

Brown, T. D., and Ferguson, A. B., Jr. (1979) Mechanical property distributions in the cancellous bone of the human proximal femur, *Acta Orthop. Scand.* (in press).

Brown, T. D., Way, M. E., and Ferguson, A. B., Jr. (1979) Mechanical property deficits in necrotic cancellous bone, *Clin. Orthop.* (in press).

Catto, M. (1976) Pathology of aseptic bone necrosis, *Aseptic Necrosis of Bone* (ed. J. K. Davidson), Chapter 2, Excerpta Medica, Amsterdam.

Curtis, B. H. (Guest editor) (1979) Symposium on Legg–Calvé–Perthes syndrome, *Orthop. Rev.*, **8,** 21–187.

Ferguson, A. B., Jr. (1975) The pathology of Legg–Perthes and its comparison with aseptic necrosis, *Clin. Orthop.*, **106,** 7–18.

Ferguson, A. B., Jr. (1978) Recent advances in understanding Legg–Perthes disease, *Orthop. Surv.*, **1,** 307–325.

Greenwald, A. S., and O'Connor, J. J. (1971) The transmission of load through the human hip joint, *J. Biomech.*, **4,** 507–528.

Hayes, W.C., and Mockros, L.F. (1971) Viscoelastic properties of human articular cartilage, *J. Appl. Physiol.*, **31,** 562–568.

Oh, I., and Harris, W. H. (1978). Proximal strain distribution in the loaded femur, *J. Bone Joint Surg.*, **60**(A), 75–78.

Petrie, J. G., and Bitenc, I. (1971) The abduction weight-bearing treatment in Legg–Perthes disease, *J. Bone Joint Surg.*, **53**(B), 54–62.

Reilly, D. T., and Burstein, A. H. (1974) The mechanical properties of cortical bone, *J. Bone Joint Surg.*, **56**(A), 1001–1022.

Rybicki, E. F., Simonen, F. A., and Weis, E. B. (1972) On the mathematical analysis of stress in the human femur, *J. Biomech.*, **5,** 203–215.

Scholten, R., Rohrle, H., and Sollbach, W. (1978) Analysis of stress distribution in natural and artificial hip joints using the finite-element method, *S. Afr. Mech. Engr.*, **28,** 220–225.

Somerville, E. W. (1979) Osteotomy in the treatment of Perthes disease of the hip, *Orthop. Rev.*, **8,** 61–63.

Valliappan, S., Svensson, N. L., and Wood, R. D. (1977) Three-dimensional stress analysis of the human femur, *Comput. Biol. Med.*, **7,** 253–264.

Finite Elements in Biomechanics
Edited by R. H. Gallagher, B. R. Simon, P. C. Johnson, and J. F. Gross

Chapter 14

A Simplified Hybrid-stress Finite Element Model of the Intervertebral Disc

R. L. Spilker

14.1 INTRODUCTION

Alterations from normal mechanical behaviour in low lumbar motion segments are usually considered to be sources of pathology. For example, intervertebral disc herniation can certainly lead to sciatica, and disc degeneration is thought to have a role in the aetiology of low-back syndromes. However, not much is understood about either normal or pathological mechanics of lumbar motion segments.

Experimental measurements can provide only limited relevant data. Experimental studies of motion segment mechanical behaviour, such as those of Nachemson *et al.*, (1979), show that it can vary widely. Mechanical behaviour seldom associates in obvious ways with the gross morphology of the intervertebral disc or with its age, sex, disc level, or degree of degeneration. Moreover, experimental studies readily provide data on the deformations and nuclear pressures that occur under load, but at best can produce only limited data on internal stresses and strains.

Structural analyses made with models of the intervertebral disc, on the other hand, can provide the desired understanding. They can show how its morphology and the properties of its constituent materials affect its response to load. Response quantities of possible clinical importance include transverse disc and longitudinal end plate bulging, intradiscal pressure, and distributions of various stresses and strains. Simple analyses of the structure of the intervertebral disc have been made by Nachemson (1960) and Sonnerup (1972), for example, but these have been limited to investigation of only a few aspects of behaviour. More sophisticated analyses have employed the finite element method; these include the report of Belytschko *et al.*, (1974) on the behaviour of a linear-elastic axisymmetric model of the disc subject to compression loading, and the follow-up report of Kulak *et al.*, (1976) which incorporated non-linear-elastic material properties. A three-dimensional linear-elastic analysis of the vertebral body/intervertebral disc has been reported by Lin *et al.*, (1978) for static loading, and a three-dimensional analysis of the vertebra has been reported by Hakim and King (1979).

The problem of stress analysis of a disc is difficult. It is not yet known how to construct a model of the disc that seems to capture the chief features of its gross structures and yet is still simple enough to enable a wide range of parameter studies at reasonable expense. No simple 'strength of materials' model seems likely to be useful. Even an elementary model at this stage of our understanding would seem to require a thick-walled short cylinder enclosing an incompressible fluid, capped on the ends by elastic plates. Attempts to capture all of these structural features in a finite element model has led to models that are too cumbersome and too expensive to use readily.

The present investigation was undertaken with this background in mind. A model of the vertebral body–intervertebral disc structure was sought that was reasonably representative of the real structure, yet simple enough to allow easy manipulation. We describe a pilot model of this kind, and show how the mechanical behaviour of that model is affected by its size and shape, and by some of the properties of its constituent materials.

The principal objective of the present simplified model and study is to determine which of the geometric and material parameters have a significant effect on the predicted mechanical response. Such information can also aid in selecting the geometric and material parameters which must be incorporated into more complex models. In the initial phase of parametric study, the model results over a range of plausible geometric and material values are compared with ranges of experimental results. The present model results are shown to be in general agreement with these experimental results. Limited correlations of individual model results with experimental results are also presented using a representation which assumes the vertebral body/endplate region to be nearly rigid compared with the annulus; good correlation is observed with appropriate choice of annulus material properties. Finally, the effects of enucleation are examined by comparing model predictions in which the nucleus pulposis is included with predictions obtained when the nucleus (fluid) is removed.

14.2 HYBRID-STRESS FUNCTIONAL AND SOLUTION PROCEDURE

The hybrid-stress finite element model, originated by Pian (1964), has been chosen for the present study because of the versatility of this model and the often-observed superiority of displacement and intra-element stress distributions compared with corresponding assumed displacement elements; e.g. Pian (1973), Spilker and Pian (1978). In this model an equilibrating stress field is assumed in the interior of the element and a compatible displacement field is assumed along the element boundary. Stress parameters are eliminated on the element level and a stiffness matrix is obtained, with nodal displacements as unknowns.

For the analysis of a vertebral body–intervertebral disc, the hybrid-stress functional must be extended to allow for the analysis of a solid enclosing an inviscid, incompressible fluid (nucleus pulposis). An energy statement based on the assumed-displacement approach has been developed for this case by Belytschko and Kulak (1973). Consider a solid of volume V, which is to be subdivided into elements, surrounding an incompressible inviscid fluid having volume V_f and let the fluid/solid interface be denoted by S_B. The hybrid-stress functional may be stated as:

$$\Pi_{mc} = \sum_n \left\{ 1/2 \int_{V_n} \boldsymbol{\sigma}^{\mathrm{T}} \mathbf{S} \boldsymbol{\sigma} \, \mathrm{d}V - \int_{\partial V_n} \mathbf{T}^{\mathrm{T}} \mathbf{u} \, \mathrm{d}S + \int_{S_{\sigma_n}} \mathbf{u}^{\mathrm{T}} \overline{\mathbf{T}} \, \mathrm{d}S + \int_{S_{B_n}} \mathbf{u}^{\mathrm{T}} (p\mathbf{n}) \, \mathrm{d}S \right\} \tag{14.1}$$

where $\boldsymbol{\sigma}$ is the vector of stress components, $\mathbf{S}$ the material compliance matrix ($\boldsymbol{\varepsilon} = \mathbf{S}\boldsymbol{\sigma}$ where $\boldsymbol{\varepsilon}$ is the vector of strain components), $\mathbf{u}$ is the vector of displacement components, $\mathbf{T}$ the vector of tractions, $\mathbf{n}$ the vector of direction cosines, p the *unknown* fluid pressure, V_n and ∂V_n are the volume and boundary, respectively, of the nth element, S_{σ_n} is the portion of ∂V_n on which tractions, $\overline{\mathbf{T}}$, are prescribed, and S_{B_n} is the portion of ∂V_n on the fluid/solid interface.

The first three terms in Π_{mc} are identical to the standard hybrid-stress model. The last term has been added and refers to the work done by the pressure, p, normal to the boundary S_B; note that only those contributions normal to S_B are included since the fluid is assumed to be inviscid and that the fluid pressure prior to application of the external loads is assumed to be zero. The Euler equations of Π_{mc} are: (1) strain-displacement in V_n; (2) continuity of $\mathbf{T}$ on interelement boundaries; (3) the mechanical boundary conditions $\mathbf{T} = \overline{\mathbf{T}}$ on S_{σ_n}; (4) continuity of tractions on S_{B_n}; and (5) incompressibility, $\varepsilon_{ii} = 0$, in V_f. The first three equations are found in the standard hybrid-stress model. Note also that only the solid is subdivided into elements.

The matrix formulation follows the usual hybrid-stress procedure. The stresses, $\boldsymbol{\sigma}$, are interpolated in V_n terms of stress parameters, $\boldsymbol{\beta}$, such that the homogeneous equilibrium equations are satisfied. The traction, $\mathbf{T}$, may also be related to $\boldsymbol{\beta}$ through the direction cosines on ∂V_n. These relations are expressed in the form

$$\boldsymbol{\sigma} = \mathbf{P}\boldsymbol{\beta}, \qquad \mathbf{T} = \mathbf{R}\boldsymbol{\beta} \tag{14.2}$$

The displacements $\mathbf{u}$ are interpolated on the element boundary, ∂V_n, in terms of nodal generalized displacements, $\mathbf{q}$, in the form

$$\mathbf{u} = \mathbf{L}\mathbf{q} \tag{14.3}$$

such that interelement continuity of the required order is guaranteed.

Equations (14.2) and (14.3) are substituted into Equation (14.1) and the following matrices are defined:

$$\mathbf{H} = \int_{V_n} \mathbf{P}^{\mathrm{T}}\mathbf{S}\mathbf{P}\,\mathrm{d}V \tag{14.4a}$$

$$\mathbf{G} = \int_{\partial V_n} \mathbf{R}^{\mathrm{T}}\mathbf{L}\,\mathrm{d}S \tag{14.4b}$$

$$\mathbf{Q} = \int_{S_{\sigma_n}} \mathbf{L}^{\mathrm{T}}\overline{\mathbf{T}}\,\mathrm{d}S \tag{14.4c}$$

$$\mathbf{N} = \int_{S_{B_n}} \mathbf{L}^{\mathrm{T}}\mathbf{n}\,\mathrm{d}S \tag{14.4d}$$

The stress parameters are eliminated on the element level (since the $\boldsymbol{\beta}$ are independent from element to element) and Π_{mc} may then be expressed in the form:

$$\Pi_{mc} = \sum_{n} \{1/2\,\mathbf{q}^{\mathrm{T}}\mathbf{k}\mathbf{q} - \mathbf{q}^{\mathrm{T}}\mathbf{Q} - \mathbf{q}^{\mathrm{T}}(p\mathbf{N})\} \tag{14.5}$$

where $\mathbf{k}$ is the element stiffness matrix, given by

$$\mathbf{k} = \mathbf{G}^{\mathrm{T}}\mathbf{H}^{-1}\mathbf{G} \tag{14.6}$$

$\mathbf{Q}$ is the element nodal force vector corresponding to applied external forces, and $\mathbf{N}$ is the nodal force vector corresponding to a unit ($p = 1$) fluid pressure. Note that $\mathbf{N}$ is calculated as the integral along S_{B_n} of the displacement component normal to S_{B_n}, and that all other matrices are identical to those defined in standard hybrid-stress analyses.

After assembly of all elements (in the solid region only), Π_{mc} can be written

$$\Pi_{mc} = 1/2\mathbf{q}^{*\mathrm{T}}\mathbf{K}\mathbf{q}^{*} - \mathbf{q}^{*\mathrm{T}}\mathbf{Q}^{*} - \mathbf{q}^{*\mathrm{T}}(p\mathbf{N}^{*}) \tag{14.7}$$

where $\mathbf{K}$ is the assembled stiffness matrix, and $(\mathbf{q})^{*}$ is the assembled vector of $(\mathbf{q})$. The governing equations for the system are obtained by setting $\delta\Pi_{mc} = 0$. For arbitrary, non-zero $\delta\mathbf{q}$ and δp, the following equations are found:

$$\mathbf{K}\mathbf{q}^{*} = \mathbf{Q}^{*} + p\mathbf{N}^{*} \tag{14.8a}$$

$$\mathbf{N}^{*\mathrm{T}}\mathbf{q}^{*} = 0 \tag{14.8b}$$

Equation (14.8a) expresses the requirement for system equilibrium and Equation (14.8b) the incompressibility assumption that no volume change occurs in the fluid region.

To obtain the unknown $\mathbf{q}^{*}$ and p, Equation (14.8a) is solved for $\mathbf{q}$ and substituted into Equation (14.8b). The fluid pressure, p, is then calculated by

$$p = \frac{\mathbf{N}^{*\mathrm{T}}\mathbf{K}^{-1}\mathbf{Q}^{*}}{\mathbf{N}^{*\mathrm{T}}\mathbf{K}^{-1}\mathbf{N}^{*}} \tag{14.9}$$

With p now known, the displacements $\mathbf{q}^*$ can be calculated from Equation (14.8a). In practice, solution of (and solution time for) the present class of problems is equivalent to static finite element solutions with two loading vectors $\mathbf{Q}^*$ and $\mathbf{N}^*$. Using the triple-factoring scheme, $\mathbf{K}$ is the first factored. The 'displacement' vectors $\mathbf{q}_1^*$ and $\mathbf{q}_2^*$ from

$$\mathbf{K}\mathbf{q}_1^* = \mathbf{Q}^* \tag{14.10a}$$

$$\mathbf{K}\mathbf{q}_2^* = \mathbf{N}^* \tag{14.10b}$$

are each evaluated by the forward and backward substitution steps. Then

$$p = \frac{\mathbf{N}^{*\mathrm{T}}\mathbf{q}_1^*}{\mathbf{N}^{*\mathrm{T}}\mathbf{q}_2} \tag{14.11a}$$

and

$$\mathbf{q}^* = \mathbf{q}_1^* + p\mathbf{q}_2^* \tag{14.11b}$$

are easily evaluated.

A physical interpretation of the solution scheme is obtained by considering the following three steps. In the first step, the fluid region is voided and the response of the structure (without fluid) subjected to the external load, $\mathbf{Q}^*$, is computed; the displacement solution for this case corresponds to $\mathbf{q}_1^*$ (Equation (14.10a)). In the second step, the external load is ignored, and the response of the structure to the fluid pressure is computed by applying loads corresponding to a *unit* fluid pressure; the resulting displacements are $\mathbf{q}_2^*$ (Equation (14.10b)). The total displacements, $\mathbf{q}^*$, are then computed by superposition of these solutions (Equation (14.11b)), once the unknown actual fluid pressure, p, is computed. The actual fluid pressure, p, is computed by requiring that $\mathbf{q}^*$ yield no volume change in the fluid region; this is accomplished by substituting Equation (14.11b) into Equation (14.8b) and solving for p; the result is given in Equation (14.11a). It should be noted that a value of p is assumed in the three-dimensional analysis of Lin *et al.*, (1978) corresponding to experimental measurements of Nachemson (1960), in contrast to the computation of p used in the present approach. Finally, it should be noted that a by-product of the present analysis is the solution for the case of no fluid (enucleation, in the context of intervertebral disc analysis), which corresponds to $\mathbf{q}_1^*$.

14.3 DESCRIPTION OF THE MODEL

The model selected is axisymmetric about a longitudinal axis, and subject to only longitudinal loads in compression. The analysis problem is then only two-dimensional; geometry is described by a radial coordinate r and a longitudinal coordinate z (Figure 14.1a). Symmetry with respect to a mid-transverse plane is

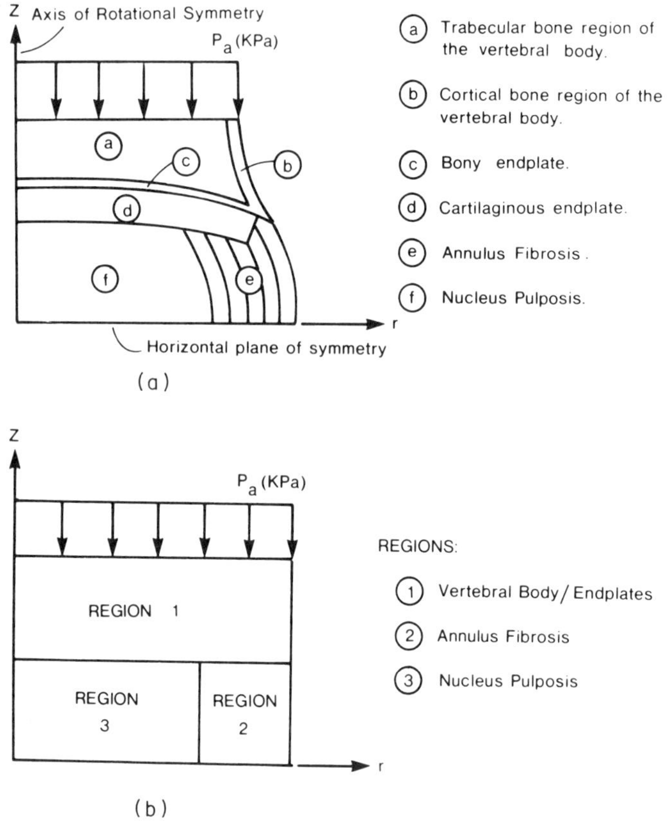

Figure 14.1 (a) Actual and (b) simplified axisymmetric representation of the vertebral body–intervertebral disc

also assumed, so that only the structure in the upper right-hand quadrant need be analysed. Moreover, every sub-structure is assumed bounded by straight lines parallel to the coordinate axes directions.

The model incorporates three homogeneous sub-structures; one that lumps the cortical and trabecular bone regions of the vertebral body together with the bony and cartilaginous end plates (region 1 of Figure 14.1b), one that represents the annulus fibrosis of the intervertebral disc (region 2), and one that represents the nucleus pulposis (region 3). The nucleus is assumed to be a hydrostatic incompressible fluid, while the annulus and vertebral body are assumed to be isotropic, linear-elastic materials. Thus, all the material properties of the model are described by four elastic constants; two for region 1 and two for region 2. Poisson ratios of $\nu_1 = 0.25$ and $\nu_2 = 0.40$ were assigned to these structures to reflect their relative hardness and softness. These ratios were not changed during the investigation. For initial parametric studies, a Young's modulus E_1 of 15.8 MPa is assigned to the vertebral body region, based on the value used by Belytschko *et al.*, (1974) and the experimental results cited in that study. Young's modulus for the annulus, E_2, is varied so that the ratio $K = E_1/E_2$ takes on values of 100, 200, 500, or 1000. Corresponding values of 200 through 500 were used in the study by Belytschko *et al.*, Region 1 is assigned a fixed height, h_v, of 0.375 cm.

With the assumptions stated, model geometry is described by three parameters: the half-height of the intervertebral disc, h_d; the total disc radius r_d; and the ratio of the area of the nucleus to the total disc cross-sectional area, A_n/A_{tot} (see Figure 14.2). Three values of h_d (0.25, 0.75, and 1.25 cm), three values of r_d (1.5, 2.0, and 3.0 cm), and four values of A_n/A_{tot} (0.16, 0.30, 0.50, and 0.64) are used in the computations, so that a wide range of lumbar vertebral disc geometries is covered.

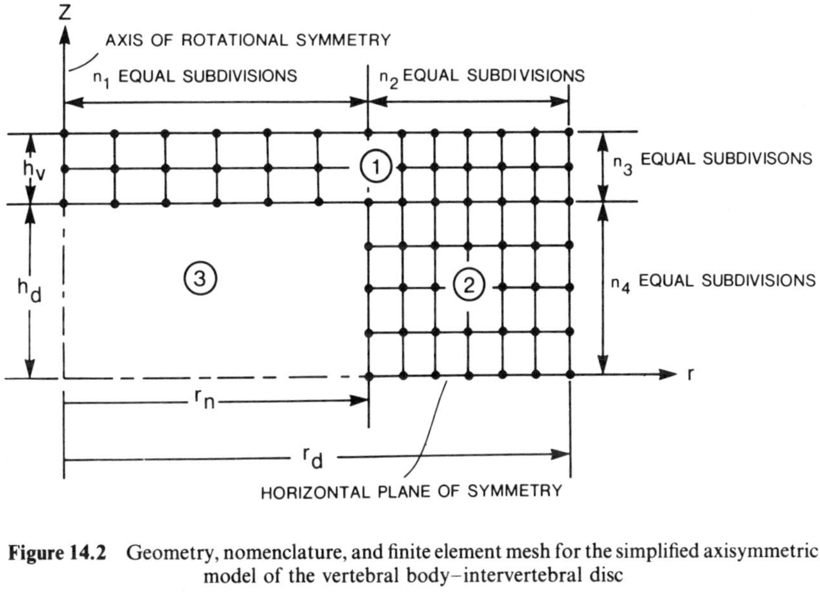

Figure 14.2 Geometry, nomenclature, and finite element mesh for the simplified axisymmetric model of the vertebral body–intervertebral disc

The finite element model is constructed by subdividing regions 1 and 2 into a series of four-noded axisymmetric ring elements as shown in Figure 14.2. Recall that no subdivision is required in the nucleus (fluid) region 3. The axisymmetric element used here corresponds to element E in the study by Spilker and Pian (1978), identified as 'best' of the alternate 4-node elements tested.

The effects of the geometric/material parameters A_n/A_{tot}, r_d, h_d, and K on the intradiscal pressure increase, P_n (p in the formulation of the previous section), disc bulge, u_B (the radial displacement at the annulus outer surface at the midsurface, $z = 0$), and the vertical deflection of the vertebral body/endplate region, u_v (the displacement, positive downward, of the loaded surface at the centre, $r = 0$), will be examined. For all cases considered, a uniformly distributed compressive pressure is applied at the upper surface of the vertebral body/endplate region (region 3); see Figure 14.1b. The magnitude, P_a, of the applied pressure is adjusted for each value of disc radius, r_d, so that a total load of 400 N is applied. Therefore, pressures of magnitude 565.88, 318.31, and 141.47 kPa are applied for disc radii values of $r_d = 1.5$, 2.0, and 3.0 cm., respectively. The choice of constant total load used here versus constant applied pressure is intended to simulate the usual experimental procedure of monitoring total applied compressive load. It should however be noted (see Equations (14.8a) and (14.9)) that for fixed geometric and material properties the ratio of nuclear pressure to applied pressure, P_n/P_a, will be independent of the applied pressure magnitude, P_a, and all displacements will vary in proportion to P_a. It should also be observed that for the range of K values selected, the endplate–vertebral body region (region 1) is relatively rigid in comparison with the annulus region (region 2). In such cases, the distribution of the compressive load on the axisymmetric model should not affect the predicted disc response; only the magnitude of the compressive load is important.

In the results which follow, the general effects of the geometric and material parameters are examined first. Next, limited comparisons of models and experimental results for individual cases are presented. Finally, the effects of enucleation are examined over a range of disc radii and half-heights. To ensure that the observed trends are not related to differences in finite element mesh refinement, a series of tests was made for each disc geometry. The number of subdivisions in each region (n_1, n_2, n_3, and n_4 of Figure 14.2) was increased gradually until convergence of the predicted pressure ratio, disc bulge, and vertical deflection were observed. All parametric studies were then conducted using these converged meshes. The final values of n_1, n_2, n_3, and n_4 used for each disc geometry (area ratio, A_n/A_{tot}) are listed in Table 14.1.

Table 14.1 Converged mesh subdivisions for the area ratios tested

Area ratio	n_1	n_2	n_3	n_4
0.16	8	10	3	8
0.30	10	8	3	8
0.50	12	6	3	8
0.64	12	6	3	8

14.4 RESULTS OF PARAMETRIC STUDIES

The results obtained for intradiscal pressure increase are shown in Figure 14.3. Pressure (recall that the pre-load intradiscal pressure is assumed to be zero; the present results must therefore be interpreted as pressure increase, and will be termed simply pressure) is plotted versus nuclear-to-total disc cross-section area ratio, A_n/A_{tot}, at various values of disc radius, r_d, disc half-height, h_d, and at

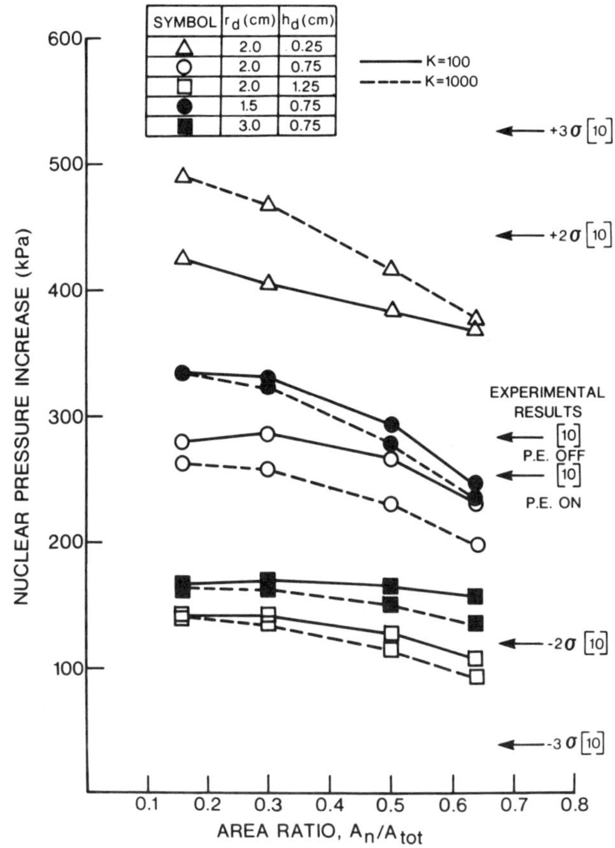

Figure 14.3 Predicted intradiscal pressure increase versus nuclear-to-disc area ratio at various disc radii and half-heights, at modulus ratios $K = 100$ and 1000

modulus ratios, $K = 100$ and 1000. The predicted pressure increases range from approximately 100 to 500 kPa. For a fixed disc radius, r_d, = 2.0 cm, increasing disc height yields a lower nuclear pressure. Similarly, for a fixed disc height, h_d = 0.75 cm, increasing disc radius yields a lower predicted intradiscal pressure. Therefore it would be expected that values of $P_n > 500$ kPa would be found for the case $r_d = 1.5$ cm, $h_d = 0.25$ cm, and values of $P_n < 100$ kPa would be found for the case $r_d = 3.0$ cm, h_d 1.25 cm. Both r_d and h_d significantly affect the value of P_n. However, the effects of K are less significant. With the exception of the case r_d = 2.0 cm, $h_d = 0.25$ cm, increasing K from 100 to 1000 causes a decrease in the pressure, but by less than 40 kPa. In the case $r_d = 2.0$ cm, $h_d = 0.25$ cm., increasing K causes an increase in pressure, but by no more than 65 kPa. Intermediate values of $K = 200$ and 500 (results not shown here), indicate that the relation between P_n

and K is non-linear. It is also observed that for r_d, h_d, and K fixed, increasing area ratio gives a decrease in pressure. The effects of area ratio on intradiscal pressure are minimal (< 30 kPa) at larger disc height and/or radius. The greatest (of the cases considered) pressure decrease (~ 85 kPa over the range $A_n/A_{tot} = 0.16$ to 0.64) is found for the case $r_d = 1.5$ cm, $h_d = 0.75$ cm.

Also shown in Figure 14.3 are results obtained in the experimental study of Berkson *et al.*, (1979). The mean value of intradiscal pressure increase (both with posterior elements intact and destroyed) and scatter in the data ($\pm 2\sigma$, $\pm 3\sigma$) are shown for an applied load of 400 N. The results obtained with the present simple model are directly within the range of experimentally measured values. Results obtained for $h_d = 0.75$ cm and $r_d = 1.5$ cm and 2.0 cm are closest to the mean experimental values. Also, the present results help to explain the scatter in experimental values due to gross disc geometry.

As shown in Figure 14.3, the effects of modulus ratio, K, on intradiscal pressure are less significant. However, effects of K on predicted disc bulge are significant, as shown in Figure 14.4 (for fixed $A_n/A_{tot} = 0.5$, and various combinations of r_d and h_d). Over the range of K values used here, bulge increases linearly with K; for

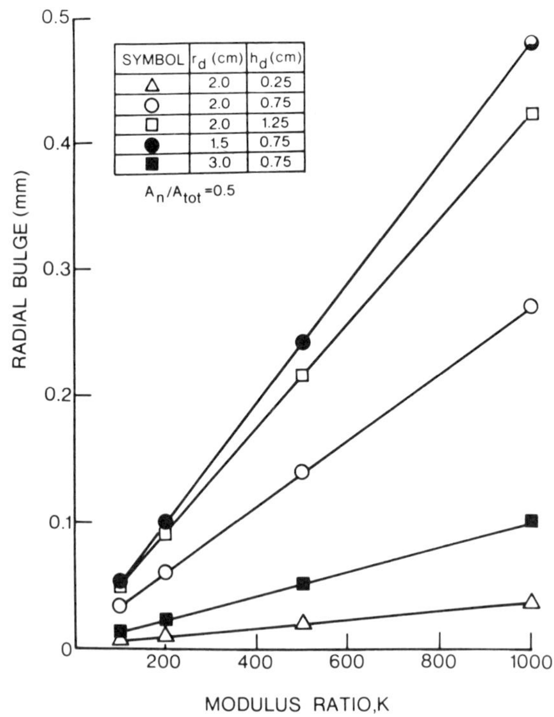

Figure 14.4 Predicted radial disc bulge versus endplate to annulus modulus ratio, K, for a nuclear-to-disc area ratio of 0.5

fixed r_d and h_d, increasing K by an order of magnitude causes a similar increase in radial bulge. On the basis of experimental measurements such as those by Lin *et al.*, (1978), a value of $K > 1000$ is necessary in order to produce predicted bulge values of the correct order of magnitude. Hence, for all subsequent results, the value $K = 1000$ will be used.

The predicted radial bulge for various combinations of r_d and h_d (with $K = 1000$) are plotted in Figure 14.5. The predicted bulges ranges from approximately 0.025 mm to 0.8 mm. Increasing disc height, h_d, causes an increase in bulge, whereas increasing radius, r_d, causes a decrease in the bulge values. For fixed r_d and h_d, increasing area ratio (increasing nucleus cross-section area relative to total disc cross-section area) increases the bulge. The effects of area ratio are more significant for discs of smaller r_d and larger h_d.

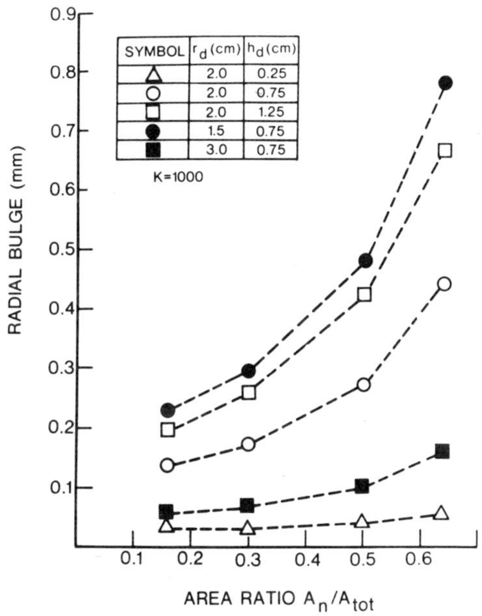

Figure 14.5 Predicted radial disc bulge versus area ratio at various disc radii and half-heights, for a modulus ratio $K = 1000$

Results of experimental measurements of disc bulge under axial loading have been reported by Lin *et al.*, (1978). The results to be compared here correspond to two lumbar disc specimens having low generation grades (i.e. normal or slightly degenerated disc). The values reported for the load 445 N (i.e. load application 0–445 N) range from 0.408 mm (right lateral bulge) to 0.482 mm (anterior bulge). If these values are scaled linearly to correspond to a 400 N compressive load, the range is 0.37 mm to 0.43 mm. Bulge values predicted for $K = 1000$ (nearly rigid endplate region compared with the annulus) are of the correct order of magnitude and for certain combinations of geometric parameters are in good agreement with the experimental measurements. For example, at $K = 1000$, area ratio of 0.5, both the cases $r_d = 2.0$ cm, $h_d = 1.25$ cm, and $r_d = 1.5$ cm, $h_d = 0.75$ predict bulge values in the range 0.4 to 0.5 mm. The experimental values cited here are for discs having an average half-height, h_d, of approximately 0.5 cm and effective radii (based on the reported disc cross-sectional area and an assumed circular cross-section) of $r_d = 2.07$ and 2.36 cm. For this range of r_d and h_d, the present model underpredicts the bulge with $K = 1000$. Better agreement would be found if values of $K > 1000$ were used.

Results obtained for the vertical deflection at the centre of the loaded surface are shown in Figure 14.6 for various r_d and h_d values (with $K = 1000$). Values in the range 0.02 mm to 0.6 mm (approximately) are found, and the effects of r_d, h_d, and area ratio on vertical deflection are the same as those observed for disc bulge. Increasing h_d causes increased deflection, increasing r_d causes decreased deflection, and increasing area ratio gives increased deflection, the effects being more significant for larger h_d and/or smaller r_d. Although not shown, vertical deflection increases linearly with increasing K (softer annulus); as in the case of bulge, the effects of K are significant.

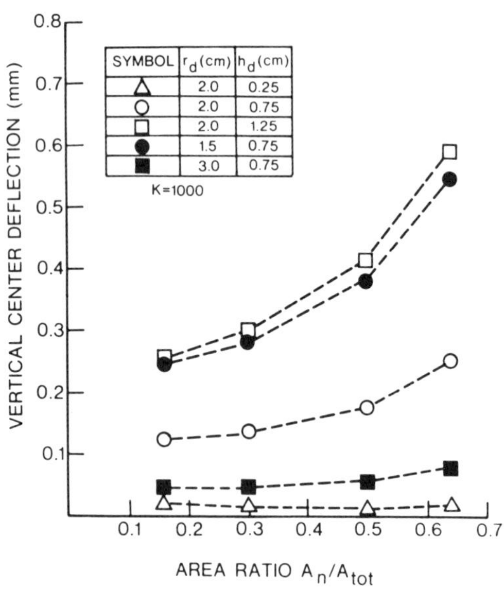

Figure 14.6 Predicted vertical deflection at the upper surface of the vertebral body-endplate region versus nuclear-to-disc area ratio for various disc radii and half-heights, for a modulus ratio $K = 1000$

Experimental results given by Berkson *et al.*, (1979) for vertical deflection under a 400 N load show a mean value of 0.51 mm with a standard deviation of 0.24 mm. For the same disc geometries described previously, Lin *et al.*, (1978) show vertical deflections of approximately 0.35 to 0.40 mm at a 400 N load. These measurements correspond to complete motion segments; in order to compare with the model predictions (based on a symmetric representation so that the predicted vertical deflection is the deflection relative to the midplane of the disc), the experimental values should be divided by 2. When this comparison is made, the predicted vertical deflection for a representative geometry is seen to be slightly lower than the mean value of Berkson *et al.*, and the values of Lin *et al.*

In view of the results and comparisons made thus far, it appears that improved correlation is possible if larger values of the modulus ratio, K, are used. As used here, increasing K implies a softer annulus relative to the endplate–vertibral body region. However, values of $K > 1000$ correspond to an endplate–vertebral body region which is essentially rigid compared with the annulus; this, in turn, suggests that for sufficiently large K, the appropriate material parameter is E_2, Young's modulus of the annulus, hereafter referred to as E_A. In a series of test cases (results not shown), E_A was fixed and E_2 increased so that $K > 1000$ was found. For values of $K > 2000$, further increases in E_2 produced no change in predicted intradiscal pressure or displacements. In view of this, a value of $K = 4000$ has been selected for all subsequent results, and E_A $(= E_2)$ is selected as the pertinent material parameter. Note that the assumption of a nearly rigid endplate region appears to be consistent with experimental measurement of endplate deformation versus overall rigid-body motion; e.g. Rolander and Blair (1975) and Reuber (1979).

With the assumption of $K > 2000$, the analysis is further simplified. The modulus E_A can be extracted from the assembled stiffness matrix (i.e. the assembled stiffness is proportional to E_A; contributions of the endplate–vertebral body region produce rigid-body motion of that region). Then, from Equation (14.9), it is observed that intradiscal pressure increase is unaffected by E_A. From Equation (14.8a), displacements will be proportional to $1/E_A$. Furthermore, strains will also be proportional to $1/E_A$, and stresses will be unaffected by E_A.

Limited case-by-case correlation of the present simplified model against experimental measurements has been attempted. As an example, consider one of the cases of Lin *et al.*, previously cited in which the effective disc radius is 2.07 cm and disc half-height is 0.5 cm. Experimental measurements for a 400 N load yield an average bulge of approximately 0.4 mm, and vertical deflection (scaled down by a factor of 2.0 to correspond to the symmetric model predictions) of approximately 0.2 mm using the above geometry, an area ratio of 0.5, $K = 4000$, and $E_A = 5100\,\text{kPa}$ yields a bulge of 0.408 mm and a vertical deflection of 0.189 mm, which are in good agreement with the experimental measurements.

With the assumption of $K > 4000$, the results of parametric studies can be more concisely presented, as shown in Figures 14.7a through 14.7c. For these

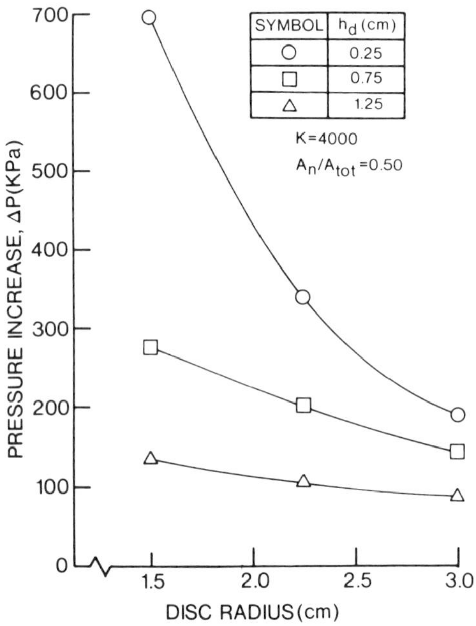
700
600
500
400
300
200
100
0
PRESSURE INCREASE, ΔP(KPa)
SYMBOL
h_d (cm)
0.25
0.75
1.25
K=4000
A_n/A_tot =0.50
1.5
2.0
2.5
3.0
DISC RADIUS(cm)

(a)

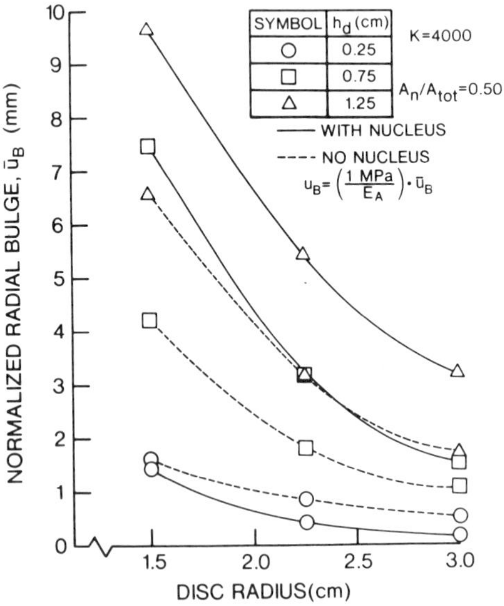
10
9
8
7
6
5
4
3
2
1
0
NORMALIZED RADIAL BULGE, ū_B (mm)
SYMBOL
h_d (cm)
0.25
0.75
1.25
K=4000
A_n/A_tot=0.50
WITH NUCLEUS
NO NUCLEUS
u_B= (1 MPa / E_A) · ū_B
1.5
2.0
2.5
3.0
DISC RADIUS(cm)

(b)

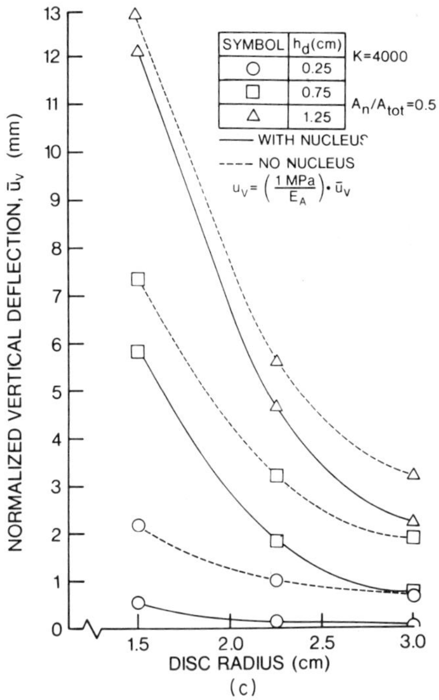

Figure 14.7 (a) Predicted intradiscal pressure increase, (b) radial disc bulge, and (c) upper-surface vertical deflection versus disc radius for an annulus modulus ratio $E_A = 1000$ kPa, modulus ratio $K = 4000$, area ratio of 0.5, at various disc half-heights, including the effects of enucleation

results, an area ratio of 0.5 is used; published values suggest that area ratios of 0.3to 0.5 are appropriate, and in this range area ratio does not have a significant influence. A value of $E_A = 1000$ kPa is used, and all 9 combinations of disc radius ($r_d = 1.5$, 2.25, 3.0 cm) and half-height ($h_d = 0.25$, 0.75, 1.25 cm) are included. Results are plotted versus disc radius for each disc half-height. Note that the disc radius of 2.25 cm is used now in place of the value of 2.0 used earlier.

Results for intradiscal pressure increase are given in Figure 14.7a. The trends observed previously are again found here. Increasing disc radius or half-height produces a lower intradiscal pressure; the effects are most pronounced for small half-height. Recall that these results are unchanged for other values of E_A provided that $K > 2000$ is maintained.

Radial disc bulge and vertical disc deflection results are given in Figures 14.7b and 14.7c. The values plotted correspond to $E_A = 1000$ kPa, and are referred to as normalized displacements; actual displacements for other values of E_A are obtained by scaling the plotted results by (1 MPa/E_A). The effects of enucleation are also shown in Figures 14.7b and 14.7c. As indicated earlier, these results correspond to the analysis of a disc in which the assumed incompressible, inviscid fluid of the nucleus is eliminated (only the solid regions remain).

As shown in Figure 14.7b, the absence of the nucleus fluid generally leads to smaller disc bulge. The effects are most significant for increasing disc height. For smaller h_d (e.g. $h_d = 0.25$ cm), the lowest bulge values are predicted, and for this h_d, the no-nucleus cases lead to a slightly larger bulge. For the cases $h_d = 0.75$ and 1.25 cm, enucleation causes decreases in bulge in the range of 30% to 45%.

In contrast, enucleation produces larger vertical deflections for all cases considered (Figure 14.7c). The percent increase in vertical deflection due to enucleation increases as disc radius is increased and/or as disc half-height is decreased. Although the absolute increase in vertical deflection is smaller than the decrease in bulge (due to enucleation), percent increases in vertical deflection are larger. For example, for the cases $h_d = 0.75$ and 1.25 cm, the percent increase in vertical deflection ranges from 7% ($h_d = 1.25$, $r_d = 1.5$ cm) to 150% ($h_d = 0.75$, $r_d = 3.0$ cm).

14.5 CONCLUDING REMARKS

A hybrid-stress-based energy functional has been presented for a solid enclosing an incompressible inviscid fluid. In this approach only the solid region must be subdivided. A simplified axisymmetric finite element model of the vertebral body/intervertebral disc has been constructed which provides an efficient tool for assessing the effects of gross disc geometry and material properties on predicted radial disc bulge, vertical deflection, and intradiscal pressure increase under compressive loading. The following observations have been made:

(a) Increasing disc radius, r_d, produces a decrease in nuclear pressure, disc bulge, and vertical deflection.
(b) Increasing disc half-height, h_d, leads to a decrease in nuclear pressure, and increasing disc bulge and vertical deflection.
(c) Increasing area ratio A_n/A_{tot} causes a reduction in the intradiscal pressure, and an increase in disc bulge and vertical deflection.
(d) In general, increasing modulus ratio, K (softer annulus), leads to a decrease in nuclear pressure, and an increase in disc bulge and vertical deflection which varies linearly with K. When $K > 2000$ is used (nearly rigid endplate–vertebral body region compared with the annulus), Young's modulus of the annulus, E_A, is found to be the significant material parameter. In such cases, displacements and strains are proportional to $1/E_A$, and intradiscal pressure increase and stresses in the annulus are unaffected by E_A.
(e) Enucleation leads to a decrease in radial disc bulge and an increase in vertical deflection. In terms of absolute change in displacements, the effects on bulge appear to be more significant.

The present simple model fulfills the principal objective of demonstrating the qualitative effects of gross disc geometry and material properties on the mechanical response, and thereby helps to explain the wide scatter often observed in raw experimental data on motion segment behaviour. Limited comparisons suggest that reasonable quantitative predictions are also possible with appropriate choice of E_A. However, further work in this area is needed. In addition, studies of the effects of enhanced geometric and material description are needed to determine if such modelling changes will produce significant improvements in accuracy; these investigations are in progress.

14.6 SUMMARY

An energy functional based on the assumed-stress hybrid model is defined for a solid enclosing an incompressible, inviscid fluid, and is applied to the analysis of a vertebral body – intervertebral disc under compressive loading. A simplified axisymmetric finite element model is constructed which employs three homogeneous substructures corresponding to the vertebral body/endplate region, the annulus fibrosis of the intervertebral disc, and the nucleus pulposis. The model is used to examine the effects of gross disc geometry and material property parameters on the predicted intradiscal pressure increase, radial disc bulge, and vertical deflection under compressive loading. The results obtained indicate that all parameters have a significant effect on disc bulge and vertical deflection. Variations of disc height and radius are found to produce significant changes in the predicted intradiscal pressure increase. Results obtained using the present simplified model are also shown to be in reasonable agreement with published experimental measurements with appropriate choice of material property parameters. Results are also presented which show the effects of enucleation on the predicted vertical deflection and radial disc bulge.

ACKNOWLEDGEMENTS

This research was supported by Public Health Service Grants OH00514 and AM15575. The helpful suggestions of Professor Albert B. Schultz as well as the technical assistance of Ms. S. Maskeri and Mr. J. Julien are gratefully acknowledged.

REFERENCES

Belytschko, T., and Kulak R. (1973) A finite element method for a solid enclosing an inviscid, incompressible fluid, *J. Appl. Mech.*, **40** (2); *Trans. ASME* **95** (E), 609–610.

Belytschko, T., Kulak, R., Schultz, A., and Galante, J. (1974) Finite element stress analysis of an intervertebral disc, *J. Biomechanics*, **7**, 277–285.

Berkson, M., Nachemson, A., and Schultz, A. (1979) Mechanical properties of human lumbar spine motion segments. Part III: Responses in compression and shear; influence of gross morphology, *J. Biomech. Engng.*, **101,** 53–57.

Hakim, N., and King, A. (1979) A three-dimensional finite element dynamic response analysis of a vertebra with experimental varification, *J. Biomechanics,* **12,** 277–292.

Kulak, R., Belytschko, T., Schultz, A., and Galante, J. (1976) Nonlinear behavior of the human intervertebral disc under axial load, *J. Biomechanics*, **9,** 377–386.

Lin, H., Liu Y., and Adams, K. (1978) Mechanical response of the lumbar intervertebral joint under physiological (complex) loading, *J. Bone Jnt. Surg.*, **60** (A), 41–55.

Lin, H., Liu, Y., Ray, G., and Nikravesh, P. (1978) Systems identification for material properties of the intervertebral joint, *J. Biomechanics*, **11,** 1–14.

Nachemson, A. (1960) Lumbar intradiscal pressure, *Acta Orthop. Scand.* (Suppl. 43)

Nachemson, A., Schultz, A., and Berkson, M. (1979) Mechanical properties of human lumbar spine motion segments. Part III: Influences of age, sex, disc level and degeneration, *Spine*, **4,** 1–8.

Pian, T. H. H. (1964) Derivation of element stiffness matrices by assumed stress distributions, *AIAA J*, **2,** 1333–1336.

Pian, T. H. H. (1973) Hybrid elements, in: *Numerical and Computer Methods and Structural Mechanics* (eds. S. H. Fenves, N. Perrone, A. R. Robinson, and W. C. Schnobrich), Academic Press, New York, pp. 59–78.

Reuber, M. O. (1979) Disc bulge in lumbar intervertebral discs, M. S. Thesis, Carleton Univ., Ottawa, Ontario.

Rolander, S. D., and Blair, W. E. (1975) Deformation and fracture of the lumbar vertebral end plate, Symposium on Lumbar Spine, *Ortho. Clinics of North America*, **6,** 75–81.

Sonnerup, L. (1972) A semi-experimental stress analysis of the human intervertebral disc in compression, *Exp. Mech*, **12,** 142–147.

Spilker, R., and Pian, T. H. H. (1978) A study of axisymmetric solid of revolution elements based on the assumed-stress hybrid model, *Computers and Structures*, **9,** 273–279.

Finite Elements in Biomechanics
Edited by R. H. Gallagher, B. R. Simon, P. C. Johnson, and J. F. Gross

Chapter 15

Finite Element Analysis for Artificial Joint Fixation Problems in Orthopaedics

R. Huiskes, J. D. Janssen, and T. J. Slooff

15.1 INTRODUCTION

Artificial joints implanted in orthopaedic surgery are usually fixated in or against the bone by using self-curing acrylic cement (polymethylmethacrylate) as space-filling material (Figure 15.1). The cement is inserted by the surgeon in a viscous phase and polymerizes *in situ* to a solid mass. During this process, heat is generated. Possible complications after such interventions include fracture or loosening of the prosthetic device. Both these phenomena can be enhanced by cement fracture, shear failure of the implant–cement and cement–bone contact regions (the 'interfaces'), and stress-related bone remodelling processes. Thus, it is evident that the stress patterns that result from the joint-loading forces play an important role in the loosening phenomena.

The cement–bone interface is a key region for the maintenance of the bone–prosthesis structural integrity. After surgery, a radioluscent line is usually found between the cement and the bone consisting of a soft-tissue layer (Charnley, 1978). The thickness of this layer varies but quite often reaches 1 to 2 mm and may still increase gradually. This increase can be a sign for forthcoming loosening (Beckenbaugh and Ilstrup, 1978). Although the mechanisms of these interface-remodelling phenomena are not yet fully understood (Charnley, 1978; Willert and Frech, 1976), there is no doubt that during the operative procedure, a small layer of endosteal bone near the cement becomes necrotic. This damage is caused by disturbance of the vascularity, the effect of toxic components within the cement, and principally, by the heat of polymerizing (Feith, 1975). The layer of dead bone is later resorbed and replaced by soft tissue in which new bone may again be formed (Vernon-Roberts and Freeman, 1977). This bone-appositioning mechanism, however, may be inhibited by stresses (or strains) at the interface. This can be shown by comparing histological findings in animal experiments (Nunamaker *et al.*, 1980) with stress analysis results of the experimental system (Huiskes *et al.*, 1980). Hence, it appears that chances of cement–bone interface

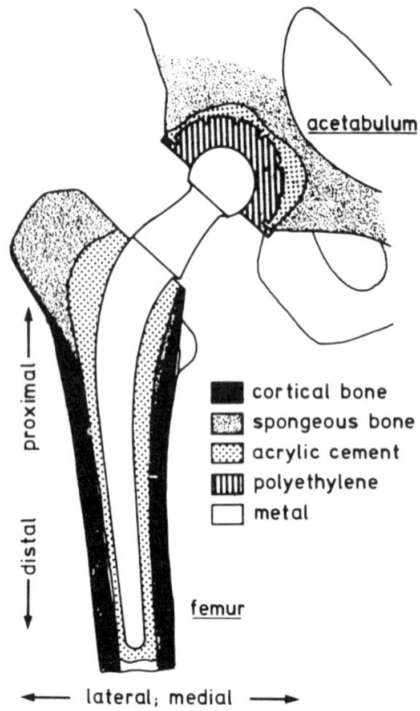

Figure 15.1 Schematic illustration of a Charnley total hip prosthesis (acetabular and femoral component) fixated in the bone with acrylic cement. Distal and proximal, medial and lateral directions are indicated

loosening might be reduced by decreasing the bone temperatures during the operative procedure and by avoiding high, unnatural stresses at the interface during the full postoperative period. It is evident that the chances for cement or prosthetic fracture will be reduced by decreasing the stresses in these materials. The occurrence of so-called 'disuse osteoporosis' (a process of bone resorption throughout the bone caused by the stress-shielding effect of an intramedullary prosthesis stem; Slooff, 1970) will be minimized by a 'natural' stress pattern in the bone. In view of these aspects, it is evident that a quantitative understanding of the heat generation and conduction processes on the one hand, and the load introduction mechanism on the other hand, are vital for increasing successful long-term performance of bone–prosthesis structures.

In the first part of this chapter, finite element analysis is used to quantify the heat generation and conduction process and to evaluate influences of geometrical and material parameters in order to provide guidelines for an optimal cement composition and fixation procedure. The second part of the chapter is devoted to stress analyses of intramedullary fixated bone–prosthesis structures in general. No doubt the finite element method (FEM) offers the best opportunity for detailed analysis of these irregular structures. Several finite element models have been described in the recent literature. Most of these models are of the hip joint, but a number of models were developed of the knee and the finger joints. These were either two-dimensional 'sandwich' models, neglecting three-dimensional bone structural integrity (Andriacchi *et al.*, 1975; McNeice *et al.*, 1975; Kwak *et al.*, 1979; Yettram and Wright, 1979; Cook *et al.*, 1980); two-dimensional models taking bone integrity into account by modelling 'side-plates' (Hampton *et al.*, 1976; Svensson *et al.*, 1977); or three-dimensional models (Scholten *et al.*, 1978; Campen *et al.*, 1979, Röhrle *et al.*, 1979; Crowninshield *et al.*, 1980; Tarr *et al.*, 1980; Valliappan *et al.*, 1980; Huiskes *et al.*, 1980). These models were all directed toward a geometrically refined representation of the real structure. It should be noted that a certain imbalance between the geometrical and other modelling aspects was often apparent. Usually, little attention was given to complicated material properties of bone and cement, conditions of continuity or discontinuity at the interfaces, and dynamic loading. Hence, even in the most refined three-dimensional models, only tendencies can be studied and the absolute accuracy of the results is not as high as suggested by the apparent visual agreement between model and reality. If it is the object to study tendencies, however, geometrical refinement often results in unnecessary complexity. By using geometrically simplified models, fundamental and general concepts can be developed owing to a reduction of parameters. These concepts might remain hidden in the complexity of more refined models.

Simplified models have been used to establish trends in material and geometrical influences (Bartel and Ulsoy, 1975; Huiskes *et al.*, 1977), to evaluate prosthesis stem designs (Calderale *et al.*, 1977), to compare finite element and experimental results in detail (Rohlmann *et al.*, 1979), and to evaluate the influences of stem–cement interface loosening (Huiskes and Schouten, 1980).

In this chapter, it will be shown that closed-form solutions can be found for a simplified, general model of intramedullary fixation structures, using a finite element model as a reference. These formulas yield a fundamental understanding of the structural behaviour of these systems in general and can be used as first-order design guidelines for prosthesis stems. It will be shown as well that quite acceptable results (compared with refined three-dimensional finite element models) can be found from finite element beam models.

15.2 HEAT GENERATION AND CONDUCTION ANALYSIS

15.2.1 Description of model

The acrylic cement dough is mixed by the surgeon from polymer powder and monomer liquid. When inserted, the monomer polymerizes (producing heat) to a solid matrix in which the original polymer is embedded. The heat of polymerization is conducted through the cement mass, the implant, and the bone.

Relatively little is known about the values for heat-generation, heat-conduction, and heat-capacity parameters in the bone–prosthesis structure. Hence, a very refined model cannot be generated. In the models presented here, it is assumed that:

(a) no material transport takes place in the system, and the materials are incompressible;
(b) the local heat loss by conduction and convection at the boundary of the system (usually the outer bone surface) is either proportional to the difference in the local boundary temperature and a constant local ambient temperature, or the boundary temperature is equal to this ambient temperature;
(c) the thermal properties of implant, cement, and bone are isotropic;
(d) the cement mixture is homogeneous;
(e) the heat conduction and capacity properties of the materials are independent of the temperature;
(f) the heat generation, as a function of time, is not affected by temperature gradients;
(g) the cement–bone interface is smooth and the local heat transport across the interface is proportional to the local difference in temperature between the two materials.

The prosthesis–cement–bone structures to be investigated are approximated as axisymmetric or plane geometry. Since the aim was to perform extensive parametric analysis and evaluate the influences of the essential system parameters in order to develop a fundamental concept of the heat-conduction process, the number of parameters in the model was kept restricted. The process of heat generation and conduction was described by the heat-conduction equation (Carlslaw and Jaeger, 1959) and Fourier's law. For example, assuming an axisymmetric construction (Figure 15.2) consisting of three materials (M_1, M_2, M_3) in contact at the interfaces (I_{12} and I_{23}) with boundaries B_1 and B_2, the time- and location-dependent temperature $T(r,z,t)$ can be described as in Table 15.1. Different solution procedures are available (e.g. Zienkiewicz, 1977). These can be divided into methods that apply finite element discretization in coordinates as well as in time, and those methods that apply finite element discretization in geometry and a finite difference method in time. The latter method was chosen because a computer program for the solution of elliptical differential equations (such as that heat-conduction equation for steady state) based on FEM was available (Brekelmans, 1975). This program was adapted to suit a step-by-step procedure using backward differences for discretization in time (Table 15.2). For each step, T_{n-1} and ϕ_n are given quantities calculated from the previous step and the source function, respectively. The time-step dimension was kept constant to suit the available program. Although this method is not the most efficient (Zienkiewicz, 1977), it is unconditionally stable (Wilson *et al.*, 1966, 1974). The

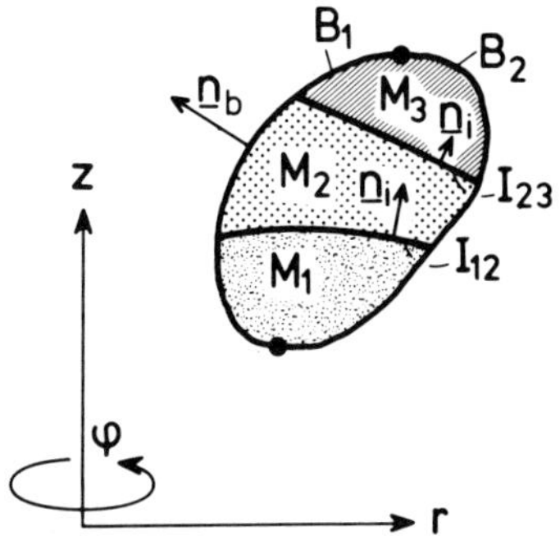

Figure 15.2 An axisymmetric structure, consisting of three materials (M_1, M_2, and M_3), two boundaries (B_1 and B_2) and two interfaces (I_{12} and I_{23}). Normal vectors on boundaries and interfaces are indicated

Table 15.1 Governing equations for the heat-conduction process in an axisymmetric and a plane structure, consisting of three materials (Figure 15.2)

$$\frac{\partial}{\partial r}\left(\lambda r\frac{\partial T}{\partial r}\right)+\frac{\partial}{\partial z}\left(\lambda r\frac{\partial T}{\partial z}\right)-\mathrm{C}r\frac{\partial T}{\partial t}+r\Phi=0 \qquad \text{in } M_1,\ M_2 \text{ and } M_3$$

$\lambda(\mathbf{n},\mathrm{grad}T) = -\alpha(T-T_a)$ on B_1

$T=T_a$ on B_2

$\lambda(\mathbf{n},\mathrm{grad}T) = -\beta(T-T_{\mathrm{adj}})$ on l_{12} and l_{23}

where:

t	time	(sec)
r, z	coordinates	(m)
$T(r,z,t)$	temperature	(°C)
$\Phi(r,z,t)$	heat source	(J/m^3 sec)
$\lambda(r,z)$	thermal conductivity	(J/m sec °C)
$C(r,z)$	heat capacity	(J/m^3 °C)
$\mathbf{n}(r,z)$	normal vector on boundary (b) or interface (i)	(−)
$\alpha(r,z)$	surface conduction and convection coefficient	(J/m^2 sec °C)
$T_a(r,z)$	ambient temperature	(°C)
$\beta(r,z)$	interface conductivity	(J/m^2 sec °C)
$T_{\mathrm{adj}}(r,z,t)$	temperature of the adjoining material	(°C)

with initial conditions: $T(r,z,0)=T_0(r,z)$ in M_1, M_2, and M_3 and on B_1, B_2, l_{12}, and l_{23}.

For plane geometry (x–z plane), assuming no heat flow in the direction perpendicular to the x–z plane, the differential equation transforms into

$$\frac{\partial}{\partial x}\left(\lambda\frac{\partial T}{\partial x}\right)+\frac{\partial}{\partial z}\left(\lambda\frac{\partial T}{\partial z}\right)-C\frac{\partial T}{\partial t}+\Phi=0$$

while all other equations remain as they are, replacing r by x.

Table 15.2 Application of the backward time-difference formula results in a transformation of the equations shown in Table 15.1, for which a step-by-step procedure can be followed

$$\frac{\partial}{\partial t}T(t) \approx \frac{1}{\Delta t}\{T(t) - T(t - \Delta t)\}$$

where Δt denotes a time step. (backward difference)

By introducing the notation: $T_n(r,z) = T(r,z,t_n)$, $T_{n-1}(r,z) = T(r,z,t_{n-1})$, $\Phi_n(r,z) = \Phi(r,z,t_n)$, etc., where $\Delta t = t_n - t_{n-1}$,

$\frac{\partial}{\partial r}\left(\lambda r\frac{\partial T_n}{\partial r}\right) + \frac{\partial}{\partial z}\left(\lambda r\frac{\partial T_n}{\partial z}\right) - \frac{C}{\Delta t}rT_n + \frac{C}{\Delta t}rT_{n-1} + r\Phi_n = 0$	in M_1, M_2, M_3
$\lambda(\mathbf{n}, \mathrm{grad}T_n) = \alpha(T_n - T_a)$	on B_1
$T_n = T_a$	on B_2
$\lambda(\mathbf{n}, \mathrm{grad}T_n) = -\beta(T_n - T_{\mathrm{adj},n})$	on l_{12} and l_{23}

solution procedure was tested on a number of problems with known solutions and proved to give excellent results provided the time-step was small enough. The element used in the FEM procedures is triangular and has three nodal points. Within an element T_n, (r,z) is linear with respect to r and z.

15.2.2 The heat source

The rate of heat generation depends directly on the rate of monomer molecule linkage to polymer chains. This was expressed as the time derivative of the polymerization function ($p(t)$, i.e. the fraction of the monomer polymerized at time t, $0 \leqslant p \leqslant 1$). Hence,

$$\phi(t) = v_m \rho_m Q_t \frac{\mathrm{d}p}{\mathrm{d}t} = Q\frac{\mathrm{d}p}{\mathrm{d}t} \tag{15.1}$$

where v_m = volume fraction of the actual polymerizing monomer in the mixture (depending on the powder-to-liquid ratio, additives, and residual monomer)

ρ_m = density of the monomer (Kg/m^3)

Q_t = total amount of heat generated per unit mass of monomer (J/kg).

The polymerization function was considered to be a function of a dimensionless time: $(t - t_r)/\tau$, where t_r denotes the retardation time and τ is the 'polymerization time' defined by $p(t = \tau) = 0.75$. Both t_r and τ depend on the composition of the cement. This $p(t)$ curve was measured by de Wijn (1977; Figure 15.3). The dimensionless form was found to be a valid approximation for acrylic cement in general.

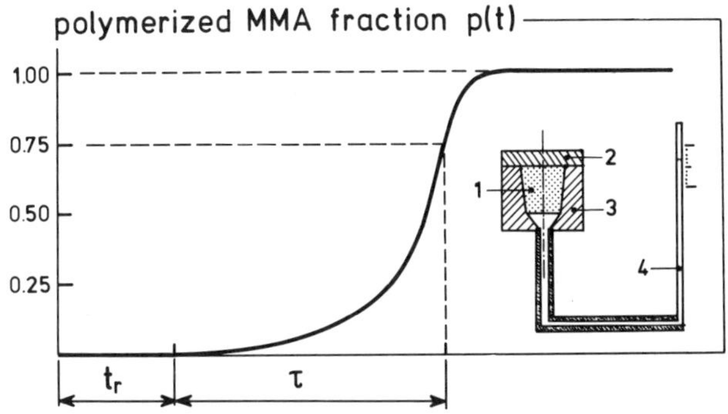

Figure 15.3 The polymerization function $p(t)$ as measured by de Wijn (1977) in shrinkage tests using the apparatus shown in the inset: 1, cement mass; 2, Delrin lid; 3, Delrin cup; 4, glass tube filled with liquid to register the amount of shrinkage

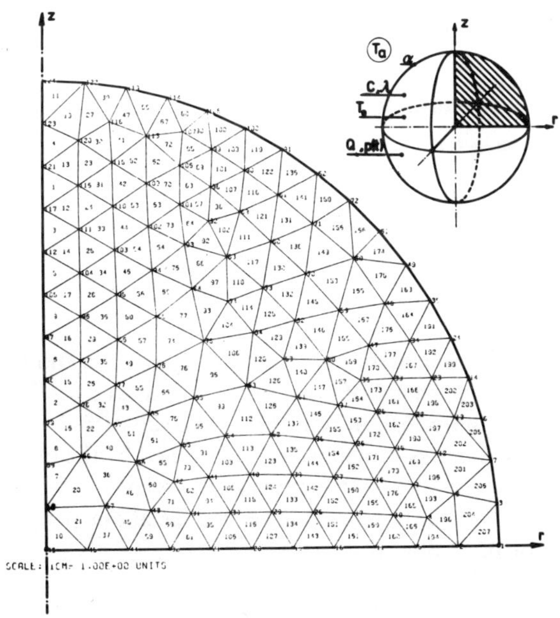

Figure 15.4 Element mesh for the acrylic-cement ball and the parameters on which the process of heat generation and conduction depends

Table 15.3 Probable and possible values of the relevant thermal parameters of acrylic cement and bone, as estimated on the basis of an extensive literature study (Huiskes, 1979); p_2 and p_3 are polymerization functions with slower and faster rates than p_1, respectively

Parameter	Unit	Probable value	Adjustable range	Possible range
Heat generation:				
ρ_m	kg/m^3	0.94×10^3	–	–
Q_t	J/kg	5.4×10^5	–	–
v_m	–	0.35	0.27–0.42	–
Q	J/m^3	1.8×10^8	1.4–2.1×10^8	–
$p(t)$	–	$p_1(t)$	$p_2(t)$, $p_3(t)$	–
t_r	sec	150	–	–
τ	sec	200–400	–	–
Heat capacity and conductivity:				
Acrylic cement:				
C_c	J/m^3 °C	1.8–2.0×10^6	–	1.7–2.1×10^6
λ_c	J/m sec °C	0.17	–	0.14–0.20
Cortical bone:				
C_b	J/m^3 °C	2.2–3.2×10^6	–	1.8–4.1×10^6
λ_b	J/m sec °C	0.3–0.5	–	0.2–0.6
Spongeous bone:				
C_b	J/m^3 °C	2.8–3.4×10^6	–	2.6–3.6×10^6
λ_b	J/m sec °C	0.5	–	0.4–0.6
Interface conductivity:				
β_{ci}	J/m^2 sec °C	1000–10000	–	500–∞
β_{cb}	J/m^2 sec °C	100–1000	–	50–∞
Bone surface heat loss:				
α	J/m^2 sec °C	500–10000	–	100–∞
Initial and ambient temperatures:				
$T_a = T_{0b}$	°C	32–35	–	30–37
T_{0c}	°C	20–25	–	–
T_{0i}	°C	20–25	–	–

15.2.3 Parameter values

An extensive review of the literature to establish probable and possible values of the relevant parameters has been carried out (Huiskes, 1979). A summary of results is shown in Table 15.3. In some cases, the values and value ranges shown are only estimates. Some parameters are determined by the manufacturer or the surgeon with the adjustable range shown in the relevant column of Table 15.3. The 'possible range' column reflects uncertainties about parameter values or a natural scatter of individual properties.

15.2.4 Evaluation of the model

The model was used to simulate experiments published by Ohnsorge and Kroesen (1969), where temperatures were measured as a function of time in the centre and at the surface of a ball of curing bone cement (ball radius: 2.2 $\times 10^{-2}$ m; initial and ambient temperature: 25 °C). Figure 15.4 shows the element mesh of ring elements. Only half of the symmetric ball was modelled and on $r = 0$ and $z = 0$ the heat flow is taken as zero by choosing $\alpha = 0$. Figure 15.5 shows a reasonable agreement between measured and calculated results. Figure 15.6 shows the temperature distribution in the ball.

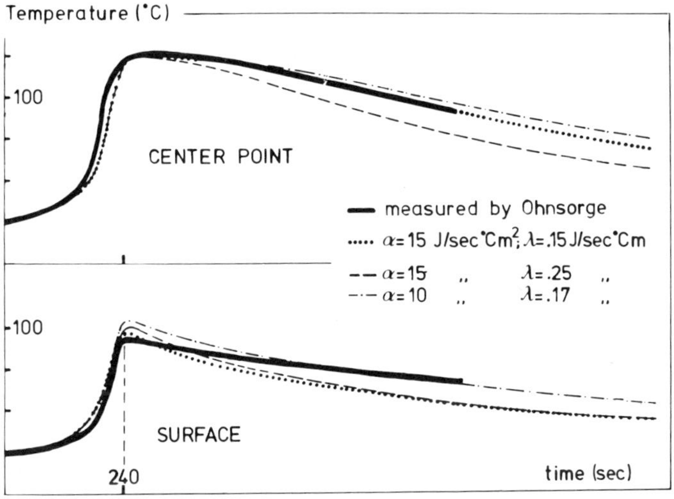

Figure 15.5 Comparison of experimental results and results calculated with different values for λ and α

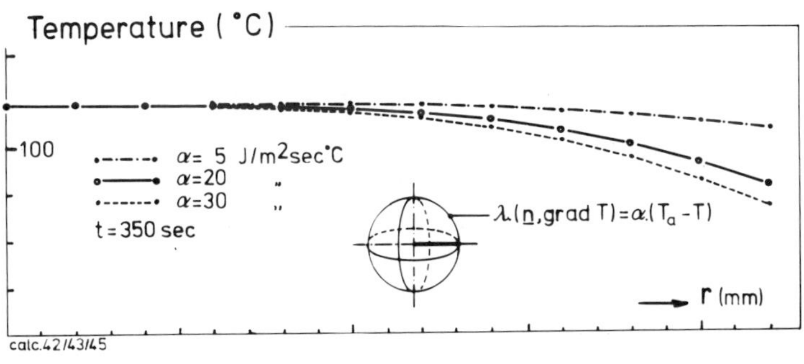

Figure 15.6 Temperature as a function of the ball radius at $t = 350$ s, for different values of the surface convection coefficient. The temperature increase of the centre point is practically adiabatic

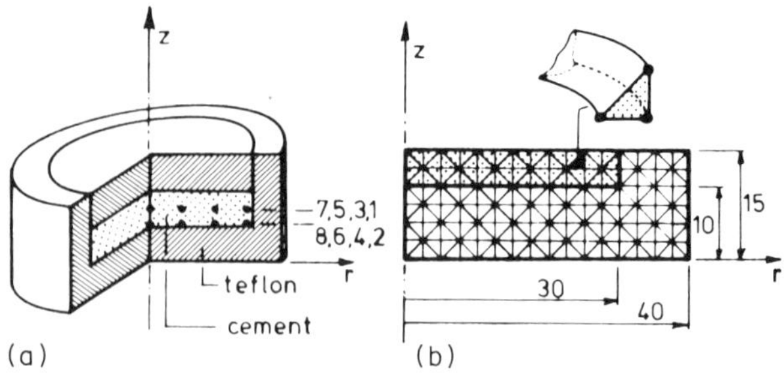

Figure 15.7 (a) The Teflon cup with the acrylic-cement mass as used by Meyer *et al.* (1973); temperature measuring points 1 through 8 are indicated; (b) FEM mesh and dimensions

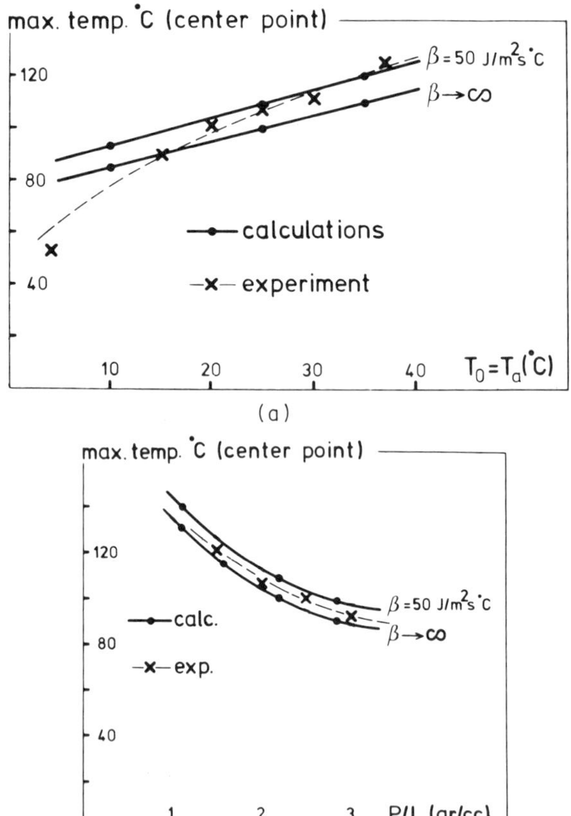

Figure 15.8 The maximal temperature of the centre point of the cement mass in the cup, as measured and as calculated (a) as a function of the initial and ambient temperature and (b) as a function of the powder-to-liquid ratio. The interface conductivity was varied in the calculations

An experiment by Meyer *et al.*, (1973) was also simulated. They measured time-dependent temperatures in acrylic cement curing in a Teflon cup (modelled in Figure 15.7). They also studied the influences of the initial and ambient temperatures and the P/L ratio (powder-to-liquid ratio). Figure 15.8 shows a comparison of measured and predicted maximal centre-point temperatures as a function of the initial temperature and P/L ratio. In this calculation, the cement–teflon interface conductivity β was varied as shown in Figure 15.8a and b.

Many uncertainties still remain with respect to the true values of the parameters. The interface conductivity is especially important but remains an unknown quantity. The model is well suited to perform parametric analyses and in doing so, possible temperature values in the bone adjoining the acrylic cement can be estimated and the influences of the important parameters on these temperature values can be established. In this respect, the model gives much more information than experiments possibly could.

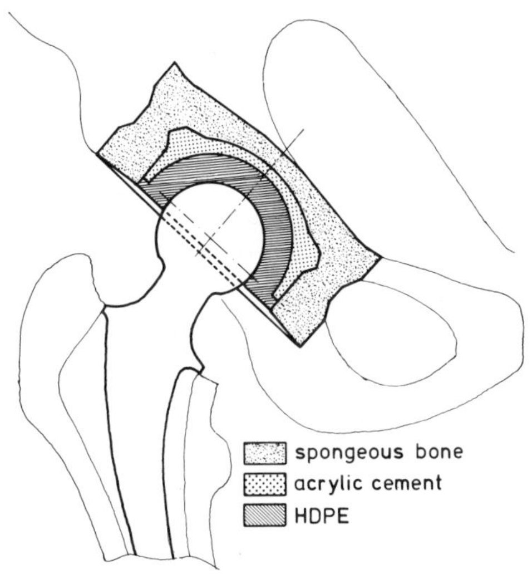

Figure 15.9 Schematic frontal section of an acetabular-cup fixation system, indicating the part of the bone that is taken into account in the model, together with the polyethyelene cup (HDPE) and the cement mass. The model is axisymmetric

15.2.5 Applications

The method described above was used to simulate the heat generation and conduction process and calculate transient temperature fields for various realistic situations (Huiskes, 1979). An example of such simulations is the acetabular cup fixation in hip joint replacement (see Figure 15.1). Figure 15.9 shows a schematic cross-section of this structure. The element mesh is shown in Figure 15.10. The

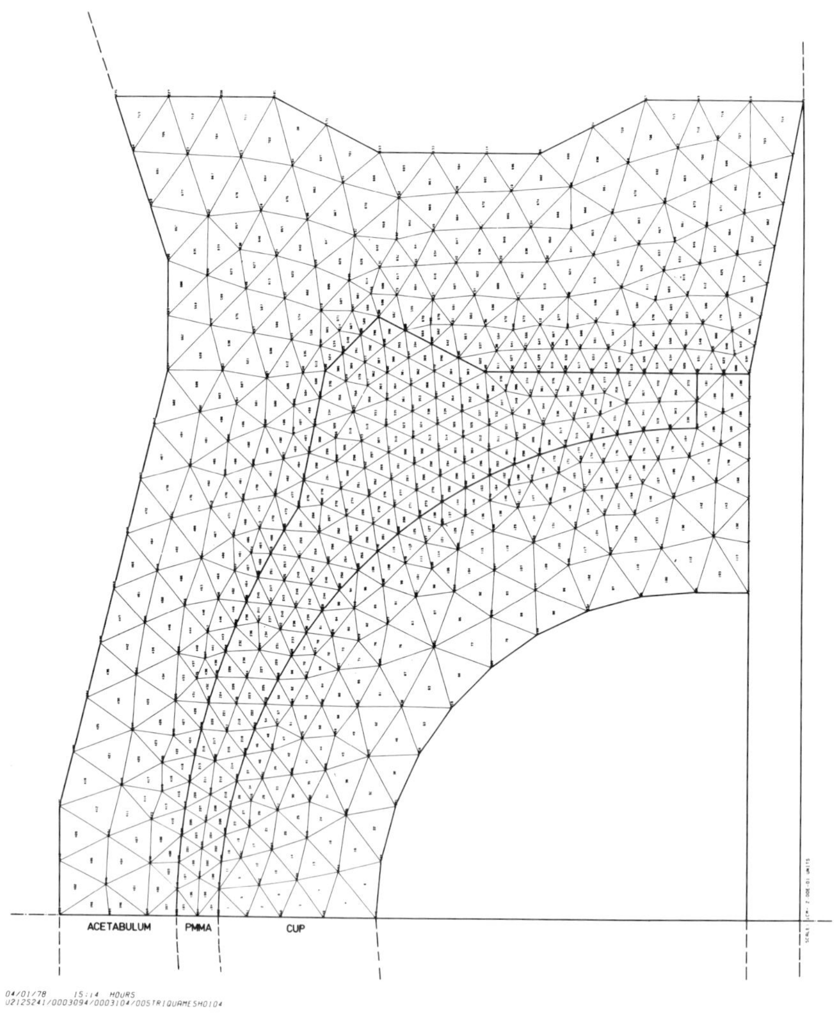

Figure 15.10 The element mesh used in the FEM calculations. At all contact regions double nodal points are generated

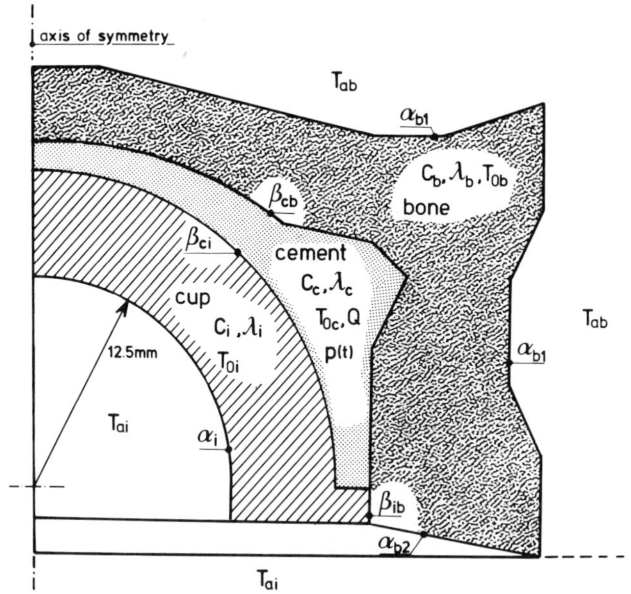

Figure 15.11 Parameters of the model

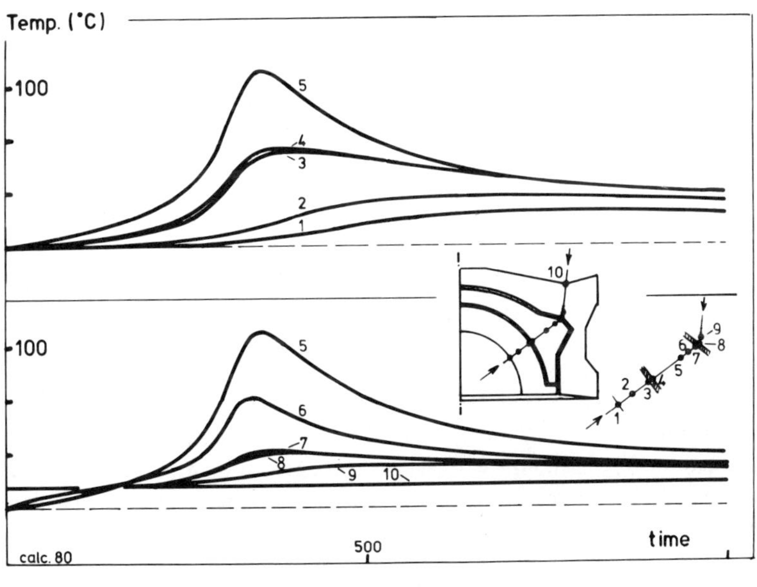

Figure 15.12 Temperatures as function of time, as calculated in several nodal points of the three materials; 1 through 3 in the cup, 4 through 7 in the cement mass, 8 and 9 in the bone

values for the parameters are given in Figure 15.11 and Table 15.4. Figure 15.12 shows an example of results. Although the temperature increase in the middle of the cement mass is substantial, the bone reaches only 57 °C in a very small zone near the cement interface. The temperatures and the exposure times have to be considered in order to judge the risks of thermal damage. Based on experiments by Moritz and Henriques (1947), it can be shown (Huiskes, 1980) that bone-cell necrosis will probably occur when:

$$\Delta T^{7.143} t_e > 3.55 \times 10^{10} \tag{15.2}$$

where ΔT = bone temperature − 37 (°C)
t_e = exposure time (sec).

Table 15.4 Values for the parameters, as used in the FE calculations concerning the acetabular cup-fixation

Parameter	Unit	Value	Parameter	Unit	Value
Heat generation			**Cement**		
Q	J/m³	1.7×10^8	C_c	J/m³ °C	1.75×10^6
$p(t)$	–	$p_1(t)$	λ_c	J/m sec °C	0.17
t_r	sec	0	T_{0c}	°C	25
τ	sec	350			
Interface conditions			**Bone (spongeous)**		
β_{ci}	J/m² sec °C	500	C_b	J/m³ °C	3.0×10^6
β_{cb}	J/m² sec °C	500	λ_b	J/m sec °C	0.4
β_{ib}	J/m² sec °C	500	T_{0b}	°C	37
Boundary conditions			**Cup (HDPE)**		
α_i	J/m² sec °C	10	C_i	J/m³ °C	2.2×10^6
α_{b1}	J/m² sec °C	10	λ_i	J/m sec °C	0.29
α_{b2}	J/m² sec °C	500	T_{0i}	°C	25
T_{ai}	°C	25			
T_{ab}	°C	37			

When these threshold levels are applied to the present results, a 'necrosis map' (Figure 15.13) can be drawn showing the regions where thermal damage is probable. Due to the variety in possible parameter values, the results of Figure 15.13 should be regarded more as indications of the possible necrosis. By systematic variation of all parameters, it was found that the cement–bone interface conductivity (β) has the greatest influence on the relevant temperature values. Taking $\beta \geqslant 500$ (J/m² s °C), a 'worst-case' situation is simulated.

When animal experiments with acrylic cements were simulated (Huiskes, 1979), it was found that the results of these experiments, as far as thermal damage is concerned, cannot be transformed to the human situation. This is due to the differences in dimensions and could not be accomplished even if all dimensional differences were subject to the same magnification factor.

The various calculations carried out showed (Huiskes, 1979) that some quite simple measures can be taken to reduce chances of thermal necrosis. These

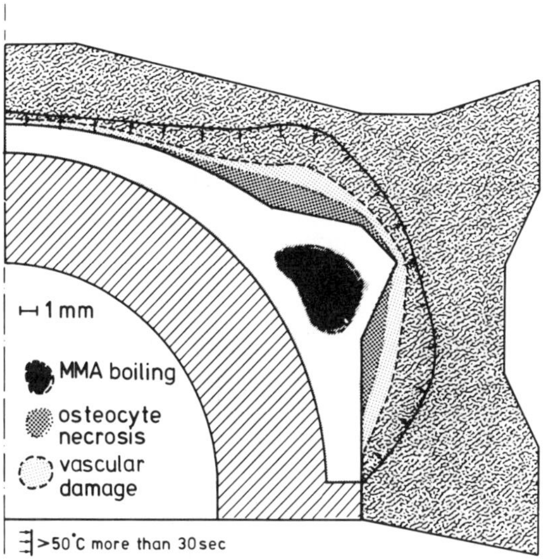

Figure 15.13 A 'necrosis map' as constructed from the calculated results, taking time-dependent threshold levels for thermal damage into account. Also shown is the region in the cement mass where gross evaporation of monomer might be expected, leading to inferior cement properties

measures were evaluated as to their order of effectiveness in a realistic range using a model simulation of the hip stem fixation.

Reduction of the cement-layer thickness has the most drastic effect on the bone temperatures. Other simple measures, in order of effectiveness, are: cooling of the operation region during and prior to cement insertion, decreasing the rate of polymerization, and increasing the P/L ratio. These measures are of special importance when acrylic cement is pressurized in trabecular bone. The penetration of the cement in the trabecular voids will then significantly increase the chances of bone necrosis (Huiskes and Slooff 1981). If the above-mentioned measures were combined during actual prosthesis fixation, chances for direct thermal damage of bone could be reduced.

15.3 STRESS ANALYSIS

15.3.1 Finite element model

A simplified model of an intramedullary fixation system is shown in Figure 15.14. It consists of a metal stem fixated in a bone cylinder with acrylic cement; the structure is axisymmetric. This model was analysed using ring elements applying Fourier series to take non-axisymmetric loading into account. Model results

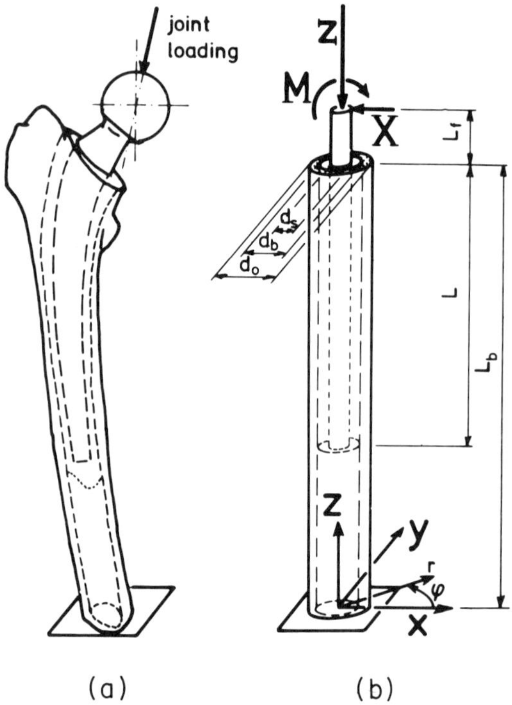

Figure 15.14 The axisymmetric general model of an intramedullary fixation system

compared reasonably well with those of experimental strain gauge analyses on a femur with an implant (Huiskes *et al.*, 1976; Huiskes, 1979). It was established that the stem and the bone in this model behave nearly in accordance with linear beam theory for all loading cases. The radial stress component was shown to be most significant in the cement layer for transverse loading, and the longitudinal shear stress component most significant for axial loading. Figure 15.15 shows an example of the results, i.e. axial stresses in the bone, radial stresses at the cement–bone interface, and axial stresses in the stem for transverse loading of the stem. It is evident that the structure can be divided into three regions: (a) a middle region that behaves in accordance with composite beam theory; (b) a proximal region where load is transferred from the stem to the bone; and (c) a distal region where load is transferred from the bone to the stem. The same trends can be found in the other stress components and other loading cases. It is clear from these results that the bone and the stem play inverse roles in the load transmission at the proximal and distal sides. Also, the proximal and distal sides behave independently provided the structure is long enough. It follows that composite beam theory can only be used to estimate stem and bone maximal stresses in the middle region.

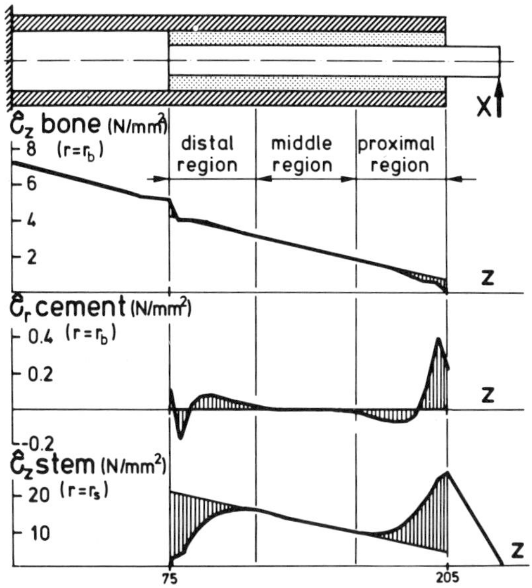

Figure 15.15 Axial stresses in the stem and the bone (near the interfaces) and radial stress at the cement – bone interface shown as deviating from composite beam theory

Cement and interface stresses reach peak values in the proximal and distal regions. Estimates for these stresses can be found using a theory of beam on elastic foundation theory.

15.3.2 Beam-on-elastic-foundation theory

The theory of beam on elastic foundation has been treated in depth by Hetény (1946). The models here differ from the usual elastic foundation models in that each beam (the stem and the bone) contributes to the other's elastic foundation. Here, only the transverse loading case will be discussed (for axial loading see Huiskes, 1979).

It is assumed that both the stem and the bone behave separately in accordance with beam theory. The transverse stiffness of the elastic layer (the cement mantle) is taken into account by assuming it as a continuous set of linear springs (Winkler hypothesis). The relevant internal loading variables of the structure are represented (Figure 15.16) by:

$M_s(z)$ bending moment in the stem (N mm)
$M_b(z)$ bending moment in the bone (N mm)
continuously distributed transverse load that the stem exerts on the bone and vice versa via the cement layer (N/mm).

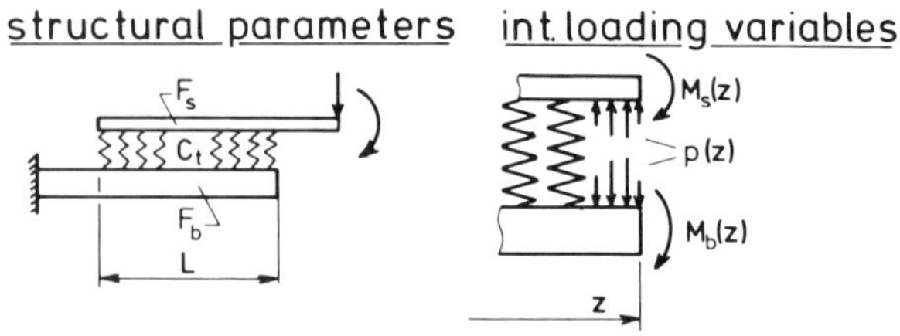

Figure 15.16 The mechanical behaviour of the beam-on-elastic-foundation model can be expressed in the lumped structural parameters F_s, F_b, C_t, and L. Internal loading variables are denoted by $p(z)$, $M_s(z)$, and $M_b(z)$

The geometrical and material properties of the structure can be lumped into the structural parameters (Figure 15.16):

F_s flexural rigidity of the stem (N mm^2)
F_b flexural rigidity of the bone (N mm^2)
C_t transverse stiffness of the elastic layer (N/mm^2)
L fixated stem length (mm)

The flexural rigidities are products of the Young's moduli and second moments of inertia; the transverse stiffness depends on the Young's modulus, Poisson's ratio, and cross-sectional geometry of the layer (Figure 15.17).

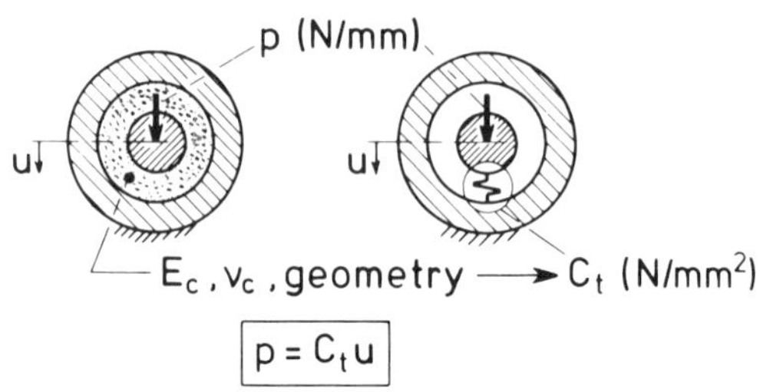

Figure 15.17 The transverse stiffness of the elastic layer depends on material and geometrical parameters, as illustrated here in the case of an axisymmetric cross-section

Table 15.5 gives the governing equations for transverse loading (formulated in the stem and bone deflections, u_s and u_b respectively), their solution, and the boundary conditions (with which the constants A_1 through A_4 and a_0 through a_3 can be calculated).

It should be noted that these formulae are valid for arbitrary cross-sectional geometry that is uniform in the z direction. The value of C_t for several shapes was found by applying analytical and finite element methods to a cross-sectional

Table 15.5 Governing differential equations, solutions, and boundary conditions for the beam-on-elastic-foundation model in transverse loading

Equations

$$p = C_t(U_s - U_b) \qquad F_s \frac{d^4 u_s}{dz^4} + p = 0 \quad \text{and} \quad F_b \frac{d^4 u_b}{dz^4} - p = 0$$

Solution	*With*
$u_s(z) = e^{\lambda_t z} D_1(z) + e^{-\lambda_t z} D_2(z) + \sum_{i=0}^{3} a_i z^i$	$D_1 = A_1 \cos \lambda_t z + A_2 \sin \lambda_t z$ $D_2 = A_3 \cos \lambda_t z + A_4 \sin \lambda_t z$
$u_b(z) = -\frac{F_s}{F_b} e^{\lambda_t z} D_1(z) - \frac{F_s}{F_b} e^{\lambda_t z} D_2(z) + \sum_{i=0}^{3} a_i z^i$	$\lambda_t = \sqrt[4]{\left\{\frac{C_t}{4}\left(\frac{1}{F_s} + \frac{1}{F_b}\right)\right\}}$
$p(z) = C_t\left(1 + \frac{F_s}{F_b}\right)(e^{\lambda_t z} D_1(z) + e^{\lambda_t z} D_2(z))$	$M_s(z) = F_s \frac{d^2 u_s}{dz^2}$ and $M_b(z) = F_b \frac{d^2 u_b}{dz^2}$

Boundary conditions

$$u_b(0) = (0); \quad \frac{du_b}{dz}(0) = 0; \quad \frac{d^2 u_b}{dz^2}(L) = 0; \quad \frac{d^3 u_b}{dz^3}(L) = 0; \quad \frac{d^2 u_s}{dz^2}(0) = 0; \quad \frac{d^3 u_s}{dz^3}(0) = 0;$$

$$\frac{d^2 u_s}{dz^2}(L) = \frac{M_L}{F_s}; \quad \frac{d^3 u_s}{dz^3}(L) = \frac{T_L}{F_s},$$

slice, and confirming the results experimentally (Huiskes and Schouten, 1980). For axisymmetric geometry, C_t can be evaluated from:

$$C_t = 5.3\, E_c g(v_c)\,(0.24 + \rho/(1 - \rho)) \tag{15.3}$$

where ρ = stem radius/inner bone radius (r_s/r_b) and $1.1 < g(v_c) < 1.5$ for $0.3 < v_c < 0.4$

Figure 15.18 shows a comparison between finite element and analytical beam-on-elastic-foundation results for the structure of Figure 15.14. In the finite element model, $p(z)$ in each cross-section can be calculated by integration of σ_r and τ_{rz} in the circumferential direction. Excellent agreement was found in $M_s(z)$ and $M_b(z)$ and reasonable agreement in $p(z)$. Deviations are principally caused by neglecting the three-dimensional stress state of the cement layer in the beam-on-elastic-foundation model.

The coefficient for transverse loading λ_t (Table 15.5) plays a key role in the mechanical behaviour of the structure. Together with the length L, it determines whether the structure is 'long' or 'short' (Hetény, 1946). If $\lambda_t L$ is small (Figure 15.19a), the structure is 'short' and in this case, both beams may be assumed as infinitely rigid. If $\lambda_t L$ is large (Figure 15.19c), the structure is 'long' and both ends behave independently while a middle region behaves in accordance with composite beam theory.

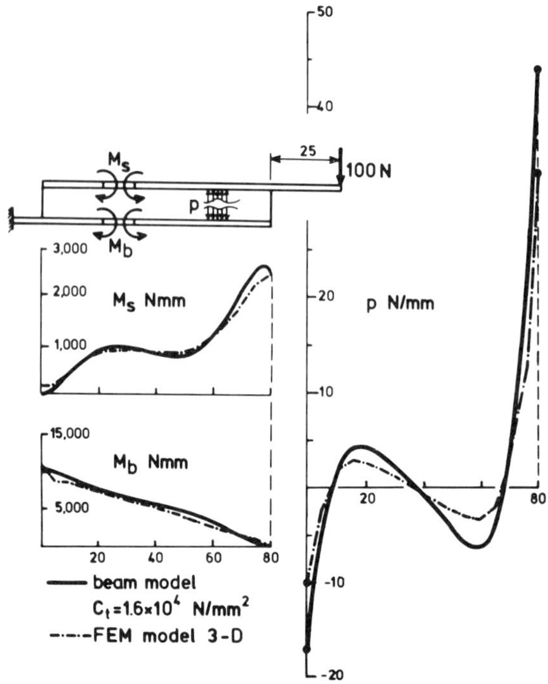

Figure 15.18 Comparison of $M_s(z)$, $M_b(z)$, and $p(z)$ as calculated with the beam-on-elastic foundation model and as derived from the FEM model results

In the latter case, when $\lambda_t L > \pi$, both sides can be treated separately by assuming the structure to be of semi-infinite length on both ends (Figure 15.20). The solutions presented in Table 15.5 are again valid in this case. Due to the different boundary conditions, however, relatively simple formulae result. For the distal side:

$$M_s(z) = \frac{F_s}{F_s + F_b}\left[M_0 - T_0 z - e^{-\lambda_t z}\left\{M_0 \cos\lambda_t z + \left(M_0 - \frac{T_0}{\lambda_t}\right)\sin\lambda_t z\right\}\right] \tag{15.4}$$

$$M_b(z) = \frac{F_s}{F_s + F_b}\left[\frac{F_b}{F_s}M_0 - \frac{F_b}{F_s}T_0 z + e^{-\lambda_t z}\left\{M_0 \cos\lambda_t z + \left(M_0 - \frac{T_0}{\lambda_t}\right)\sin\lambda_t z\right\}\right] \tag{15.5}$$

$$p(z) = \frac{1}{2}\frac{C_t}{\lambda_t^2 F_b} e^{-\lambda_t z}\left\{M_0 \sin\lambda_t z - \left(M_0 - \frac{T_0}{\lambda_t}\right)\cos\lambda_t z\right\} \tag{15.6}$$

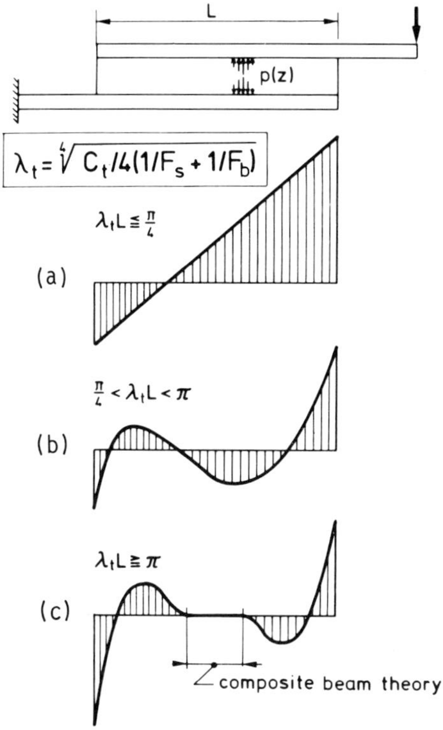

Figure 15.19 The coefficient for transverse loading λ_t represents a first-order characterization of the behaviour of the structure. For small $\lambda_t L$ ('short' structure, (a)) $p(z)$ will be linear by approximation, stem and bone behave rigidly relative to the cement deformation. For large $\lambda_t L$ (a 'long' structure, (c)) both ends behave independently from each other and a middle region behaves according to composite beam theory. The effects are somewhat exaggerated in (a) and (c), to illustrate the tendencies

For the proximal side, the above formula give the solution by replacing z by $\xi(=L-z)$, F_s by F_b and vice versa, M_0 by M_L and T_O by $-T_L$. It is evident that in the structure $M_0 = M_L + T_L L$ and $T_0 = T_L$. For larger z (or larger ξ), the exponential expressions diminish and the formulae approximate those of composite beam theory.

15.3.3 Applications

The formulae derived in the preceeding paragraph are of great use in gaining a better understanding of the structural behaviour of intramedullary fixated artificial joints and can be applied as first-order design guidelines for prosthesis

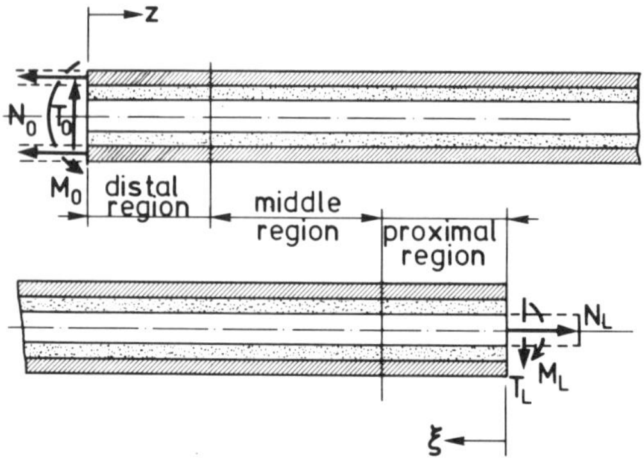

Figure 15.20 On assuming the structure to be of semi-infinite length, the proximal and distal sides are treated separately. The loading system for these two cases are shown

stems. Stresses in the stem reach maximal values that are proportional to $\varepsilon_m = F_s/(F_s + F_b)$ in the middle region. Stresses in the cement layer and at the interfaces reach maximal values at the proximal and distal sides that are proportional to α_m and β_m (Figure 15.21) respectively. The lengths of the proximal and distal load-transmission regions are approximately equal to $\pi/2\lambda_t$.

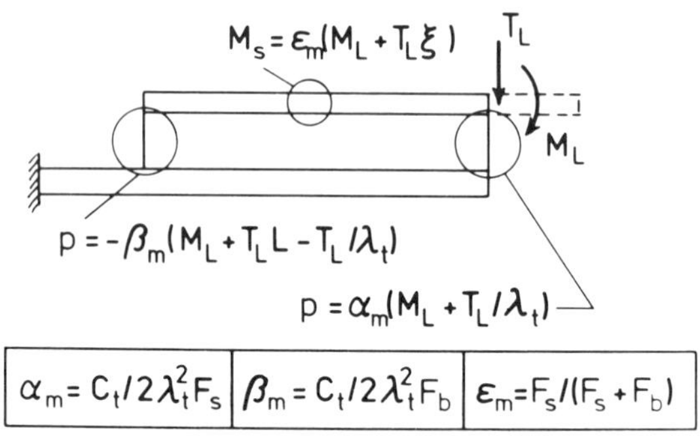

Figure 15.21 Relatively simple formulae can be derived to approximate stem loading (M_s) in the middle region, interface and cement loading (p) at the proximal and distal sides. Stresses in these regions can in turn in turn be calculated from M_s and p (Figure 15–22)

The following conclusions (assuming given bone properties and dimensions) may be drawn from the formulae for arbitrary cross-sectional geometry:

(a) The length of the stem should be restricted to the length of the proximal load-transmission region plus the length of the distal region since the distal cement stresses increase with the factor $T_L L$ (Figure 15.21) and the middle region serves no purpose in the load transmission. This 'optimal' length can be approximated using λ_t.

(b) The stem bending moment (M_s) will increase with increasing F_s, which in turn increases with increasing stem modulus and thickness following directly from ε_m (Figure 15.21). The stem stresses then increase in proportion to the stem modulus. An increase in stem thickness, however, also affects its moment of resistance (Figure 15.22), hence, the stresses might decrease, depending on the geometry.

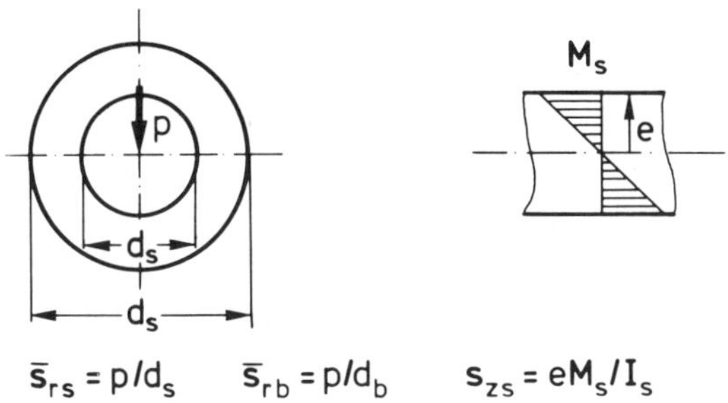

Figure 15.22 From the internal loading variable p, average transverse stresses at the stem–cement interface ($\bar{s}_{rs}$) and the cement–bone interface ($\bar{s}_{rb}$) can be calculated, as shown, in this case, for an axisymmetric cross-section. Stem stresses can be calculated from M_s, using its moment of resistance

(c) The proximal and distal cement and interface stresses are directly proportional to C_t/λ_t^2 (Figure 15.21), hence, to $\sqrt{C_t}$. From this, it follows that these stresses (1) decrease with decreasing layer modulus proportional to $\sqrt{E_c}$; (2) decrease with the layer Poisson's ratio (for an axisymmetric structure, as shown by Equation (15.3), proportional to $\sqrt{g(v_c)}$); and (3) decrease with increasing layer thickness (for an axisymmetric structure proportional to $\sqrt{(0.24+\rho/(1-\rho))}$).

(d) For a given bone, the layer thickness depends on the stem thickness. This parameter affects the stem flexural rigidity as well. The expressions for α_m and β_m (Figure 15.21) indicate that the proximal stresses decrease and the distal stresses increase with increasing F_s. Thus, on the distal side the stresses will be minimal for a flexible stem (thin, low modulus) and on the proximal side the stresses will be small for a stem with a high modulus. However, on the proximal side a thicker stem will increase the stresses through the influence of C_t while decreasing them through the influence of F_s. Hence, an optimal thickness can be expected to exist. This is illutrated for axisymmetric cross-sectional geometry in Figure 15.23, which shows α_m and β_m as functions of the

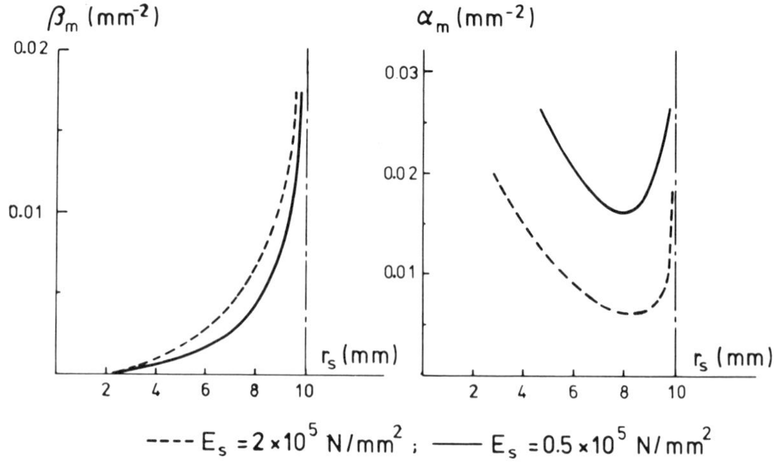

Figure 15.23 The factors α_m (proximal side) and β_m (distal side), defined in Figure 15–21, as functions of the stem thickness, as shown, in this case, for an axisymmetric cross-section, for two different stem Young's moduli

stem thickness for two different stem moduli. The optimal cement layer thickness on the proximal side (~ 2 mm shown in Figure 15.23) was found to be rather insensitive to other structural parameters (Huiskes, 1979). Hence, it appears that an intramedullary prosthesis should have a good fit on the proximal side, be as thick as possible, and leave a gap of around 2 mm. This indicates the significance of the 'custom-fit' approach.

(e) Much speculation can be found in the literature with respect to the possible influence of interface loosening on the stress distribution. With respect to 'loosening upon tension', note that acrylic cement is not an adhesive but rather is a filler. It is evident then that this phenomenon can only affect C_t. It can be shown that for a loose stem, C_t is reduced by a factor of 2 to 3 (Huiskes and Schouten, 1980). Thus it follows that when considering the gross mechanical behaviour of the structure (with lumped internal loading variables $M_s(z)$, $M_b(z)$, and $p(z)$), loosening has the same effect as a reduction in layer modulus by a factor of 2 to 3. This effect was confirmed with a two-dimensional finite element model taking loosening upon tension into account (Huiskes and Schouten, 1980). It should be noted here that this stress-smoothening effect of a lower C_t value, and thus a lower value of $p(z)$, does not imply lower stress values in the cement and the interfaces in this case. When loosening upon tension occurs, roughly half the interface is available for load transmittance. Hence, the same value of p will result in twice the stress values.

(f) Based on the results presented here, a two-dimensional model will be a reasonable representation of a three-dimensional structure with respect to its gross mechanical behaviour if the structural parameters C_t, F_s, F_b, and L are modelled correctly. Comparable conditions can be formulated for axial loading and it follows (Huiskes, 1979) that unless separate models are applied for transverse and axial loading, and a method of scaling is introduced, the results of two-dimensional models will be unrealistic in the absolute sense.

The above guidelines can be used for a first-order evaluation of stem designs. It follows that an optimal stem should have a taper and the next step would be to optimize the taper shape.

15.3.4 Extension to non-uniform cross-sectional geometry

The models based on beam-on-elastic-foundation theory can also be simulated using FEM beam analysis. In this case, no closed-form solutions result but non-constant cross-sectional geometries of the stem and the bone can be taken into account. Also, variations in elastic properties (incomplete cement mantles, bone resorption, spongeous bone layers of varying thickness, etc.) can be included. The stem and bone are represented by a number of beam elements and the intermediate layer by special spring elements (elastic properties calculated using local values for C_t and C_a for axial loading; Huiskes, 1979).

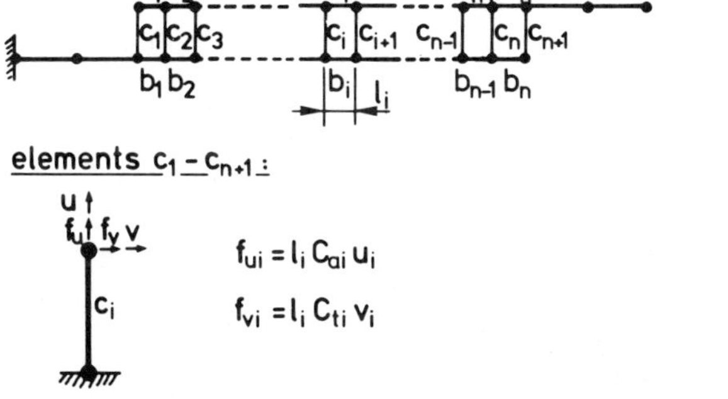

Figure 15.24 Characterization of the FE-beam model; the elements s_i and b_i are beam elements representing the stem and the bone, respectively. The intermediate layer is represented by 'spring-elements' (c_i)

The model characteristics are shown in Figure 15.24. Cross-sectional geometrical and material properties as a function of z are used as input and the computer program automatically generates the mesh and the element parameters. The lumped internal loading variables $M_s(z)$, $M_b(z)$, p(z), and comparable variables for axial loading are calculated. Since these lumped internal loading variables are not uniformly proportional to the stresses (for this case with non-uniform cross-sectional geometry), stresses will have to be calculated as well. [Stem stresses (S_{zs}) and average cement–interface stresses ($\overline{S}_{rs}$ and $\overline{S}_{rb}$) are calculated from $M_s(z)$, $M_b(z)$, and $p(z)$ shown in Figure 15.22 and assuming axisymmetric geometry (although this is not essential for the method).

Examples of applications of this model are shown in Figures 15.25 through 15.27. In these examples, the bone is assumed to have constant cross-sectional geometry but this again is not essential. Figure 15.25 shows stem–cement transverse stresses and stem stresses for different distal taper lengths. The length and thickness of the proximal stem have been optimized for this 'bone' using the formulae discussed in the previous paragraph.

Figure 15.26 shows the same stresses for three taper shapes. A small difference in shape significantly affects the distal cement stresses. Both figures illustrate the independence of the proximal stresses from the distal geometry. In Figure 15.27, the same stresses are shown for the undercut taper shape (shape No. 3 in Figure 15.26) and three different stem Young's moduli (simulating titanium, stainless steel, and ceramics). Note that the effects found in this figure can be predicted in a qualitative sense by the guidelines mentioned in the preceeding paragraph.

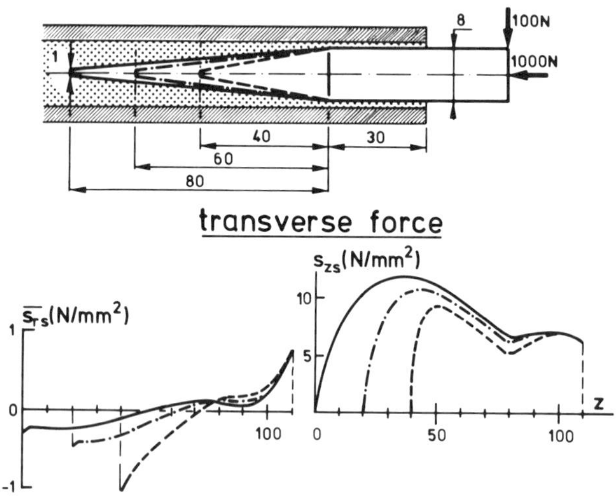

Figure 15.25 Stem–cement interface stresses ($\overline{s}_{rs}$) and axial stem stresses (s_{zs}) as functions of z, for three different taper lengths (FE-beam model)

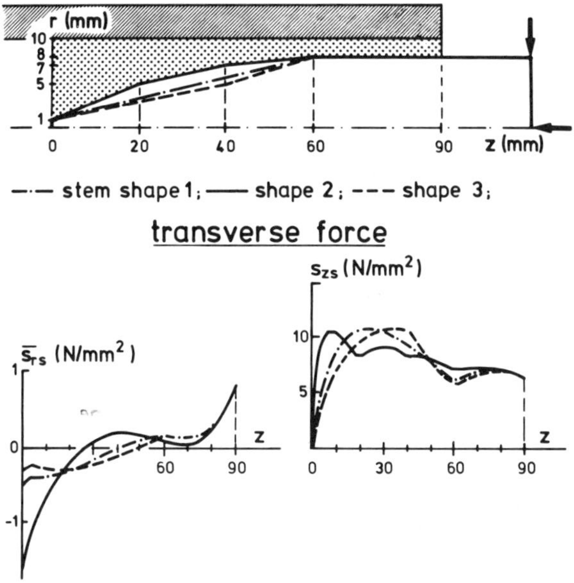

Figure 15.26 Stem–cement interface stresses ($\bar{s}_{rs}$) and axial stem stresses (s_{zs}) as functions of z, for three different taper profiles. Shape no. 1 equals the moderate taper of Figure 15.24 (FE-beam model)

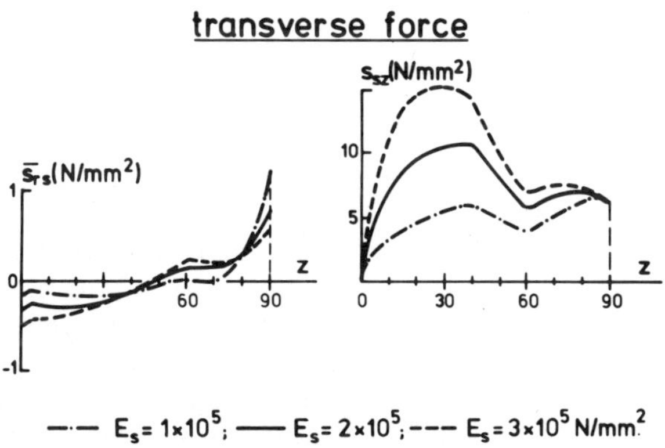

Figure 15.27 Stem–cement interface stresses ($\bar{s}_{rs}$) and axial stem stresses (s_{zs}) as functions of z, the taper shape no. 3 of Figure 15.26, and three different stem Young's moduli

15.4 CONCLUSION

It is evident that the methods presented here are suited only for rough approximations and first-order evaluations. They cannot replace detailed FE modelling but serve to amplify it. They can be used for a first estimation of design dimensions and materials while a three-dimensional model would be better suited for an ultimate (relative) test. Likewise, they can be used for rough comparisons of different designs and for the choice of a best suitable design for a specific bone. These are specific questions for which refined models would be too expensive and time consuming. Above all, they are of great use for a better fundamental understanding of refined FE model results and for placing these results in a general setting.

When comparing the guidelines mentioned previously with the results of anatomic three-dimensional FE models (Röhrle *et al.*, 1979; Crowninshield *et al.*, 1980; Tarr *et al.*, 1980), it is found that they are compatible so that many of these results can now be understood in a more fundamental and general sense. Moreover, good qualitative agreement is found between these three-dimensional results and the stresses calculated with the FE beam model discussed above. These last results show qualitative agreement with strain-gauge measurements made on the outside of the femur (Huiskes *et al.*, 1976; Huggler *et al.*, 1978) as well as *in situ* on femoral stems (Huggler *et al.*, 1978).

The closed-form solutions discussed here are valid only for those intramedullary fixated prostheses that have no 'shoulder' to rest on the bone. However, these theories can easily be extended to take this feature into account. Also, formulae comparable to those discussed here for transverse and axial loading can be derived for torsional loading (Huiskes and Chao, 1980). Transverse and axial loading can also be applied to the FE beam models.

Although it has been noted that '(geometrically simplified models) can hardly be expected to define the relative merits of an actual prosthesis' (Clarke *et al.*, 1980), it should be evident that this, if true, applies to other model aspects as well and that an apparent visual agreement between model and reality does not guarantee accuracy in itself. It can hardly be expected that models will be developed that are complete and refined in every sense (i.e. with realistic geometry and realistic descriptions of the non-linear material properties, conditions of interface continuity, and the loading) and still yield interpretable results. Hence, the simplified models described here can be helpful in producing boundary conditions for detailed local models and in investigating consequences of certain simplifying assumptions. Such a combined step-by-step fundamental approach can give a basis for future success in modelling bone–prosthesis structures.

REFERENCES

Andriacchi, T. P., Galante, J. O., Belytschko, T. B., and Hampton, S. (1975) A stress analysis of the femoral stem in total hip prostheses, *J. Bone Jt. Surg*, **58**(A), 618–624.

Bartel, D. L., and Ulsoy, G. A. (1975) The effect of stem length and stem material on stresses in bone-prosthesis systems, *21st Annual Meeting, Orthopaedic Research Society*, San Francisco.

Beckenbaugh, R. D., and Ilstrup, D. M. (1978) Total hip arthroplasty, a review of three hundred twenty-three cases with long follow-up, *J. Bone Jt. Surg.*, **60**(A), 306–313.

Brekelmans, W. A. M. (1975), A method for the numerical solution of a certain type of elliptical differential equations (in Dutch), *Report WE-75-03*, Dept. Applied Mechanics, Eindhoven University of Technology, The Netherlands.

Calderale, P. M., Gola, M. M., and Gugliotta, A. (1977) Calculation of load distribution between bone and prosthesis stem by an analytical method; comparison of several stem types, *6th International Conference of Biomechanics*, Copenhagen.

Campen, v. D. H., Croon, H. W., and Lindwer, J. (1979) Mechanical loosening of knee endoprostheses with intramedullary stems; influence of dynamic loading, *25th Annual Meeting, Orthopaedic Research Society*, San Francisco.

Carlslaw, H. S., and Jaeger, J. C. (1959) *Conduction of Heat in Solids*, Oxford University Press.

Charnley, J. (1978) *Low-friction Arthroplasty of the Hip*, Springer-Verlag, Berlin, Heidelberg, New York.

Clarke, I. C., Gruen, T. A. W., Tarr, R. R., and Sarmiento, A., (1980) Finite element analyses studies of total hip versus clinical reality, *Proc. of the International Conference on Finite Elements in Biomechanics*, Tucson, AZ (ed. B. R. Simon), The College of Engineering, University of Arizona, Tucson.

Cook, S. D., Klawitter, J. J., and Weinstein, A. M. (1980) The influence of design parameters on calcar stresses following femoral head arthroplasty, *J. Biom. Mat. Res.*, **14,** 133–144.

Crowninshield, R. D., Brand, R. A., Johnston, R. C., and Milroy, J. C. (1980) An analysis of femoral component stem design in total hip arthroplasty, *J. Bone Jt. Surg.*, **62**(A), 68–78.

Feith, R. (1975) Side-effects of acrylic cement implanted into bone, *Acta Orthop. Scand.*, Supp. No. 161.

Hampton, S. J., Andriacchi, T. P., Galante, J. O., and Belytschko, T. B. (1976) Analytical approach to the study of stresses in the femoral stem of total hip prostheses. *29th ACEMB*, Boston, MA, Paper No. 32.1, November.

Hetény, M. (1946) *Beams on Elastic Foundations*, The University of Michigan Press, Ann Arbor, Michigan.

Huggler, A. H., Jacob, H. A. C., and Schreiber, A. (1978) Biomechanische Analyse der Lockerung von Femurprosthesen, *Arch. Orthop. Traumat. Surg.*, **92,** 261–272.

Huiskes, R. (1979) Some fundamental aspects of human joint replacement, analyses of stresses and heat conduction in bone prosthesis structures, *Acta Orthop. Scand.*, Supp. No. 185.

Huiskes, R. (1980) Criteria for thermal damage of bone, (to be published).

Huiskes, R., and Chao, E. Y. (1980) Optimal stem dimensions for intramedullary fixated 'custom-fit' prostheses. *Report of Orthopedic Biomechanics Lab.* Mayo Clinic, Rochester, MN.

Huiskes, R., Elangovan, P. T., Banens, J. P. A., and Slooff, T. J. J. H. (1977) Finite element computer methods for design and fixation problems of orthopaedic implants, In: *Biomechanics VIB,* Int. Series on Biomechanics (eds. E. Asmussen and K. Jørgensen), University Park Press, Baltimore, MD

Huiskes, R., Heck, J. v., Walker, P. S., Green, D. J., and Nunamaker, D. M. (1980) A three-dimensional stress analysis of a new finger-joint prosthesis fixation system, *Proc. Int. Conf. on Finite Elements in Biomechanics*, Tucson, AZ (ed. B. R. Simon), The College of Engineering, Univ. Arizona, Tucson, AZ.

Huiskes, R., v. Heugten, P. C. M., and Slooff, T. J. J. H. (1976) Strain-gauge measurements on a loaded femur, intact as well as provided with prostheses, *29th ACEMB*, Boston, MA, November.

Huiskes, R., and Schouten, R. Y. (1980) The effect of interface loosening on the stress distribution in intramedullary fixated artificial joints. Presented at the 1980 Centennial Winter Annual Meeting of the ASME, Chicago, November 17–21 (To be published in *ASME 1980 Advances in Bioengineering*).

Huiskes, R., and Slooff, T. J. (1981) Thermal injury of cancellous bone following pressurized penetration of acrylic cement, *Trans. 27th Annual Meeting, Orthopaedic Research Society*, Las Vegas, Nev., February.

Kwak, B. M., Lim, O. K., Kim, Y. Y., and Rim, K. (1979) An investigation of the effect of cement thickness on an implant by finite element stress analysis, *Int. Orthop. (SICOT)* **2,** 315–319.

McNeice, G. M., Eng, P., and Amstutz, H. C. (1975) Finite Element studies in hip reconstruction, *5th Int. Conf. of Biomechanics*, Jyväskylä, Finland. Published (1976) in: *Biomechanics VA* (ed P. V. Komi), Univ. Park Press, Baltimore, M. D.

Meyer, J. R., Lautenschlager, E. P., and Moore, E. K. (1973): On the setting properties of acrylic bone cement, *J. Bone Jt. Surg.*, **55**(A), 149–156.

Moritz, A. R., and Henriques, F. C. (1947) Studies of thermal injury, *Am. J. Path.*, **23,** 695.

Nunamaker, D. M., Walker, P. S., van Syckle, P., and Green, D. (1980) A one year in vivo study of a new total finger-joint prosthesis, using biological ingrowth for stabilization in the knee of the cat, *Trans. 26th Annual Meeting of the Orthopaedic Research Society*, Atlanta, GA, p. 279.

Ohnsorge, J., and Kroesen, A. (1969) Thermoelektrische Temperaturmessungen des abhärtender Knochenzementes 'Palacos', *Z. Orthop.* **106,** 476–482.

Rohlmann, A., Kölbel, R., and Bergmann, G. (1979) Ueber den Aussagewert der Finiten-Element-Methode bei orthopädische Problemen, Pauwels-Symposium, Berlin, January 19–20 (to be published).

Röhrle, H., Scholten, R., and Sollbach, W. (1979) Der Kraftflusz bei neuartigen Hüftendoprothesen, Pauwels-Symposium, Berlin, Febr. 18–19 (to be published).

Scholten, R., Röhrle, H, and Sollbach, W. (1978) Analysis of stress distribution in natural and artificial hip joints, using the finite-element-method, *The South Afr. Mech. Eng.*, **28,** 20.

Slooff, T. J. J. H. (1970) De invloed van het acryl-cement bij de fixatie van de heupprothese, Dissertation, Univ. of Nijmegen, The Netherlands.

Svensson, N. L., Valliappan, S., and Wood, R. D. (1977) Stress analysis of human femur with implanted Charnley prosthesis, *J. Biomechanics*, **10,** 581–588.

Tarr, R., Lewis, J., Ghassemi, F., Sarmiento, A., Clarke, I., and Weingarten, V. (1980) Anatomic three-dimensional FE model of the proximal femur with total hip prosthesis, *Proc. Int. Conf. on Finite Elements in Biomechanics*, Tucson, AZ (ed. B. R. Simon), The Coll. of Eng., Univ. of Arizona, Tucson, AZ.

Valliappan , S., Kjellberg, S., and Svensson, N. L. (1980) Finite element analysis of total hip prosthesis, *Proc. Int. Conf. on Finite Elements in Biomechanics*, Tucson, AZ (ed. B. R. Simon), The Coll. of Eng., Univ. of Arizona, Tucson, AZ.

Vernon-Roberts, B., and Freeman, M. A. R. (1977) The tissue response to total joint replacement prostheses, In: *The Scientific Basis of Joint Replacement* (eds. S. A. V. Swansson and M. A. R. Freeman), Pitman Med. Publ. Comp., Tunbridge Wells.

Willert, H. G., and Frech, H. A. (1976) Tissue damage caused by bone cement, In: *Advances in Artificial Hip and Knee Joint Technology* (eds. N. Schaldach and D. Hohmann), Springer-Verlag, Berlin, Heidelberg, New York.

Wilson, E. L., and Nickell, R. E. (1966) Application of the FEM to heat conduction analyses, *Nucl. Eng. and Des.*, **4,** 276–288.

Wilson, E. L., Bathe, K. J., and Peterson, F. E. (1974) Finite element analysis of linear and nonlinear heat transfer, *Nucl. Eng. and Des.*, **29,** 110–124.
Wijn, de J. R. (1977) Private communication.
Yettram, A. L., and Wright, K. W. J. (1979) Biomechanics of the femoral component with particular reference to the stress in the bone cement *J. Biomed. Eng.*, **1,** 281.
Zienkiewicz, O. C. (1977) *The Finite Element Method* (3rd edn), McGraw-Hill, Ltd. London.

Finite Elements in Biomechanics
Edited by R. H. Gallagher, B. R. Simon, P. C. Johnson, and J. F. Gross

Chapter 16

Predictions of Cement-bone Failure Criteria: Three-dimensional Finite Element Models Versus Clinical Reality of Total Hip Replacement

R. R. Tarr, I. C. Clarke, T. A. Gruen, and A. Sarmiento

16.1 INTRODUCTION

The relationship between the biomechanical behaviour of hip prostheses and the physiological behaviour of the femur is poorly understood, even after 18 years of clinical experience. Loosening of the prosthesis is still one of the major problems associated with this procedure (Beckenbaugh and Ilstrup, 1978; Charnley, 1979). While detailed sequential radiographic analysis of failing prostheses have described their modes of failure (Gruen *et al.*, 1979), the mechanism that initiates failure is unknown. Speculation exists: (1) that the cement is the weak link and must be protected (Crowninshield *et al.*, 1980); (2) that the bone interfaces reject the prosthesis–cement composite as unphysiological (Huiskes, 1979); or (3) that the stress-shielding on the bone by a 'stiff' prosthesis results in detrimental bone re-modelling (Tarr *et al.*, 1979). There are obviously many complex parameters to consider, including clinical, surgical, mechanical, and biological aspects (Clarke *et al.*, 1980).

Recent radiographic studies (Gruen *et al.*, 1979) have suggested that one of the first visible signs of loosening is a radiolucent line at the proximal-lateral aspect of the stem–cement interface or a transverse crack of the cement at the tip of the stem. The former represents a separation of the stem away from the cement, which must involve a longitudinal fracture of the cement. Since longitudinal fractures of the cement cannot be seen on standard radiographs, this raises the question of what type of cement stresses are produced during normal activities. Was the first visible crack the result of failure of proximal-medial ('calcar') bone support or an overload due to failure elsewhere, for example, other undetected cement cracks? It is therefore important to analyse the type and magnitudes of stresses present in bone and cement and how they are influenced by the design and mechanical behaviour of the stem. It is implicitly assumed, for the purpose of this discussion, that the stem alloy material is strong enough to eliminate stem fracture as anything but a secondary problem which is always preceded by the loosening phenomenon (Muller, 1974; Gruen *et al.*, 1979).

Many studies have compared the predicted overall performance of a femur implanted with one or more stem types with varied results. Weightman (1976) studied five stems of varied sizes and alloys (stainless steel and cobalt-chromium-molybdenum) and concluded that the stiffer stems provided a significant stress-shielding effect. In contrast, Oh and Harris (1978) conducted similar tests on a range of six stem types and concluded that the variations in stem size and shape, i.e. structural stiffness, had little effect on the strains measured in the bone. Svensson *et al.*, (1977) apparently confirmed the latter results when they varied the material properties in a two-dimensional computer model. They found the bone strains apparently unchanged. Since considerable difference of opinion exists as to the state of stress in the proximal femur following total hip replacement, it may well be that detailed three-dimensional models of proximal femur are required to provide a closer approximation to the stresses in the bone and cement structures, particularly where prosthesis design changes are modelled (Tarr *et al.*, 1979, 1980; Crowninshield *et al.*, 1979, 1980). The objective of the study described in this chapter was to compare, using three-dimensional finite element methods, examples of very stiff (large cobalt-chromium-molybdenum alloy prosthesis), moderately stiff (small cobalt-chromium-molybednum alloy prosthesis), and flexible (small titanium, 6-aluminum, 4-vanadium prosthesis) femoral components cemented in the proximal femur.

16.2 MATERIALS AND METHODS

16.2.1 Finite element analysis

Four models were generated for the computer analysis. The first two duplicated the exact geometry of a natural femur inplanted with stem-type prostheses. One model included a small stem with a relatively low-modulus alloy (Class I; Ti-6Al-4V) design as typified by the STH (Sarmiento *et al.*, 1979) or the Charnley higher-modulus alloy (Class II, cobalt-chromium-molybdenum) types, and the second model included a much larger stem reminiscent of the CAD (Class III) type. These finite element meshes constituted the models used for parametric design comparisons (Figure 16.1). Each mesh contained 1215 nodes and 896 eight-noded, three-dimensional solid, isoparametric, hexahedron elements.

Mesh 3 was a model of a proximal femur without total hip replacement and was used to predict stress distribution for the natural intact femur. Mesh 4 replicated the exact geometry of a proximal femur with an STH prosthesis used for an experimental strain-gauge validation study. This mesh contained 1395

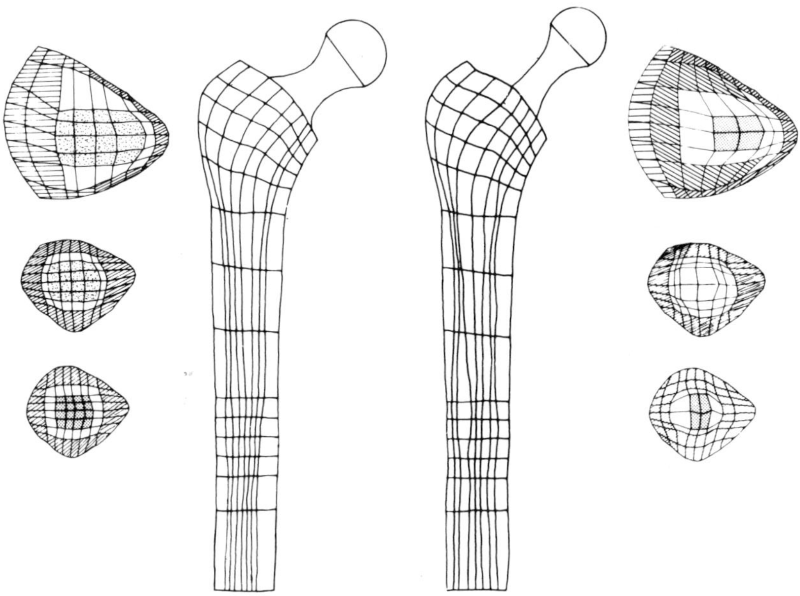

Figure 16.1 Finite element representations

nodes and 1032 elements. The actual femoral geometry for all meshes was determined by sectioning femurs and inputting internal and external boundaries as well as the cancellous and cortical bone variations.

Bone cement completely surrounded the prosthesis and extended approximately 15 mm below the tip of the stem. The distal bone was constrained well below the prosthesis (near midshaft). All interfaces were assumed perfectly bonded. This is not a realistic simulation, but necessary because of the available element library and since a linear analysis was carried out. All materials were assumed isotropic and linearly elastic with values determined from the literature (Table 16.1).

Table 16.1 Comparison of Young's modulus of elasticity and Poisson's ratio for various materials included in the finite element computer model

Material	Young's modulus E (kN/mm^2)	Poisson's ratio v
Stem (Ti-6A1-4V)	110.0	0.30
Stem (Co-Cr)	220.0	0.30
Cortical Bone	15.5	0.28
Cancellous Bone I	0.6	0.32
Cancellous Bone II	0.3	0.32
Bone Cement (PMMA)	2.1	0.35

Two normalized load cases were simulated for each run of the parametric analysis. Load case 'A' had a 1 N load applied to the femoral head in a medio-lateral plane, 18° from vertical. Load case B had the same load as case A' with the addition of a resultant abductor muscle force (one-third times the resultant hip force) applied to three elements in the greater trochanteric region. Analysis was performed using the SAP finite element computer program. Normal and shear stress results are presented for a load of 2000 N which represents a typical peak hip-load for normal walking.

16.2.2 Experimental verification

To verify the results of the finite element model, a human adult femur was implanted with an STH prosthesis and instrumented with strain gauges. The distal third of the femur was sectioned and the remaining proximal portion potted vertically in a steel cylindrical base support (75 mm dia. × 60 mm). Seven strain gauges (WA-06-125AD-120) were bonded to the external bone surface, four medially and three laterally. An STH curved stem femoral component was also prepared routinely for strain gauging. Six strain gauges (EA-06-125BB-120) were bonded to the metal stem, three medially and three laterally (Figure 16.2a). Gauge bonds were cured overnight.

The following day the femoral neck was osteotomized and the medullary canal prepared according to operative guidelines for insertion of the prosthesis. Since the computer model assumes a complete bond at all interfaces, epoxy was used as the cement layer to simulate this condition.

Three static load cases were tested. For load directions of 18° (Load Case 2), 40° (Load Case 1), and 74° (Load Case 3) to the prosthetic neck, incremental loads were applied to the femoral head and strain values recorded for all thirteen gauges.

The bone with prosthesis was radiographed and sectioned so that the exact geometry could be duplicated and the location of all gauges would correspond to a specific finite element in the computer model. The material properties of the epoxy which are similar to acrylic cement were used in the analysis. Data from the fine-mesh model (Mesh 4) described earlier was then compared to the experimental results.

16.3 RESULTS

Comparative results of the experimental and theoretical models showed satisfactory correlation of strain data in both magnitude and direction for all three load cases (Figure 16.2b). Comparisons for cortical bone strains were found to be more accurate than metal stem strains, probably owing to the lower mesh density of the stem (only two elements wide). It was therefore concluded that comparisons of finite element solutions for similar meshes would be accurate for this study of bone and cement stresses.

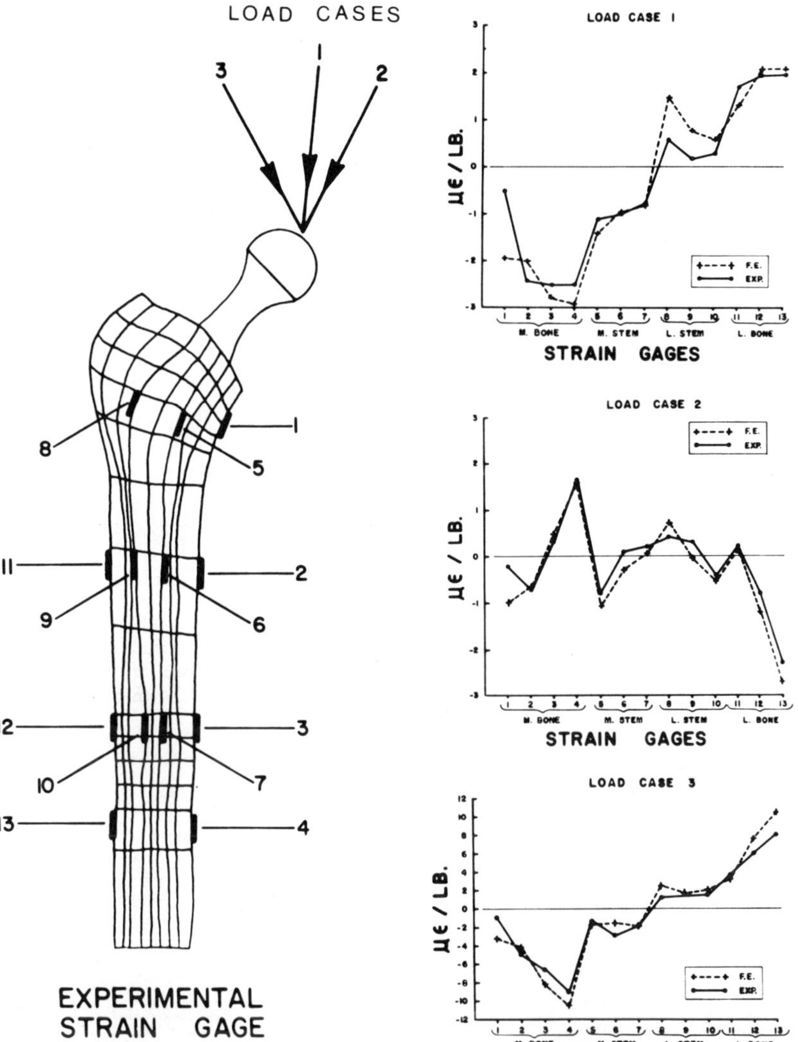

Figure 16.2 Experimental strain gauge locations

Comparison of computed stresses for the various materials of the bone-cement-prosthesis system are listed in Table 16.2 for a load of 2000 N (approximately three times body weight). The largest prosthesis naturally had the lowest ultimate compressive, tensile and shear stress in the stem. However, the titanium alloy design had the next lowest stresses due to the reduction in the metal's modulus of elasticity. Failure stresses for stem alloys, cement, and bone are listed in Table 16.3. Thus for the highest predicted stem stresses (Class II) the metal is only between 10–20% of its fatigue limit. Titanium alloy stem stresses were only 5–10% of reported fatigue limits.

Table 16.2 Predicted stresses (N/mm^2) for load case one with a three times body weight joint force (2000 N). Numbers in parentheses are percentages relative to Class I predicted stress levels. Class I refers to small femoral component made of titanium alloy, Class II the same stem geometry as Class I made of cobalt alloy, and Class III a large stem, cobalt alloy prothesis

Stresses Location and direction	Class I		Stem type Class II		Class III	
Max stem tension, axial	9.80	(100)	14.22	(145)	8.20	(84)
Max stem compression, axial	– 40.00	(100)	– 56.00	(140)	– 18.70	(47)
Max stem shear	18.30	(100)	22.40	(122)	6.50	(36)
Max cement tension, axial, distal-medial	1.30	(100)	1.54	(118)	2.48	(191)
Max cement compression, axial, distal-lateral	– 4.82	(100)	– 5.10	(106)	– 6.02	(125)
Max cement compression, axial, proximal-medial	– 5.14	(100)	– 3.64	(71)	– 1.00	(19)
Bone calcar, axial	– 7.64	(100)	– 5.34	(70)	– 1.60	(21)
Bone calcar, hoop	2.58	(100)	1.78	(69)	0.64	(25)

Table 16.3 (a) Comparison of failure criteria for metal alloys used in total joint replacements. (b) Comparison of failure criteria for bone and acrylic cement. All stresses shown in N/mm^2

(a)

Material	Yield	Fatigue
Colbalt chrome, cast	470	270
Titanium-6A1-4V	780	550

(b)

Material	Tension	Compression	Bending
Acrylic cement	25	90	
Cortical bone	90	140	175

Bone stresses in the proximal-medial calcar support area were reduced following total hip arthroplasty. Of the three stem types, the titanium-alloy stem transferred more axial stress to this region. As stem stiffness increased, due to increased modulus of elasticity and/or increased cross-sectional geometry, the calcar consequently was required to carry less of the load in the system (Table 16.2 and Figure 16.3). Compared to a femur without prosthetic replacement, axial

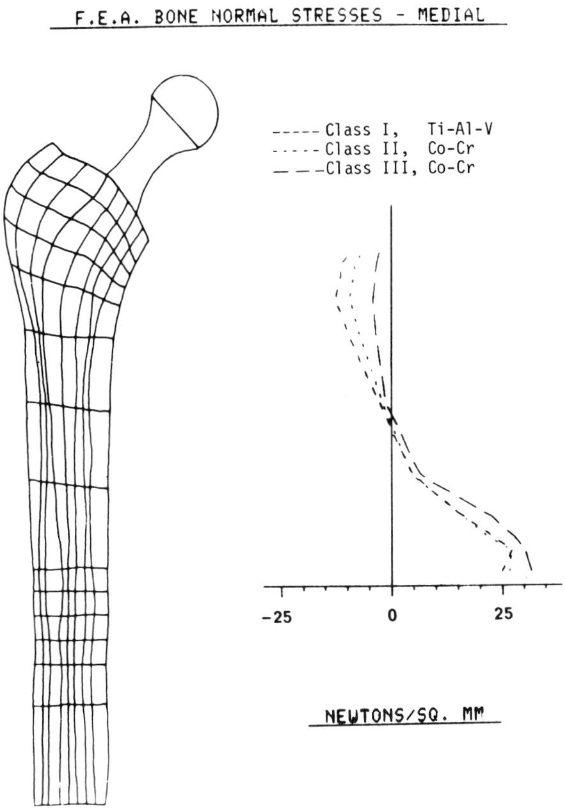

Figure 16.3 Finite element analysis bone normal stresses–medial

bone stresses in the calcar region for the titanium stem are approximately 30% of normal. For Class II stems these stresses are approximately 21% of normal, and for the largest stem (Class III) these stresses are only approximately 6% of normal. At the distal end of the prosthesis, bone stresses were tensile on the medial border (Figure 16.3) and compressive on the lateral side. Compared to Class I and II stems, the stiffer Class III stem produced the highest axial bone stresses in the region of the tip. These stresses were 39% higher on the medial cortical wall and 20% higher on the lateral cortical wall than the stresses produced by either the Class I or II stems.

When the abductor muscle force was combined with femoral head loading, the resultant load axis was shifted leaving the entire medial cortical bone in axial compression (Figure 16.4). Comparing the same stem design with titanium alloy and cobalt-chromium alloy showed a similar pattern as in load case A of

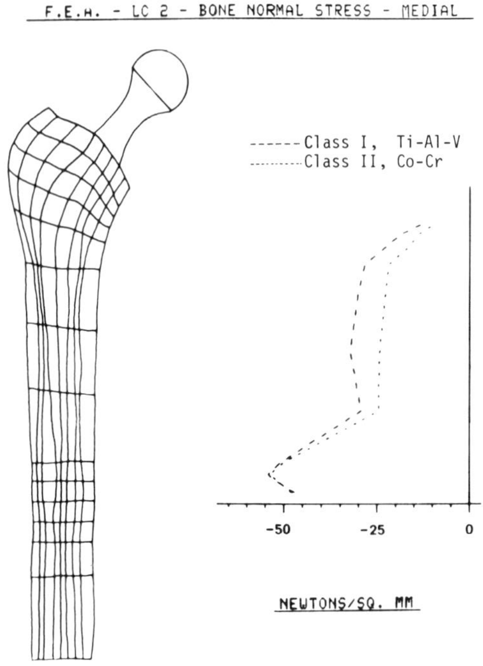

Figure 16.4 Finite element analysis. LC2. Bone normal stress – medial

increased load transfer to the proximal medial cortical bone. The titanium alloy stem transferred approximately 25 % higher axial compression stress to the bone in this region than the Class II stem and approximately 60 % higher compressive stress than the Class III stem.

The distribution of stresses throughout the cement sheath also showed differences between the three prostheses analysed Figure 16.5). The titanium stem produced the highest axial compressive cement stresses in the proximal-medial region. For a load of three times body weight, the axial compressive stress in the cement in the proximal medial (calcar) region was 6 % of the ultimate failure strength with the titanium alloy stem, compared to 4 % of ultimate strength with the cobalt-chromium stem. The large stem showed the lowest cement stress of all in the proximal-medial region. Proximal-lateral cement (tensile) stresses were negligible in all cases.

The cement stresses near the stem tip were of considerable interest. The large stem design produced the highest axial stresses in both compression (-6.02 N/mm^2) laterally and tension (2.48 N/mm^2) medially (Figure 16.5 and Table 16.2). For a 2000 N load these values were 7 % and 10 % of reported failure criteria for cement. Class I stems (i.e. with the titanium alloy) gave the lowest

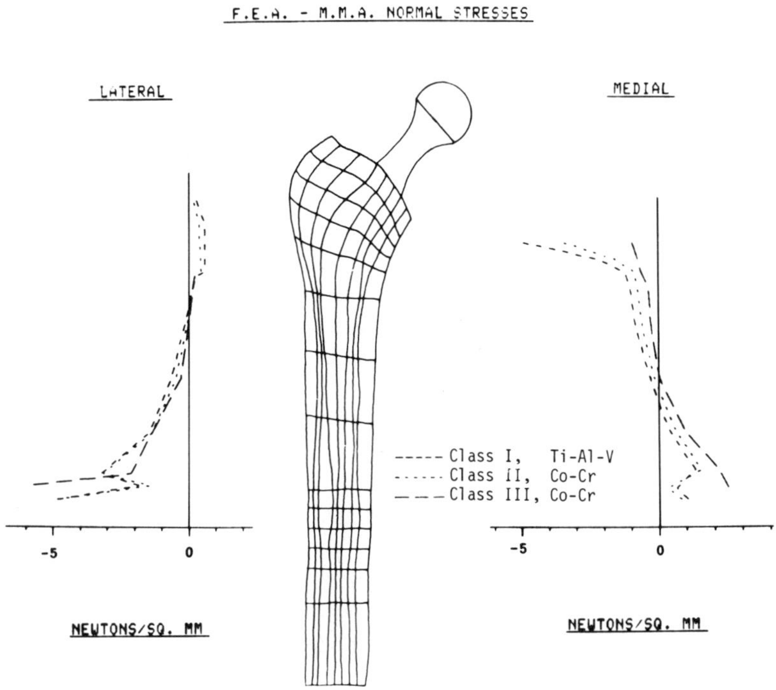

Figure 16.5 Finite element analysis. M.M.A. normal stresses

predicted cement stresses distally, $-4.82\ \mathrm{N/mm^2}$ laterally and $1.30\ \mathrm{N/mm^2}$ medially, which are both 5 % of reported failure strength of cement. Since it was already shown that the larger, stiffer stems transferred load distally, by-passing the proximal bony support and loading the distal cortex, it also follows that the cement in the distal regions would be subjected to increased stresses as predicted.

16.4 DISCUSSION

There is now considerable information in general as to how the stresses are transferred from the femoral prosthesis to the cement and than the adjacent bone. What is not clear is the influence of prosthesis design on the success or failure of the composite system. Huiskes (1980) and others have stated only that we should 'minimize' the stresses in the cement sheath and 'optimize' the bone stresses. Here we have attempted to determine, for prostheses typical of current clinical usage, whether the stem design affected the resultant cement stresses, and if so, were these stresses in the critical range for cement failure.

For the tensile stress criteria (Table 16.3b), the cement has been reported to fail at approximately 25 N/mm^2. The predicted peak tensile cement stresses (at the stem tip) in the model (Table 16.2) only varied from 1.3 to 2.5 N/mm^2, increasing from the flexible to the stiffer stems. In the proximal lateral region, the cement stresses in all cases were also tensile but so small as to be negligible. This does not therefore appear to be a significant enough change to justify any concern for this model.

Adjacent to the stem collar, the maximum compressive cement stresses decreased from 5.1 N/mm^2 for the flexible to 1.0 N/mm^2 for the stiff stem. This is a large relative change but the compressive cement stresses required to cause failure are as high as 90 N/mm^2, so this again does not offer much insight as to a possible failure criteria to distinguish clinical success or failure of these designs. In fact the cement may not be the weak link in a well-fixed stem. It may not be prejudiced until bone remodelling in time creates loosening at the cement–bone interface. Then and only then do the cement stresses rise significantly to prejudice the integrity of the cement sheath.

In assessing the magnitudes of the predicted cement stresses (Table 16.2) and their failure criteria (Table 16.3) there are several other points to consider. In the clinical situation, the effects of poor cement technique at surgery and possible degradation effects in its hostile environment may have a marked effect on its mechanical properties. However, there are no laboratory data for comparison. It may also be argued that the cement may fail at much lower stresses in the fatigue mode (Crowninshield *et al.*, 1980). This is no doubt true, but the magnitudes of the tensile cement stresses in the stem designs selected are relatively close so that both designs would probably undergo the same failure mechanism, rather than one failing preferentially to the other.

On the other hand it may be argued that the applied loads are unrealistically low. In this study we selected 2000 N peak load, which corresponds to most modelling studies (Clarke *et al.*, 1980) and represents a value of three times body weight as measured in hip-joint patients (Rydell, 1966; Paul *et al.*, 1975). Therefore the magnitudes of the applied loads appear appropriate.

Alternatively, it could be argued that the critical load case was not selected. Many different loading conditions have in fact been used (Clarke *et al.*, 1980; Berme and Paul, 1979) but there is no indication as yet as to their relative significance or clinical relevance.

Another possible criterion for failure is the stress shielding of the bone under the collar of the prothesis. If this bone remodels or reabsorbs due to the stress protection offered by the stem, it may well prejudice the intergrity of the composite system (Charnley, 1975; Bechtol 1975). The flexible STH stem provided the most loading of the bone (2.6 N/mm^2, hoop; 7.6 N/mm^2, axial) compared to the stiffer cobalt-chromium design (0.6 N/mm^2; hoop; 1.6 N/mm^2 axial). Currently there is no way of assessing the clinical relevance of the predicted values. The actual significance of prosthesis design on bone remodelling will only come from animal models or detailed clinical and radiographic studies.

The theoretical analyses have demonstrated that a decrease in the elastic modulus of the stem, as with titanium alloy, generally increases the stresses within the proximal part of the cement and decreases stress within the cement at the stem tip. Consequently there is more transfer of stresses onto the proximal-medial ('calcar') femoral cortex and less into the bone around the stem tip.

It is interesting to contrast these theoretical results with reported observations from radiographic analysis of total hip replacements (Tables 16.4 and 16.5). It should be noted that comparison of results from the various radiographic evaluations is frequently difficult due to varying patient selection, patient population, surgical and cement techniques, as well as varying criteria for identifying and interpreting radiographic changes, variations in radiological techniques, and insufficient serial X-ray films.

Table 16.4 Comparison of severe calcar resorption incidences of Class I flexible stems with Class II rigid femoral components of similar structural geometry

	'Calcar' resorption		
Reference	THR type	Maximum X-ray follow-up mos.	'Severe' resorption (criteria %)
Nicholson (1973)	Charnley (Co-Cr-Mo)	48	> 5 mm, 5.6
Beckenbaugh and Ilstrup (1978)	Charnley (Co-Cr-Mo)	84	> 2 mm, 13.7
Sarmiento *et al.*, (1979)	Charnley (316L steel)	84	> 3 mm, 13.3
Gruen *et al.*, (1980)	STH (Ti-6A1-4V)	52	> 3 mm, 2.1

Table 16.5 Comparison of radiographic evidences of cortical hypertrophy associated with Class I flexible titanium alloy stems and Class II rigid femoral components

	Cortical hypertrophy		
Reference	THR type	Maximum X-ray Follow-up mos.	Incidence
Nicholson (1973)	Charnley (Co-Cr-Mo)	48	10.8
Salvati *et al.*, (1976)	Charnley (Co-Cr-Mo)	84	31.0
Beckenbaugh and Ilstrup (1978)	Charnley (Co-Cr-Mo)	84	1.2
Gruen *et al.*, (1980)	STH (Ti-6A1-4V)	52	0.8

The incidence of 'severe' (i.e., greater than 2 or 3mm) resorption for the Co-Cr-Mo Charnley prosthesis (Class II) ranges from 5.6% (Nicholson, 1973) to 13.7% (Beckenbaugh and Ilstrup, 1978); on the other hand, the incidence of 'severe' resorption for the titanium alloy STH prostheses (Class I) is reduced to 2% (Gruen *et al.*, 1979). It remains to be seen if the incidence will or will not increase with longer X-ray follow-ups. It is acknowledged that many other factors may contribute to the mechanism of 'calcar' resorption such as pre-operative aetiology, patient sex, weight, and activity, stem alignment (varus versus valgus), and cement techniques (such as thicker proximal-medial acrylic support and/or 'optimal' fixation of the distal stem). It is quite possible that the resorption phenomenon is related to various combined interactive parameters rather than a single factor. This needs to be carefully evaluated with reasonable X-ray interpretation as well as proper statistical analysis considering the multifactorial problem.

Another radiographic phenomenon of bone remodelling often associated with apparently well-fixed stem-type femoral components is cortical hypertrophy (i.e., thickening) at the level of the stem tip. This has been attributed to be 'a manifestation of Wolff's law caused by artificially raised stress at the point of sudden transition from the flexible distal femur to the artificially stiffened proximal femur containing prosthesis and cement' (Charnley, 1979).

The incidence of cortical hypertrophy with Co-Cr-Mo Charnley prostheses as reported in the literature ranges from 1.2% (Beckenbaugh and Ilstrup, 1978) to 31% (Salvati *et al.*, 1976). The low incidence in the Mayo series is a reflection of the reported incomplete filling of the distal intramedullary canal at the stem tip, which consequently led to loosening at the cement–bone interface rather than being transferred into the femoral cortex as in the case of well-fixed stems.

A recent detailed radiographic survey of a titanium alloy Charnley-type prosthesis with lower flexural stiffness (Class I) indicated a negligible incidence (less than 1%) of cortical hypertrophy even in the presence of well-fixed components (Gruen *et al.*, 1979).

Therefore this clinical assessment would tend to support the theoretical conclusions that the lower flexural stiffness of the titanium alloy prosthesis does provide a more even load distribution into the femoral cortex as indicated by the lower incidence of 'calcar' resorption and cortical hypertrophy.

16.5 CONCLUSION

A comparison of large, stiff; small, moderately stiff; and small, flexible femoral total hip components cemented in the proximal femur has been investigated using a three-dimensional finite element model, and validated using experimental strain-gauge techniques. A load of three times body weight (2000 N) 18° from vertical with and without abductor forces was utilized to compute maximum stresses in the bone, cement, and metal.

For a well-fixed total hip replacement, the results show stem stresses to be between 5 and 20 % of the reported fatigue limits for the metals analysed. Cement stresses in the proximal lateral region were negligible for all stems tested. In looking at the failure stress for cement in compression, the proximal-medial cement stresses were only 4–6 % of reported failure criteria, and in tension distally only between 5 and 10 % of failure criteria.

The proximal-medial bone stress was 30 % of normal for titanium alloy, small stems; 21 % of normal for moderately stiff, small stems; and only 6 % of normal for large, stiff stems. Distal bone stress on the medial cortex was 39 % higher for the stiff stem compared to the flexible stem and laterally 20 % higher (stiff compared to flexible stem).

Viewing these results in the light of stem design or femoral component failure criteria, the cement does not appear to be the weak link for the well-fixed total hip replacement regardless of stem design. If failure initiation does not begin within the cement in a well-fixed hip, do physiological changes in the bone reverse this situation until loosening results at the cement–bone interface and the cement is then prejudiced? Some advantages of load transfer to bone are gained with the use of flexible stems, but the significance of these changes can only be confirmed or disputed with detailed clinical and radiographic follow-up.

ACKNOWLEDGEMENTS

Thanks are due to the Orthopaedic Foundation of Los Angeles and the Department of Orthopaedics, University of Southern California, for their financial support.

REFERENCES

Bechtol, C. O. (1975) Failure of femoral implant components in total hip replacement operations, *Orthop. Rev.*, **4,** 23–29.

Beckenbaush, R. D., and Ilstrup, D. M. (1978) Total hip arthroplasty. A review of three hundred and thirty-three cases with lons follow-up, *J. Bone Joint Surg.* **60**(A), 306–313.

Berme, N., and Paul, J. P. (1979) Load actions transmitted by implants, *J. Biomed. Eng.*, **1,** 268–272.

Bonfield, W., and Datta, P. K. (1974) Young's modulus of compact bone, *J. Biomech.* **7,** 147–149.

Charnley, J. (1975) Fracture of femoral prostheses in total hip replacement. A clinical study, *Clin. Orthop.*, **111,** 105–120.

Charnley, J. (1979) *Low Friction Arthroplasty of the Hip. Theory and Practice*, Springer-Verlag, New York.

Clarke, I. C., Gruen, T. A. W., Tarr, R. R., and Sarmiento, A. (1980) Finite element analyses studies of total hips versus clinical reality, In: *International Conference Proceedings of Finite Elements in Biomechanics* (ed. B. R. Simon) Tucson, pp. 487–510.

Crowninshield, R. D., Brand, R. A., and Johnston, R. C. (1979) An analysis of femoral stem design in total hip arthroplasty, *Trans. Orthop. Res. Soc.*, **4,** 33.

Crowninshield, R. D., Brand, R. A., Johnston, R. C., and Milroy, J. C. (1980) An analysis of femoral component stem design in total hip arthroplasty, *J. Bone Jt. Surg.* **62**(A), 68–78.

Curry, J. D. (1970) The mechanical properties of bone, *Clin. Orthop.*, **73,** 210–231.

Gruen, T. A., McNeice, G. M., and Amstutz, H. C. (1979) Modes of failure of cemented stem-type femoral components – a radiographic analysis of loosening, *Clin. Orthop.*, **141,** 17–27, 1979.

Huiskes, R. (1980) Some fundamental aspects of human joint replacement. Analyses of stresses and heat conduction in bone-prosthesis structures, *Acta Orthop. Scand.*, Suppl. No. 185.

Kummer, B. K. F. (1972) Biomechanics of bone: Mechanical properties, functional structure, functional adaptation, In: *Biomechanics: Its Foundation and Objectives* (eds. Y. C. Fung and M. Anliker), Prentice-Hall, Inc. Englewood Cliffs, pp. 237–271.

Lautenschlager, E. D., Marshall, G. W., Marks, K. E., Schwartz, J., and Nelson, C. L. (1976) Mechanical strength of acrylic bone cements impregnated with antibiotics, *J. Biomed. Mater. Res.*, **10,** 837–845.

Lee, A. J. C., Ling, R. S. M., and Wrighton, J. D. (1973) Some properties of polymethylmethacrylate with reference to its use in orthopaedic surgery, *Clin. Orthop.*, **95,** 281–287.

Muller, M. E. (1974) Late complications of total hip replacement, In: *The Hip* (Proceedings of the Hip Society), C. V. Mosoy, St. Louis, pp. 319–327.

Nicholson, O. R. (1973) Total hip replacement. An evaluation of the results and techniques, 1967–1972, *Clin. Orthop.*, **95,** 217–223.

Oh, I., and Harris, W. H. (1978) Proximal strain distribution in the loaded femur, *J. Bone and Joint Surg.*, **60**(A), 75–85.

Paul, J. P., McGrowther, D. A., and Clarke, I. C. (1975) Forces translated at the hip and knee joint of normal and disabled persons during a range of activities, *Acta Orthop. Belg.*, **41,** supp. 1, 78–88.

Pope, M. H., and Outwater, J. O. (1974) Mechanical properties of bone as a function of position and orientation, *J. Biomech.*, **7,** 61–66.

Reilly, D. T., Burstein, A. H., and Frankel, V. H. (1974) The elastic modulus for bone, *J. Biomech.*, **7,** 271–275.

Rohlmann, A., Bergmann, G., and Kolber, R. (1980) The relevance of stress computation in the femur with and without endoprostheses, *Proc. Int. Conf. on Finite Elements in Biomechanics* (ed. B. R. Simon), pp. 549–566.

Rydell, N. W. (1966) Forces acting on the femoral head-prosthesis. *Acta Orthop. Scand.*, **37,** Suppl. 88.

Salvati, E. A., IM, V. C., Alglieti, P., and Wilson (1976) Radiology of total hip replacements, *Clin. Orthop.*, **121,** 74–82.

Sarmiento, A., Turner, T. M., and Latta, L. L. (1979) Factors contributing to lysis of the femoral neck in total hip arthroplasty, *Clin. Orthop.*, **145,** 208–212.

Sarmiento, A., Zych, A. G., Latta, L., and Tarr, R. R. (1979) Clinical experiences with the titanium alloy total hip prosthesis, *Clin. Orthop.*, **144,** 166–173.

Svensson, N. L., Valliappan, S., and Wood, R. D. (1977) Stress analysis of human femur with implanted Charnley prosthesis, *J. Biomech.*, **10,** 581–588.

Tarr, R. R., Lewis, J., Ghassemi, F., Sarmiento, A., Clarke, I. C., and Weingarten, V. (1980) Anatomic three-dimensional finite element model of the proximal femur with total hip prosthesis, *Proc. Int. Conf. on Finite Elements in Biomechanics* Tucson, pp. 511–525.

Tarr, R. R., Lewis, J. L., Jaycox, D., Sarmiento, A., Schmidt, J., and Latta, L. L. (1979) Effect of materials, stem geometry and collarcalcar contact on stress distribution in the proximal femur with total hip, *Trans. Orthop. Res. Soc.*, **4,** 34.

Weightman, B. (1976) The stress in total hip prosthesis femoral stems: a comparative experimental study, In: *Advances in Artificial Hip and Knee Joint Technology* (eds. M. Schaldach and D. Hohmann), Springer-Verlag, New York, pp. 138–147.

Yamada, H., and Evans, F. G. (1970) *Strength of Biological Materials*, Williams and Wilkins, Baltimore.

Finite Elements in Biomechanics
Edited by R. H. Gallagher, B. R. Simon, P. C. Johnson, and J. F. Gross

Chapter 17

The Relevance of Stress Computation in the Femur with and without Endoprostheses

A. Rohlmann, G. Bergmann, and R. Koelbel

17.1 INTRODUCTION

Loosening of an implant is one of the common causes of failure of joint replacement. It seems that this is brought about when bone is subjected to stresses above (or below) a range which is favourable to the maintenance of bone mass and structure. In order to avoid this, endoprostheses should be designed and fixed in such a way that stresses in the bone are kept within the favourable range. Calculations related to this objective can, at the present time, best be performed using the finite element method (FEM). Several groups have used the FEM for stress analyses of the femur with and without endoprostheses. Three-dimensional models of the proximal end of the intact femur were investigated by Scholten (1975), Wood (1975), Olofsson (1976), and our group (Rohlmann *et al.*, 1979). The stresses in the femur and the implant were investigated by Roehrle *et al.*, (1977), Huiskes *et al.*, (1977), Crowninshield *et al.*, (1979), and Tarr *et al.*, (1979) using simplified spatial models.

Unlike most problems in engineering, stress calculations in bone require a number of additional assumptions about the determination of the border between cortical and cancellous bone and the material properties and their distribution as well as the support and loading of the model. These assumptions can influence the results considerably, as was shown in another work (Rohlmann *et al.*, 1979). We have included here experimental measurements of the strains to support the assumptions made in our FEM investigations.

17.2 MATERIALS AND METHODS

A three-dimensional model was developed of a complete femur. Both femora of the same individual obtained at post mortem were macerated. Comparison of X-ray pictures confirmed uni-axial symmetry in the femora.

Figure 17.1 Femur with 34 strain gauges . (Reproduced by permission of Fachverlag Schiele of Schön GMBH)

17.2.1 Experimental model

The right femur was used for the experimental investigation. The condyles of the right femur were embedded in aluminum box. Thirty-four strain guage rosettes were attached to the surface of the femur. In the area of the diaphysis (shaft) three strain guages each were attached to the circumference at several levels (Figure 17.1). The loading rig (Figure 17.2) was designed to allow loading of the femoral head (with the resultant hip joint force R) and loading of the greater

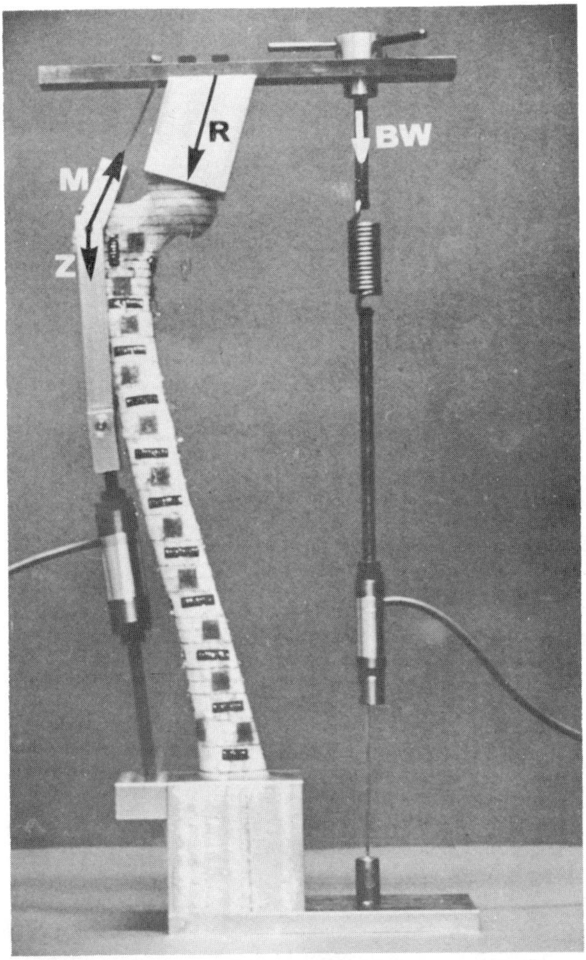

Figure 17.2 Femur with loading rig and force transducers. (Reproduced by permission of Ferdinand Enke Verlag)

trochanter (with the force *M* of the abductor muscles) corresponding to the stance-phase of walking. The rig acts as a two-sided lever: the body weight minus the weight of the standleg acts on one arm, the abductor force on the other. The femoral head then acts as the fulcrum.

In addition, variable tension can be applied between the greater trochanter and the femoral condyles. This allows simulation of the tension banding action of the tractus iliotibialis on the femur. This effect was shown by Pauwels (1965) in his photoelastic studies of plane femoral models.

Measurements were taken on a multichannel measurement system (Figure 17.3). The model was loaded with a 'body weight' (BW) of 150 N corresponding

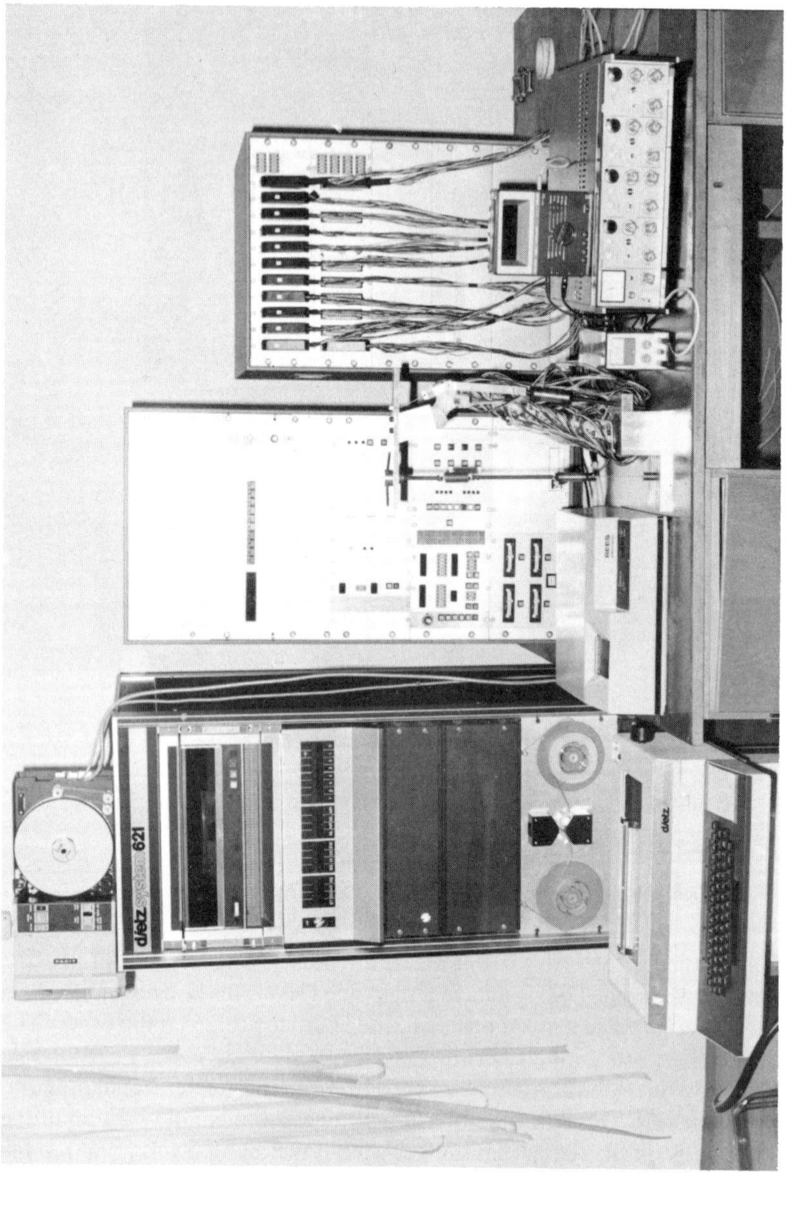

Figure 17.3 Test set-up with multichannel measuring system. (Reproduced by permission of Fachverlag Schiele of Schön GMBH

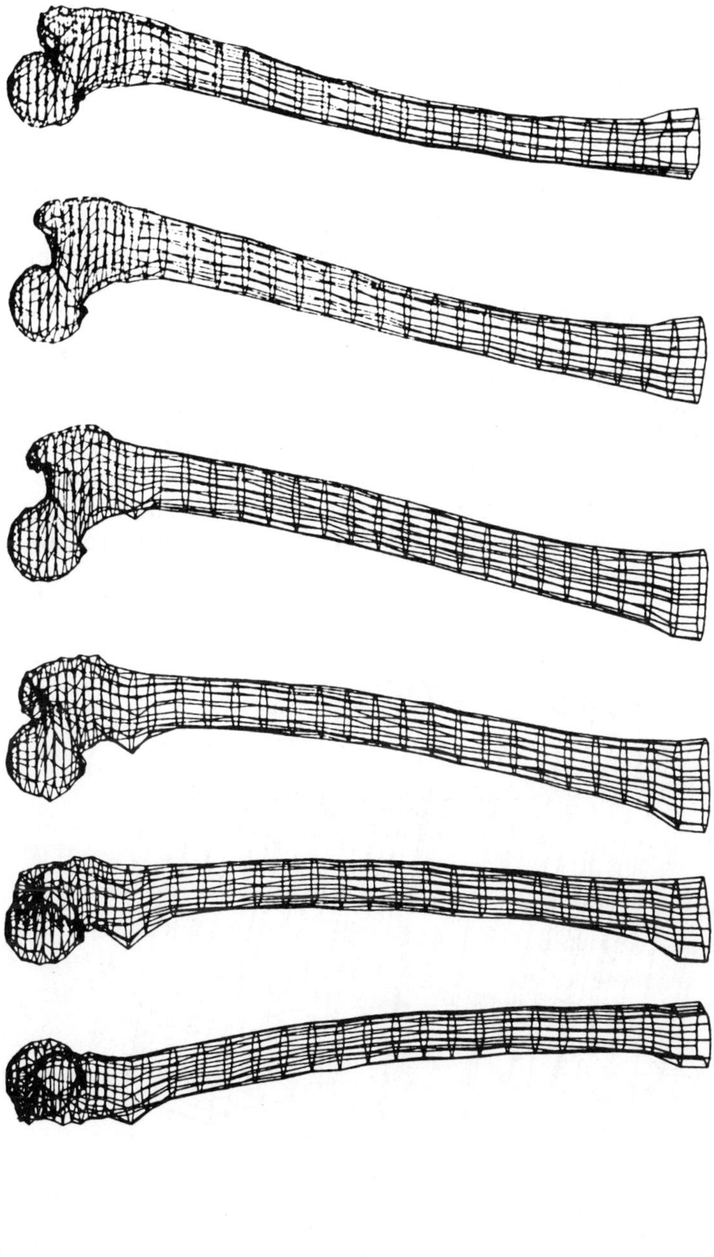

Figure 17.4 Outer net of elementation from different points of view. (Reproduced by permission of Ferdinand Enke Verlag

to a resultant hip joint force of 534 N and an abductor force of 394 N. While the load was applied, 102 strain measurements were taken and recorded. These measurements allowed the principal strains to be calculated at 34 points on the surface of the femur. Measurements were repeated with a tension banding force Z in the iliotibial tract of between 1 and 4 times BW. This corresponds to the possible *in vivo* range of this force.

17.2.2 Finite element model

The left femur was used for the FEM. This femur was completely embedded in epoxy resin and was saw-cut horizontally into 38 slabs. Each slab was radiographed and areas of equal density were determined using Konermann's method (Konermann, 1971). A stepped aluminum wedge was used to allocate material properties to each level of density. The FE model was composed of 1950 eight-noded isoparametric elements (using linear displacement functions) with 2532 nodal points and 7188 degrees of freedom. Calculations were performed on a CYBER 175 and required an average CPU-time of 20 minutes.

Figure 17.4 shows the surface net of elements viewed from different angles. The femur was rotated 30° in each view. The elements in one cross-sectional slab of the femur is shown in Figure 17.5. Forty-eight elements and 128 nodal points (64 each at the top and 64 at the bottom with three rings at 16 elements each) were used to model each slab.

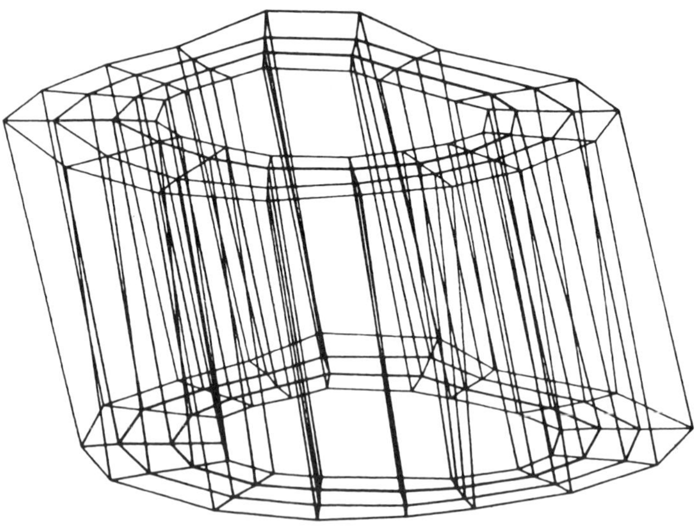

Figure 17.5 Elementation of a cross-sectional slab. (Reproduced by permission of Ferdinand Enke Verlag)

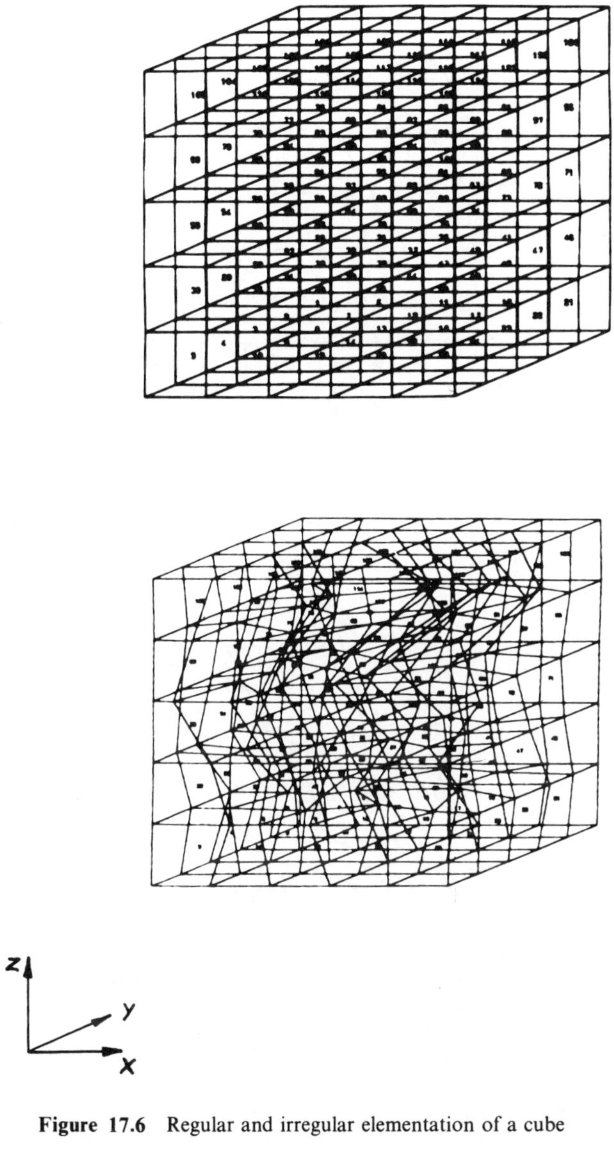

Figure 17.6 Regular and irregular elementation of a cube

The three-dimensional FE models were composed of volume elements with eight nodal points. In order to study the effect of elemental shape on calculated stress results, a cube was modelled. The cube was first modelled using cuboid elements (Figure 17.6a) and then modelled with elements in which the top and bottom faces were parallel, but of varying quadrangular shape (Figure 17.6b). The same number of elements (125) was used in both models. The models were

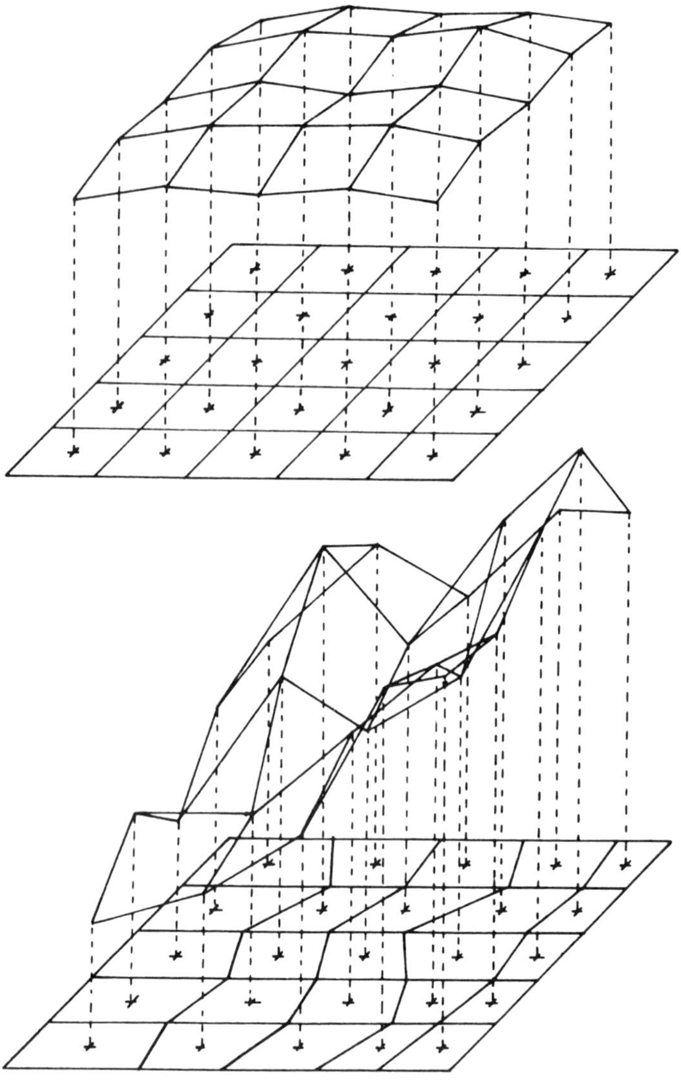

Figure 17.7 Normal stresses σ_z in the middle plane of Figure 17.6; left = for regular elementation, right = for irregular elementation.

loaded with a uniformly distributed load at the top surface and the nodal points of the bottom surface were fixed. For the first computation we used non-conforming eight-noded volume elements including linear displacement shape functions with internal variables. Figure 17.7 shows the distribution of the normal stresses in the third plane of nodes (viewed from below). When a uniform element mesh was used, the resulting stresses were nearly uniform as expected. In the second finite element model, the calculated stresses varied from 50 % to 180 % of the expected stresses (Figure 17.7). After incompatible modes were suppressed, the stresses in the model composed of irregular elements were found to be between 96 % and 104 % of the expected values. Thus the choice of element form used in the calculation of elemental stiffness is of great importance in the femoral models where a uniform or rectangular element mesh is not possible.

17.3 RESULTS

17.3.1 Principal Strains

Figure 17.8 shows a comparison of the principal strains on the surface of the femur determined experimentally and by FE analysis. Strain guage rosettes were attached to the circumference of the femur at 11 levels (at locations A, B, and C as indicated by the interrupted lines 1–11 in Figure 17.8). Loadings applied to the femur were the hip joint force R, the abductor muscle force M, and a tension banding force Z, equal to BW. The principal strains for these 11 levels are shown on the left-hand side of Figure 17.8 Maximum tensile strains are plotted to the right and minimum compression strains are plotted to the left. FE strain values are shown as dotted lines and the experimental strain values are plotted as solid lines. The difference in compression strain is greatest on the medial aspect and the difference in tensile strain is greatest on the anterior surface of the femur.

Figure 17.9 compares experimental and analytical values for the principal strains for the two-legged stance, i.e. the femoral head was subjected to a vertical load of one-third BW. The principal strains show a similar trend to Figure 17.8; however, the calculated strain values are smaller than the measured strain values at most levels.

The former results (Figure 17.8) indicate that in the shaft area of the femur, either the border between cortical and cancellous bone or the distribution of the material properties were not properly modelled. Therefore another computation was made in which the elastic modulus for cortical bone was changed. The modulus was decreased in direction of the medullary space. The results are shown in Figure 17.10, where the measured principal strains and the strains obtained by the FEM are much closer. The average of the difference between the measured results and the FE results (when related to the maximum value) is now only 8.5 % whereas it had been 14.8 % before. Making allowances for the uncertainty in the assumptions, the precision of the FE model is very good.

Possible reasons for variance between the measured results and the FE calculations were the following:

1. Insufficient knowledge of the inner architecture of the bone. The transition between cortex and cancellous bone is gradual. In the FE model a borderline had to be chosen more or less arbitrarily within the area of transition of

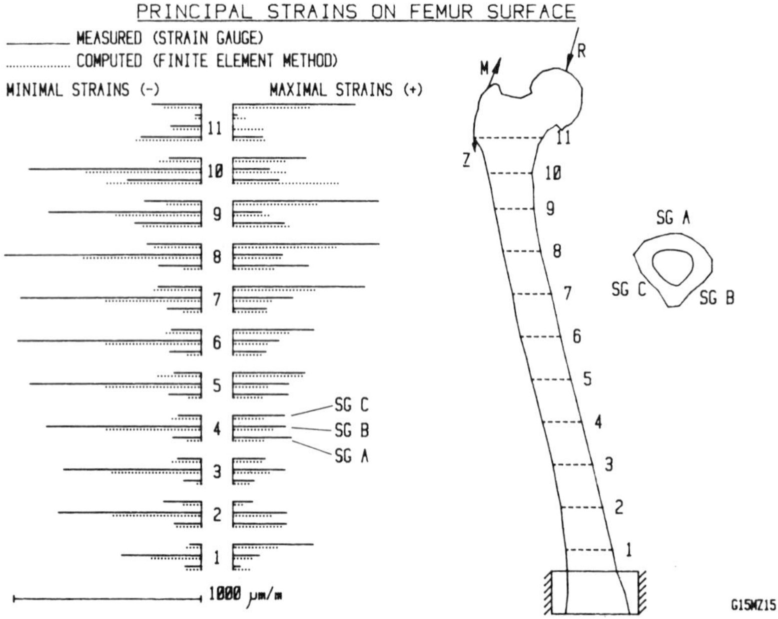

Figure 17.8 Comparison between measured and computed principal strains on femur surface for 11 levels for one-legged stance.

R = resultant hip joint force
M = abductor muscle force
Z = tension banding force
SG = strain gauge

densities. Small errors in the transfer of the geometry to the FE model have great effects, since the bone interface diameter enters the computation of bending stress to the power of 3.

2. Inaccuracy in material properties. Bone is non-homogeneous, anisotropic, and to some extent time-dependent in its behaviour. These properties and their distribution can only roughly be transferred to the FE model.

17.3.2 Deflections

Figure 17.11 shows the deflections of the nodal points in the frontal plane for various magnitudes of the force Z in the iliotibial tract. The interrupted outline in Figure 17.11 shows the model in its loaded state. Tension banding forces Z are from left to right $Z = 0 \times$ BW, $Z = 1 \times$ BW, $Z = 3 \times$ BW, $Z = 4 \times$ BW. The effect of the tension banding on the deflection, and hence the bending stresses of

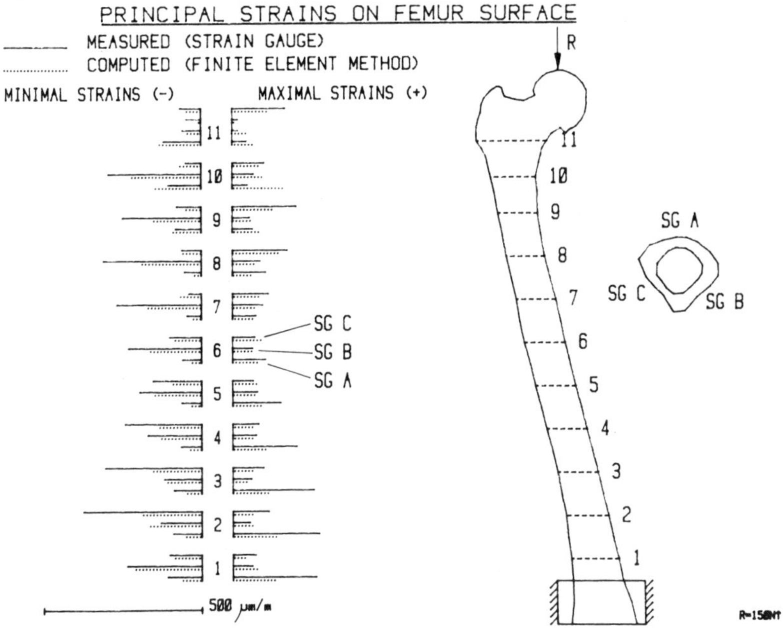

Figure 17.9 Comparison between measured and computed principal strains on femur surface for two-legged stance.

R = resultant hip joint force

SG = strain gauge

(Reproduced by permission of Ferdinand Enke Verlag)

the model, is quite pronounced. For small *Z* values, the upper end of the femur is deflected medially whereas for large *Z* it is deflected laterally.

Figure 17.12 shows the deflection of the nodal points in the sagittal plane. The loads were the same as in Figure 17.11. The deflections in the sagittal plane were much smaller than the deflections in the frontal plane.

17.3.3 Stresses in the Femur

Principal stresses on the femur surface in the frontal plane for the stance-phase of walking are shown in Figure 17.13. The calculated FE values were plotted perpendicular to the contour line. Positive values represent tension and point outward, whereas negative values point inward and represent compression. Variation of the tension banding force *Z* in the possible *in vivo* range causes a fundamental change of the principal stresses in the femur. The principal stresses in the shaft area are fairly constant when *Z* is zero and the femur is loaded in

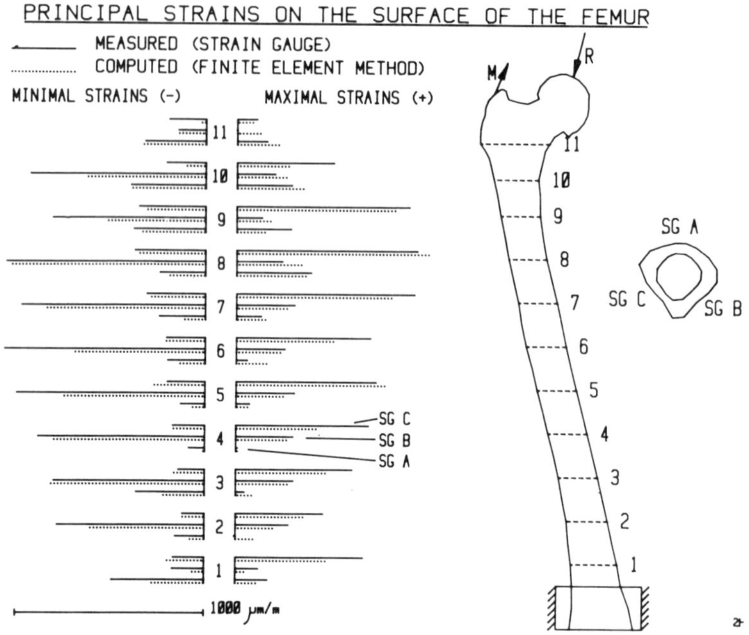

Figure 17.10 Comparison between measured and computed principal strains on femur surface for one-legged stance with changed elastic modules for cortical bone. (Reproduced by permission of Ferdinand Enke Verlag)

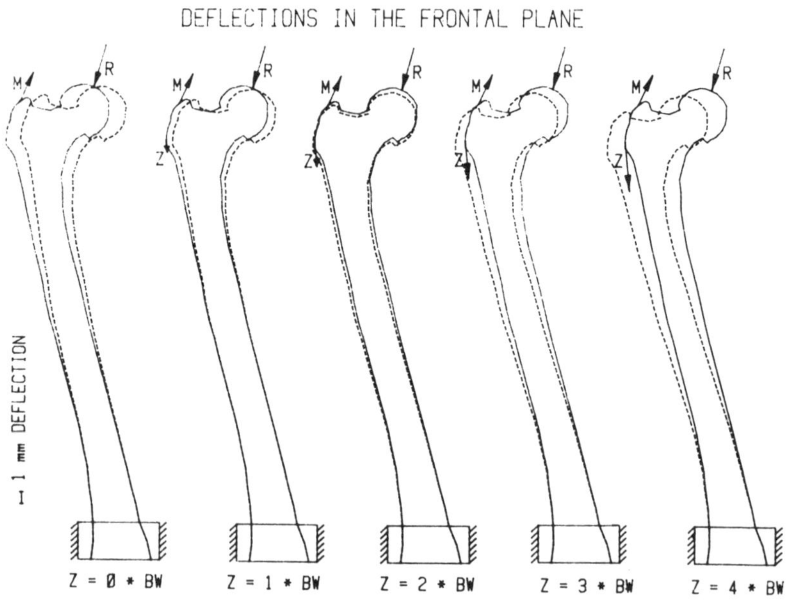

Figure 17.11 Deflections of the nodal points in frontal plane as a function of the force Z in the iliotibial band. (Reproduced by permission of Ferdinand Enke Verlag)

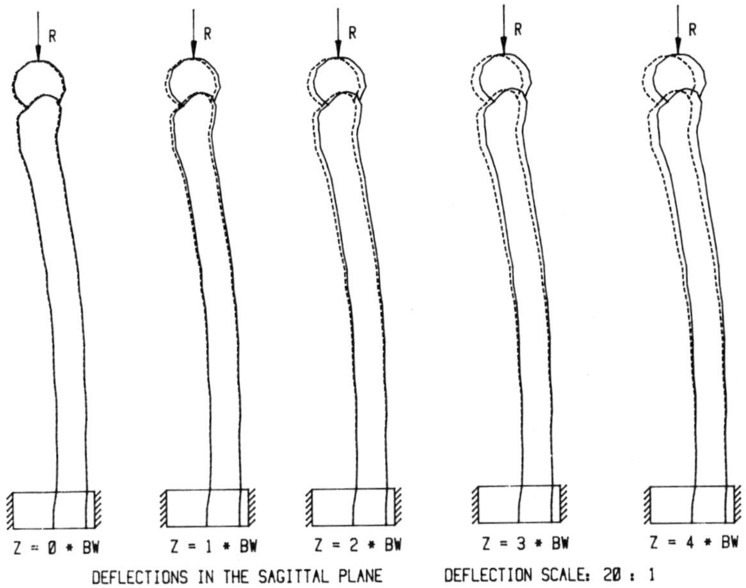

Figure 17.12 Deflections of the nodal points in the sagittal plane as a function of the force Z. (Reproduced by permission of Ferdinand Enke Verlag)

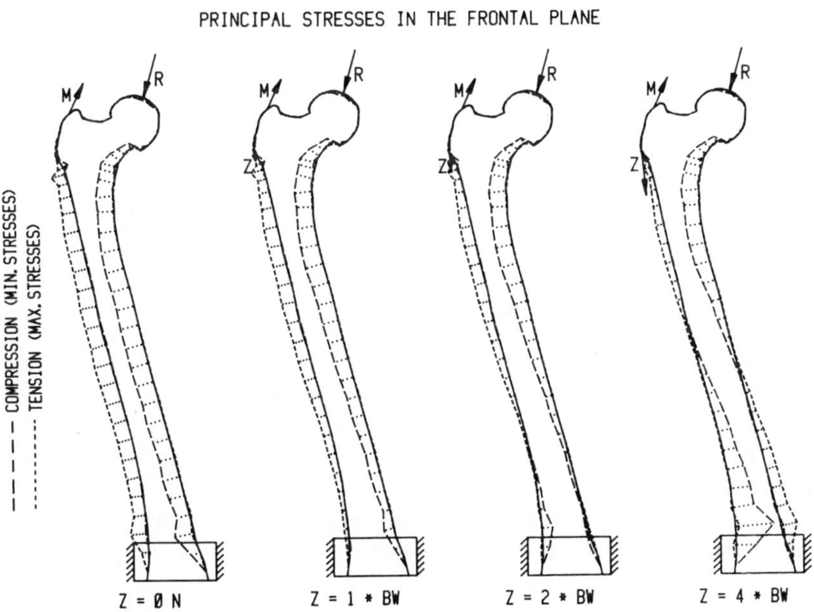

Figure 17.13 Calculated inflence of the tension banding force Z on principal stresses. Calculated values are charted at right angles to the contour line

bending. There are compressive stresses medially and tensile stresses laterally. Under increasing tension banding force Z, the principal stresses change in the shaft area. In the distal part the principal stresses even change sign when the iliotibial pull reaches $1.5 \times$ BW.

Figure 17.14 compares the stresses in the frontal plane for different loads of the femur. The comparison stresses are always positive. They are plotted at right angles to the contour line. Figure 17.14 (left) represents the femur loaded with only the resultant hip joint force R (stance-phase of walking) and the muscle forces M were neglected. The influence of the force Z in the iliotibial tract on the comparison stresses is seen in Figure 17.14. The force R ranges from zero, two times body weight, and four times body weight.

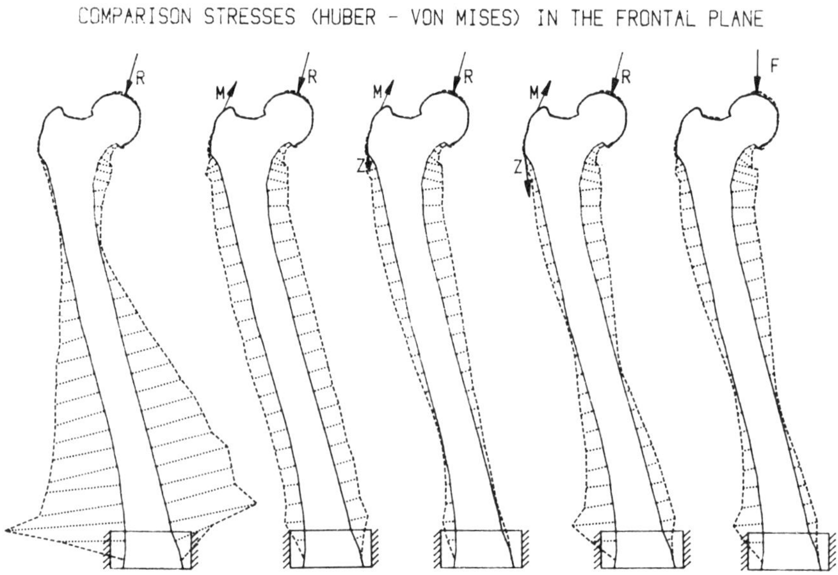

Figure 17.14 Comparison stresses in the frontal plane for different loads of the femur

On the right-hand side of Figure 17.14 the femur head was loaded vertically with a force F. The effect of these two load situations on the bone stresses indicates that attempts at optimization of endoprosthetic fixation will only make sense, if a variety of loading conditions that occur during human activities are simulated.

A Moore-type endoprosthesis with stem length of 160 mm was inserted next, again without cement and the best possible collar contact. Figures 17.15 and 17.16 show the strains for the 160 mm stem implant as dotted lines, and the strains for the 135 mm stem implant are shown as solid lines. The difference between the strains at the femoral surface is small.

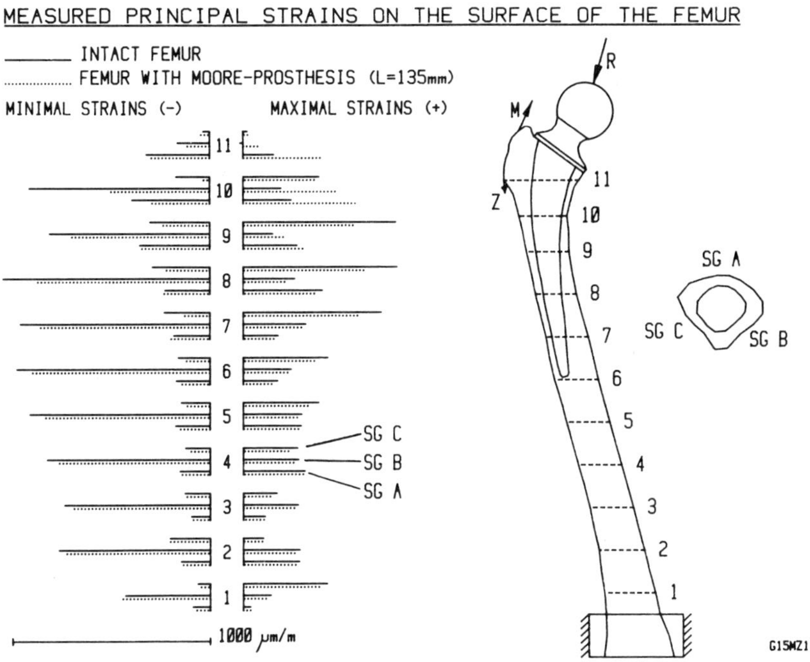

Figure 17.15 Comparison of measured principal strains on femur surface with lines) and without (solid lines) endoprosthesis

17.4 CONCLUSION

Changes in the principal strains on the surface of the femur may not be the ultimate criterion for evaluation of the effect of implant configuration. Stresses at the implant–bone interface are also important. The FEM is a practical method for determining these stresses. Investigation of bone-prosthesis models are currently under way in our laboratory.

When the FEM is used for stress analyses in orthopaedics (e.g., an analysis of the state of stress in a bone, with or without implant), the following points require particular attention:

1. Determination of bone geometry and its representation in the FE model.
2. Determination of a boundary between cortical and cancellous bone and its representation in the FEM.
3. Simulation of natural behaviour of the material bone.
4. Loading of the model.
5. Support of the model.
6. Elementation of the model.
7. Bonding between the elements at the implant–bone–cement interface.

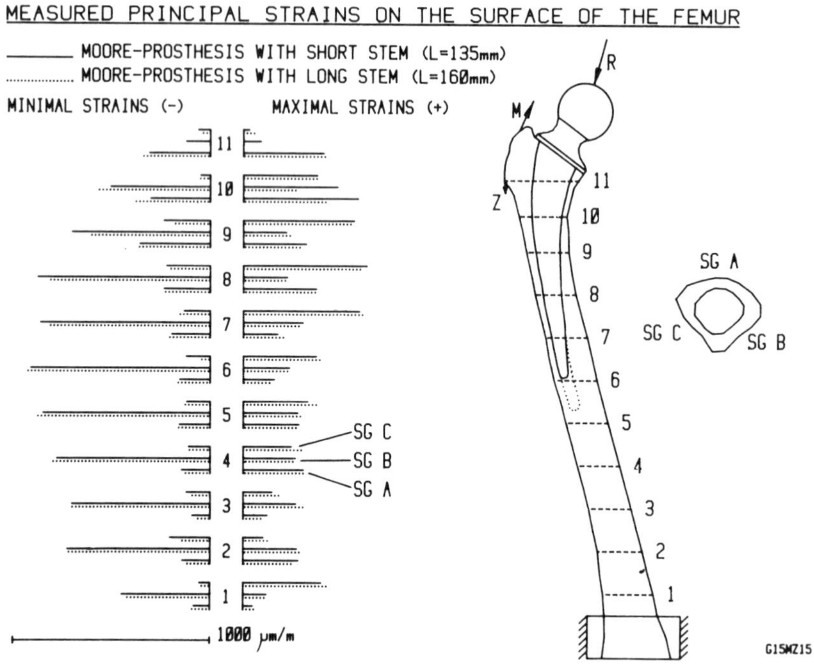

Figure 17.16 Comparison of measured principal strains on femur surface for Moore-prostheses with different length of stem

It appears that the FEM is a very appropriate method for comparative investigations. FE results alone, however, will not be sufficient if quantitative values are required. The structures in bone and joint replacement are highly complex, and it appears mandatory, therefore, that FE stress and strain results in bone will only be valuable if complemented by reliable experimental measurements.

ACKNOWLEDGEMENTS

We would like to express our gratitude to Dipl.–Ing. A. Rubert (Institut fuer allgemeine Bauingenieurmethoden, Technische Universitaet Berlin) for adaptating and running the FE-program, Dipl.–Ing. J. Hoffmann (Institut fuer allgemeine Bauingenieurmethoden, Technische Universitaet Berlin), and Dipl.–Ing. H. Herrmann (Institut fuer Luft– und Raumfahrt der Technischen Universitaet Berlin) for their help in the elementation of the models. The multichannel measuring system of the Institut fuer Tragwerkslehre, Technische Universitaet Berlin (Prof. Dr.–Ing. K. Dierks) was kindly put at our disposal for the experimental investigation.

Access to the SAP computer program was made possible by the Institut fuer Allgemeine Bauingenieurmethoden, Fachgebiet Statik der Baukonstruktionen (Prof. Dr.-Ing. G. Hees).

This work was financially supported by the Deutsche Forschungsgemeinschaft.

REFERENCES

Crowninshield, R. D., Brand, R. A., and Johnson, R. C. (1979) An analysis of femoral stem design in total hip arthroplasty, *Trans. ORS*, **4**, 33.

Huiskes, R., Slooff, T. J. J. H., Elangovan, P. T., and Banes, J. P. A. (1977) Finite element computer methods for design and fixation problems of orthopaedic implants, *Biomechanics VI-B*, (eds. E. Asmussen and K. Joergenson), Univ. Park Press, Baltimore, pp. 229–238.

Konermann, H. (1971) Quantitative Bestimmung der Material = verteilung nach Roentgenbildern des Knochens mit einer neuen fotografischen Methode, *Z. Anatomie.*, **134**, 13–48.

Olofsson, H. (1976) Three-dimensional FEM-calculation of elastic stress field in femur, Doctoral dissertation, Uppsala University.

Pauwels, F. (1965) *Gesammelte Abhandlungen zur funktionellen Anatomie des Bewegungsapparates*, Springer, Berlin.

Roehrle, H., Scholten, R., Sollbach, W., Ritter, G., and Gruenert, A. (1977) Der Kraftfluss bei Hueftendoprothesen, *Arch. orthop. Unfall-Chir.*, **89,** 49–60.

Rohlmann, A., Bergmann, G., and Koelbel, R. (1980) Aussagewert und Grenzen der Spannungsberechnung mit der Finiten-Elemente-Methode bei orthopaedischen Problemen, *Z. Orthop.*, **118,** 122–131.

Rohlmann, A. Bergmann, G., and Koelbel, R. (1979) Three-dimensional finite elemente analysis and experimental investigation of strain in the femur with and without prosthetic replacement, *Second Meeting of European Society of Biomechanics*, Strasbourg.

Scholten, R. (1975) Kraftflussberechnungen an Knochenstrukturen und Prothesen, *Med.-Orth. Technik*, **6,** 130–138.

Tarr, R. K., Lewis, J. L., Jaycox, D., Sarmiento, A., Schmidt, J., and Latta, L. L. (1979) Effect of materials, stem geometry and collar-colar contact on stress distribution in the proximal femur with total hip, *Trans. ORS*, **4,** 34.

Wood, R. D. (1975) Stress analysis of the femur, Ph.D. thesis, University of New South Wales, Australia.

Finite Elements in Biomechanics
Edited by R. H. Gallagher, B. R. Simon, P. C. Johnson, and J. F. Gross

Chapter 18

A Homeomorphic Finite Element Model of the Human Head and Neck

Ronald R. Hosey and Y. King Liu

18.1 INTRODUCTION

The mechanics of head injury has been, and still is, of special interest in biomechanics research. Although the head injury mechanism is inherently complex, the implicit assumption of most investigators is that a few parameters can give a good characterization of the biomechanical result. Numerous experimental studies have attempted to establish the functional relationship between the impact parameters and the resultant stress, strain, pressure, linear and angular acceleration, etc., in the tissues of the head and neck. The impact force history as an impact parameter has, with a few exceptions, seldom been measured or calculated in these studies. Many investigators have chosen to measure the resultant head acceleration and then use it as an index of impact severity. However, the early experimental work of Gurdjian *et al.*, (1961) has shown that the intracranial pressure in head impact is caused by both the acceleration of the brain/skull ensemble and the associated compression of the skull, i. e., a blow delivered to the head yields two quantities simultaneously: (a) an acceleration of the ensemble, and (b) a propagation of the skull deformation around the skull and into the intracranial space.

Ignoring the impact force and measuring only rigid-body acceleration implies that both the local and remote effects of impact are unimportant, i. e., that contact impact injuries are equivalent to accelerating and decelerating the brain in a rigid body, which is obviously not true. Granted that in a 'soft blow', where the amount of skull deformation is neglibible, the increase in intracranial pressure is primarily due to the acceleration pulse; however, when a 'hard blow' is delivered to the skull, the inbending and subsequent restitution of the skull produces a deformation wave, which is rapidly dispersed and attenuated around the skull. Simultaneously, the transient volume change in the epidural content as a result of the deformation wave is propagated into the cerebrospinal fluid/brain space. In spite of a fairly large number of experiments, the effects of bone deformation on the dynamic intracranial pressure distribution is still, at present, not well understood.

In 1966, the Head Injury Conference Proceedings were published in book form under the editorship of Drs. Caveness and Walker. In Chapter 29, Goldsmith suggested that the analytic solution of a fluid-filled elastic spherical shell under impulsive loading would provide some insight into the problem of fluid–solid interaction which characterizes the head injury problem. The challenge was met in two papers by Engin (1969) and Engin and Liu (1970), in which the problem of fluid-filled spherical shells was examined for axisymmetric vibrational response as well as the axisymmetric response to a Dirac delta impulse. Subsequent to the publication of these two papers, many other investigators have picked up on the theme; and, as of last count, there are over forty papers which are essentially variations on the same theme. Each one examined the importance of some additional aspects in the material-filled spheroid problem as it relates to the impact phenomena.

One of the shortcomings of the spheroid brain injury models is that they define a closed system, and hence the pressures computed from them have been much higher than those obtained experimentally. The high pressures in a closed system can be easily explained in terms of the relationship between the change in pressure ΔP, the dilatation (volumetric strain) $\Delta V/V$, and the bulk modulus B of the brain material, i.e., $\Delta P = B(\Delta V/V)$. The *in vitro* bulk modulus of the brain has been determined experimentally by McElhaney *et al.*, (1973) to be approximately 2.07 GPa (300 000 psi). Hence, a 1 % change in the volumetric strain would produce a change of pressure of about 20.7 MPa (3000 psi), i.e., of about 204 atmospheres of pressure. However, the *in vivo* experimental impact studies by Ommaya *et. al.*, (1966) on the rhesus monkey and the impact studies by Nahum *et. al.*, (1977) on human cadavera indicate that the maximum intracranial pressure is below 203 kpa (29 psi), i.e., is below 2 atmospheres of pressure, and this value is surely reasonable since cavitation can occur when the negative pressure drops below the vapour pressure of the liquid, i.e., below −95 kPa (−13.8 psi). Löfgren *et al.*, (1973) determined the pressure − volume ($P - V$). relationship of the cerebrospinal fluid (CSF) space in dogs, and their results show that the actual *in vivo* system is not nearly as incompressible as indicated by the high bulk modulus found from brain-specimen tests. Undoubtedly, potential spaces or regions exist within the head and spinal cavity, e.g., collapsed dura and venous vasculature, which are softer and more compressible than the surrounding tissues. These potential spaces and low-pressure sites serve as an expansion cavity to moderate the intracranial pressure during changes of volume. Therefore, the central nervous system (CNS) does *not* behave as a closed, incompressible system. A comprehensive review of the $P - V$ relationship within the CNS and the role it plays in the biomechanics of closed head impact is provided by Liu (1978).

Another obvious shortcoming of the spheroid head injury models is their inability to simulate the intricate geometry of the head with its numerous soft tissue interconnections. However, the severe idealizations of the spheroid models can be overcome by the use of the finite element modelling technique. Hardy and Marcal (1971) gave a faithful reproduction of the human skull by using a finite element approach. They analysed the static behaviour of the skull by incorporating the 'exact' skull geometry and experimentally obtained bone properties to calculate the stress and deformation. Shugar (1975, 1977) constructed a three-dimensional model of the skull and in addition filled it with fluid. The usefulness of this model apparently lies in the study of skull deformation and fracture.

Dr. Carley Ward and her collaborators have, over the past few years, formulated a 'morphologically' exact model of the brain using finite element techniques. The model incorporates the three-dimensional geometry of the brain together with its various investing membranes, e.g., the dura, the falx cerebri, and the tentorium. The soft tissue and the contained fluid are simulated by isoparametric brick elements. Membrane elements represent the tentorium and the falx. To avoid modelling the complex geometry of the skull, Ward *et al.*, (1974, 1975, 1978) used the measured head rotation and/or translational accelerations instead of impact force, as the input to the model. By doing so, they were able to obtain a system of linear differential equations in terms of the relative displacement coordinates, i.e., the displacement of the brain relative to the skull. The dynamic response is calculated in terms of the stress, strain, and intracranial pressure distribution. The model was revised a number of times in order to correlate better with experimentally derived head impact data in cadavera.

18.2 A HOMEOMORPHIC MODEL OF THE HEAD AND NECK

A finite element model of the head and neck, as a system, has recently been formulated by Hosey and Liu. The model incorporates the 'exact' geometry of the skull, dura, cerebrospinal fluid (CSF) space, brain, and spinal cord. The brain geometry from Ward and Thompson (1975), skull elements similar to those used by Hardy and Marcal (1971), and a structural formulation of the cervical column analogous to that of Pontius (1975) were used in the construction of the model. Continuity of the CSF space and brain material in the region of the foramen magnum was maintained in order that the proper distribution of pressures and 'flow' of material between the head and spine would take place during the excitation of the model. The geometry of the brain, spinal cord, CSF space, vertebra, cervical disks, jaw, and the floor of the cranial cavity is defined by means of eight-node, isoparametric brick elements. The frontal and parietal bones of the skull and the superior portions of the occipital and temporal bones of the skull are represented by thin shell elements. Figures 18.1 through 18.5 give some sagittal, transverse, and frontal (coronal) views of the model.

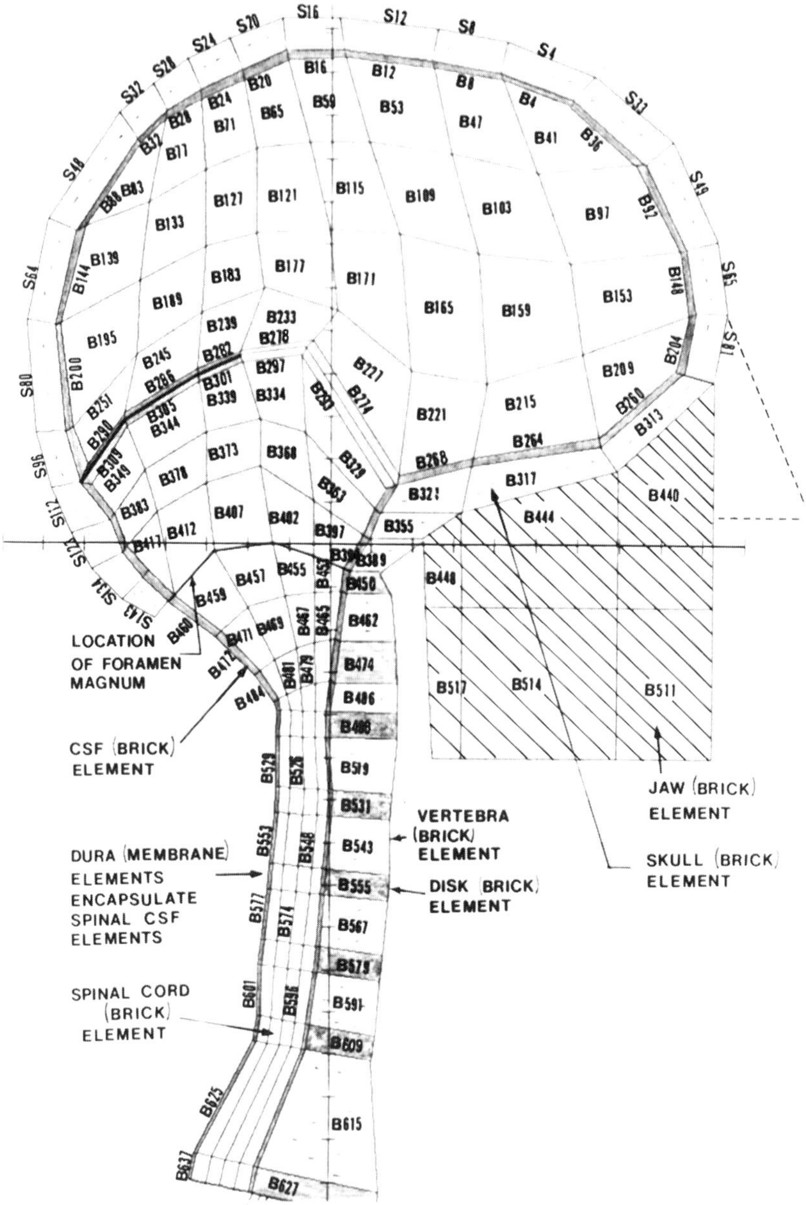

Figure 18.1 Sagittal section J–J, elements illustrated

Figure 18.1 is a sagittal section of the model illustrating lateral faces of the eight-node, isoparametric brick elements. The brain elements shown in this view are adjacent to the CSF space, e.g., B65, B109, B189, B153. The brain elements in the region of the corpus callosum are B159, B165, B171, B177, B227, and B233. CSF elements separate the brain elements from the skull elements, e.g., B20, B148. The CSF elements immediately superior to the tentorium are B282, B286, B290, and the CSF elements immediately inferior to the tentorium are B301, B305, B309. All of the elements inferior to the tentorium are adjacent to the mid-sagittal plane.

Figure 18.2 is the same sagittal section as Figure 18.1 except that the nodal points are illustrated. As an example, the lateral face shown for cerebellum element B378 in Figure 18.1 is defined by nodes 477, 483, 533, 527 in Figure 18.2.

Figure 18.3 is a cephalic, transverse section of the head illustrating the transverse faces of the elements. This view is defined by the section cut through nodes 228, 223, 217, 210, 203, 196, 189, 182, 175, 169, and 164 of Figure 18.2. This section shows the CSF elements between and adjacent to the brain elements and falx cerebri elements. Compare, as an example, brain element B115, which is shown in both Figures 18.1 and 18.3, with CSF element B116 shown in Figure 18.3. Also, this view illustrates the CSF elements between and adjacent to the brain and skull elements, e.g., the contiguity of CSF element B105 and the brain and skull (thin shell) elements B106 and S55, respectively.

Figure 18.4 is a caudal, transverse section of the head showing the tentorium and the floor of the anterior cranial cavity. The tentorium is defined by plane stress (membrane) elements. This view is defined by a section through nodes 427, 350, 355, 360, 365, as well as through the nodes between 317 and 397, 322 and 402, 327 and 407, 332 and 412, 337 and 417, and, finally, through node 390 of Figure 18.2.

Figure 18.5 is a posterior coronal section of the head and spinal cord. This view is defined by the section through nodes 25, 70, 127, 196, 265, 322, 370, 402, 465, 515, 586, 641, etc. of Figure 18.2. The CSF space, which is located between the two cerebral hemispheres and adjacent to the falx, is represented by brick elements B66 and B122. Brick elements B177, B178, B232, B233, B234, B277, B278, B296, B297, B333, and B334 represent the region of the corpus callosum third ventricle, diencephalon, and the superior portion of the pons. This view also shows brain elements, e.g., B62 and B118, surrounded by CSF elements, e.g., B17, B61, and B117, which are adjacent to the skull (thin shell) elements S17, S41, and S57. Brain elements B231 and B230 are adjacent to CSF elements B276 and B275, and these CSF elements are immediately superior and adjacent to the tentorium cerebelli. Immediately inferior to the tentorium lie CSF elements B294 and B295. At CSF element B294 and cerebellum element B331, however, the view is that of a more posterior sectional cut so that the CSF elements in the region of the cerebellum, e.g., B294, B335, and B369, and the cerebellum elements, e.g., B331 and B365, lie posterior to those brain and CSF elements which are superior

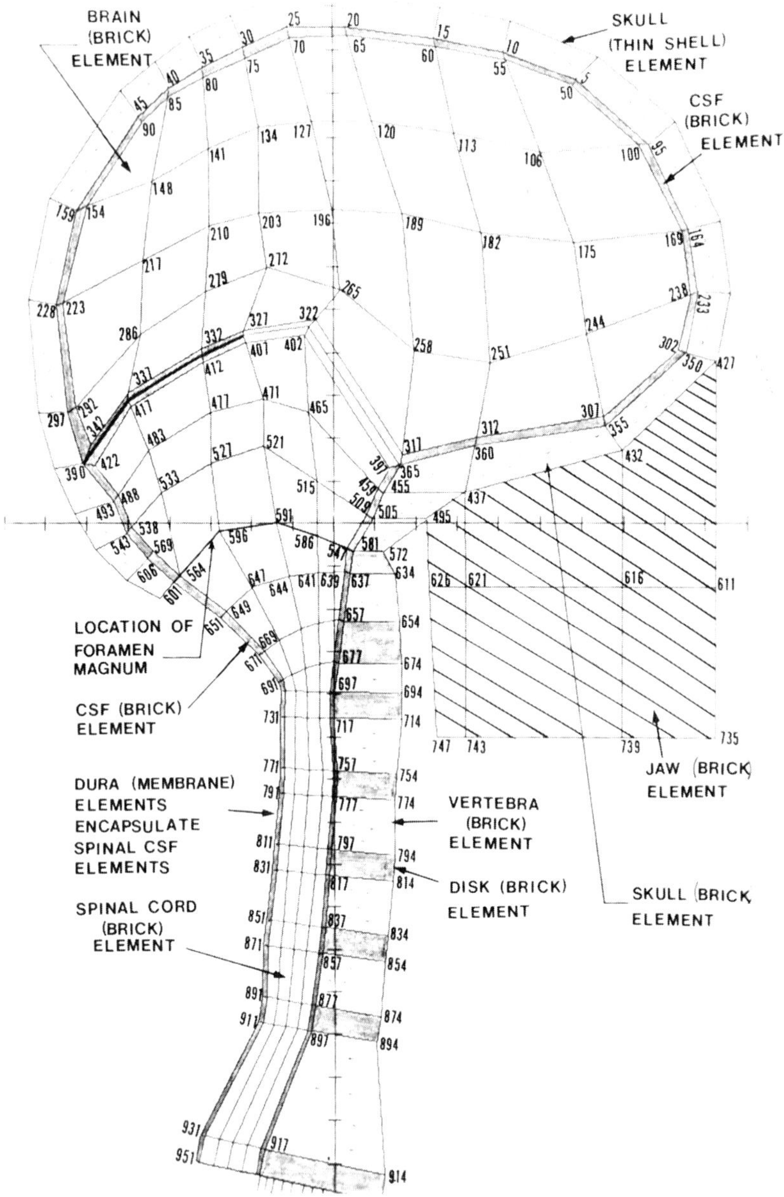

Figure 18.2 Sagittal section J–J, nodes illustrated

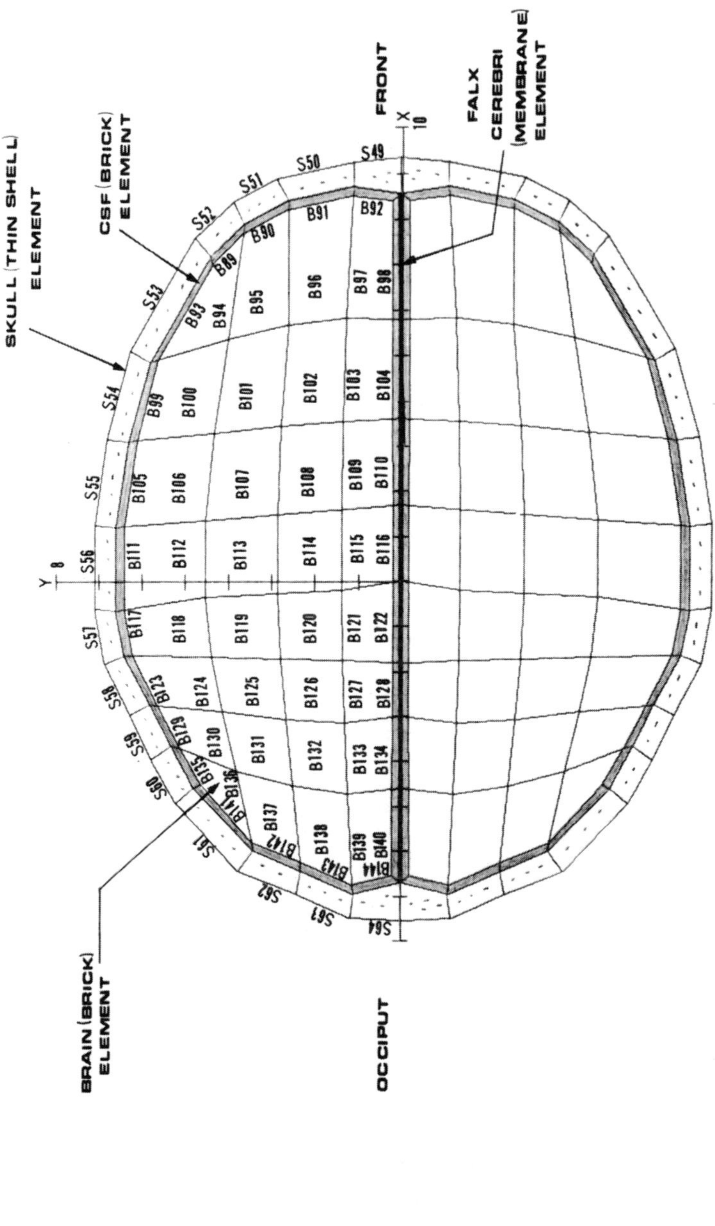

Figure 18.3 Transverse section A–A, elements illustrated

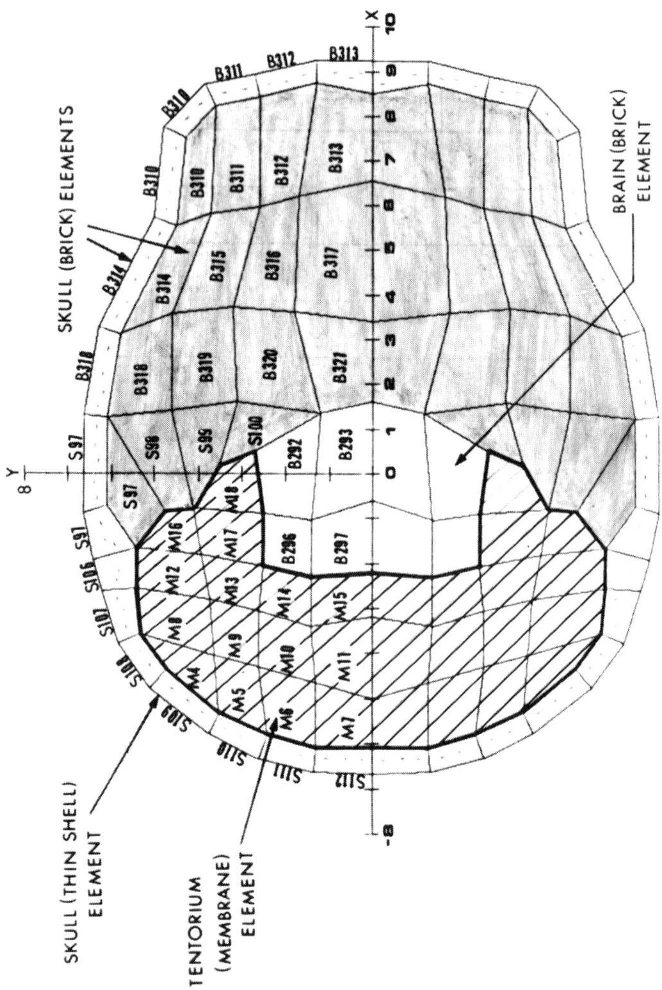

Figure 18.4 Transverse section C–C, elements illustrated

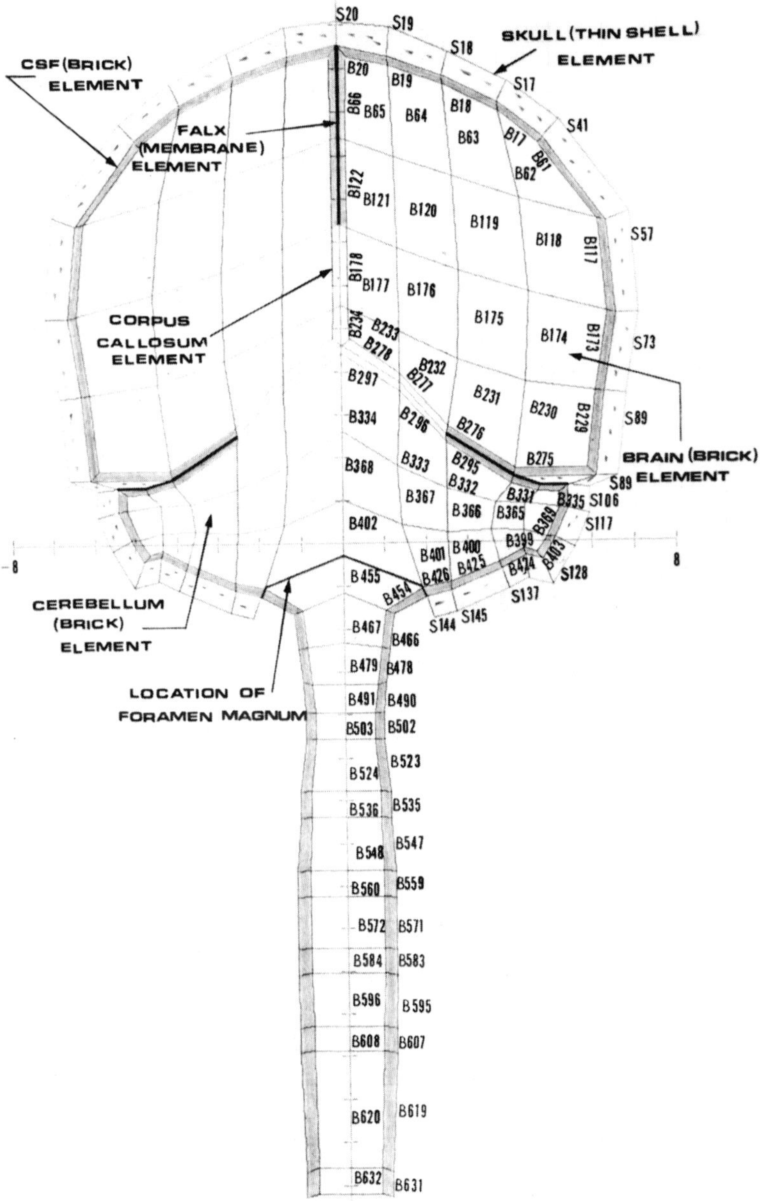

Figure 18.5 Frontal section H–H, elements illustrated

to the tentorium. The tentorium does not extend into the region of brain elements B227 and B296. Figure 18.5 also illustrates the cervical cord elements, e.g., B524, B536, and B548, surrounded by CSF elements, e.g., B523, B535, and B547. Dura (membrane) elements sheathe the CSF and cord elements of the cervical spine. The anterior edge of dura (membrane) element M40 is shown in Figure 18.5 as the outer edge of CSF element B523. Spinal cord element B467 and CSF element B466 are located at the Cl level of the cervical spine, and elements B491 and B490 at the C2 level.

Mid-sagittal symmetry, i.e., symmetry about the X–Z plane, is assumed for the model. This assumption, while quite reasonable anatomically, also reduces the number of nodes and elements by a half. As it is, 922 nodal points, 89 membrane elements, 637 brick elements, and 149 thin shell elements are required to define just the initial configuration of the model.

Accelerations usually exceed 100 g during a head impact. Hence, inertial forces will cause large changes in the pressure gradients within the central nervous system (CNS) during the contact phase of the impact. Hence, a head injury model should be reasonably equivalent, inertially, to the anatomical specimen it seeks to represent, i.e., the total mass, centre of mass, and the mass and product moments of inertia should agree reasonably well with experimentally determined values.

Becker (1972) reported the total mass, centre of mass relative to a coordinate system defined by the auditory meatuses and the orbital notches, and the mass and product moments of inertia with respect to the mass centre for 5 human head segments and 3 human head and neck segments. Walker *et al.*, (1973) performed a similar set of inertial data experiments using a sample size of 20 human cadavera, but their moment of inertia calculations were restricted to the rotary of inertia $\overline{I}_y$ (the y-axis is normal to the mid-sagittal plane). The mean values of the experimentally determined inertial data are given in Table 18.1. The mass centre is referenced to the system coordinates of the model. Only the one product moment of inertia $\overline{P}_{xz}$, where $(\overline{X}, 0, \overline{Z})$ denotes the mass centre, is listed in Table 18.1 since $\overline{P}_{xy} = \overline{P}_{yz} = 0$ by definition if mid-sagittal symmetry is assumed. In fact, the experimental data collected by Becker shows that the product moments of inertia $\overline{P}_{xy}$ and $\overline{P}_{yz}$ are an order of magnitude less than the value of $\overline{P}_{xz}$.

In order to render the model inertially equivalent, at least to a reasonable

Table 18.1 Inertial data with respect to mass centre

	Mass	Mass centre		Moments of inertia			
Head	M (g)	$\overline{X}$ (cm)	$\overline{Z}$ (cm)	$\overline{I}_x$ (g cm^2)	$\overline{I}_y$ (g cm^2)	$\overline{I}_z$ (g cm^2)	$\overline{P}_{xz}$ (g cm^2)
Becker	4059	1.33	2.09	189 500	235 100	163 800	−39 900
Walker *et al.*	4376	0.88	2.15		233 000		
Head & neck							
Becker	5827	−0.36	0.01	354 900	451 100	263 000	22 000
Walker *et al.*	6001	−0.48	−0.32		446 000		

approximation, to the experimental data reported by Becker and by Walker *et al.*, point masses were optimally located at the surface nodes of the skull and cervical column. Specifically, point masses m_i were located at surface nodes (X_i, Y_i, Z_i), $i = 1, 2, \ldots, N$, as a 'solution' to the system of inertial equations

$$\sum_{i=1}^{N} m_i + M_0 = M_e \tag{18.1}$$

$$\sum_{i=1}^{N} X_i m_i + \overline{X}_0 M_0 = \overline{X}_e M_e \tag{18.2}$$

$$\sum_{i=1}^{N} Z_i m_i + \overline{Z}_0 M_0 = \overline{Z}_e M_e \tag{18.3}$$

$$\sum_{i=1}^{N} (Y_i^2 + Z_i^2) m_i + (I_x)_0 = (\overline{I}_x)_e + M_e \overline{Z}_e^2 \tag{18.4}$$

$$\sum_{i=1}^{N} (X_i^2 + Z_i^2) m_i + (I_y)_0 = (\overline{I}_y)_e + M_e (\overline{X}_e^2 + \overline{Z}_e^2) \tag{18.5}$$

$$\sum_{i=1}^{N} (X_i^2 + Y_i^2) m_i + (I_z)_0 = (\overline{I}_z)_e + M_e \overline{X}_e^2 \tag{18.6}$$

$$\sum_{i=1}^{N} X_i Z_i m_i + (P_{xz})_0 = (\overline{P}_{xz})_e + M_e \overline{X}_e \overline{Z}_e \tag{18.7}$$

where N is the number of selected surface nodal points in the half-space $Y_i \geqslant 0$ (mid-sagittal symmetry), subscript 0 identifies the mass, mass centre, and the mass and product moments of inertia of the finite element model prior to the point mass distribution, subscript e represents experimentally determined values available in the literature, and the 'barred' inertial terms denote the experimentally determined mass and product moments of inertia with respect to the experimentally determined mass centre $(\overline{X}_e, 0, \overline{Z}_e)$. The 'unbarred' inertial terms I_x, I_y, I_z, P_{xz} are calculated with respect to the coordinate system origin of the model. Observe that there are only 7 inertial equations instead of the possible 10 due to the mid-sagittal symmetry.

Since there are many more unknowns than equations, the system of inertial equations was solved as an optimization problem, one with many 'peaks' and 'valleys'; and hence, a best solution was chosen only after much trial and error. The final total mass and mass/product moments of inertia of the model with respect to the coordinate system origin are represented by the left-hand sides of Equations (18.1), (18.4), (18.5), (18.6), and (18.7). These final values for the head segment of the model along with the experimental values of Becker (1972) and Walker *et al.*, (1973) referenced to the coordinate system origin of the model are given in Table 18.2. Notice that only the mean, minimum, and maximum values of Becker's data are provided in Table 18.2 since the sample size used by Becker was very small, but the 'mean $\pm$ one standard deviation' values are reported for the experimental data of Walker *et al.*

Table 18.2 Comparison of inertial data of head, with respect to global coordinate system

		Mass M (g)	Mass $\overline{X}$ (cm)	Centre $\overline{Z}$ (cm)	I_X (g cm^2)	Moments of inertia I_Y (g cm^2)	I_Z (g cm^2)	P_{XZ} (g cm^2)
	Min	3533	1.02	1.73	160 676	219 205	137 420	−44 729
Becker	μ	4059	1.33	2.09	207 500	260 524	171 212	−28 628
	Max	4420	1.82	2.46	241 119	306 243	193 601	−16 326
	$\mu-\sigma$	3785	0.17	1.65		221 438		
Walker	μ	4376	0.88	2.15		260 064		
et al	$\mu+\sigma$	4967	1.59	2.65		298 690		
Model		4305	1.54	2.76	167 189	233 728	150 109	−17 836

The inertial data for the head segment of the model are within the minimum and maximum values of the data of Becker (1972) and within a standard deviation to the head segment data of Walker *et al.*, (1973). The only exception is the Z component of the mass centre, i.e., $\overline{Z}$, which is within $1\frac{1}{4}$ standard deviations.

During the construction of the model, the material space occupied by hair, facial and neck muscles, and other surrounding soft tissues of the head and neck were not included in the finite element representation since it would have only demanded additional computer storage. Therefore the point masses along the cervical column could not be spaced sufficiently to obtain the inertial data for the head and neck system to the same degree of accuracy as was obtained for the head segment. However, the head and neck inertial data is within 2 to 3 standard deviations of that reported by Becker (1972). The integrity of the pressure and displacement calculations obtained from the model should still be maintained, even with the absence of neck mass, since head motion is surely the predominant motion which results when an occipital contact force of millisecond duration is applied to the head.

Jacob and Francone (1965) state that the average weight of the human brain is about 1300 g for man and 1250 g for woman, and that the total cerebrospinal fluid volume ranges from 80 to 200 ml; 15 ml of CSF are found in the ventricular cavities, and the remainder is in the subarachnoid space. The actual brain weight of the model is 1245 g, and the designated CSF space is 158 ml. Hence, the brain of the model corresponds in size to that of a woman, and it is protected by a surrounding 'water jacket' which is a little above average in content.

18.3 SYSTEM PARAMETERS AND RESULTS

The head injury model described in Section 18.2 has much potential and flexibility as a research tool. It can be exercised using one of the general-purpose, commercially available finite element programs such as SAP, EASE, NONSAP, and ADINA. The version of the finite element program used to exercise the model must, however, have an element formulation capable of representing the nearly incompressible tissues of the head and neck, i.e., brain, spinal cord, intervertebral discs, and cerebrospinal fluid (CSF). The program will also require an allocation of computer storage (primary or virtual) sufficient to process a dynamic run with approximately 2500 degrees of freedom. The material properties of the brain, CSF, bone (skull and vertebra), intervertebral disc, and dura mater elements are assigned during the data set input phase of program execution, and hence they may easily be changed from computer run to computer run in order to utilize or verify experimental data. Boundary conditions can be assigned to constrain a set of nodes or to specify initial nodal point displacements or velocities, depending upon the finite element program used. A contact force can be applied to the head in order to simulate a head impact, or the classical pilot injection problem can be modelled by applying an acceleration pulse to the inferior nodal points at the C7 level. The level of sophistication incorporated into the definition of the contact force used to approximate a very short-duration impact depends also upon the finite element computer program used.

The results reported here were obtained using the linear finite element program EASE2 with the split energy option. Use of the split energy option during program execution invokes a special numerical routine to calculate the stiffness matrix of nearly incompressible elements. The boundary conditions consisted of: (1) fixing the inferior nodal points of the C7 vertebral elements in order to simulate the effect of the larger torso mass on the head/neck system; (2) constraining the nodal points in the mid-sagittal plane against out-of-plane displacement; and (3) applying a half-sine wave contact force of 4 ms duration to ten occipital nodes of the skull. Since mid-sagittal symmetry is assumed in the construction of the model, the resultant input force has a peak value of 6000 N (1349 lb) and corresponds to an occipital contact force which is applied to the occiput over an area of 26.7 cm^2 (4.14 in^2) in a posterior-anterior direction. The material properties of the brain, dura, and skull are those reported in the literature by McElhaney *et al.*, (1973). The anatomically small volume of CSF space was assigned the material properties of the brain in order to satisfy the requirements of the EASE2 program. A Poisson's ratio of 0.499 was chosen to represent the essential incompressibility of the brain material. The modulus of elasticity of the cervical disc was taken from Pontius (1975), and the density values are the same as those used by Ward (1974). A summary of the material properties assigned to the finite elements is provided in Table 18.3. A 0.2 ms time step was chosen for the integration routines, and program execution was carried out to 6 ms.

The results of the program output gave a peak negative pressure in the contre-coup (frontal element 153) region of the brain of approximately −100.1 kPa (−14.5 psi) at 1.6 ms while the peak positive pressure in the coup region (occipital element 195) was 93.5 kPa (13.6 psi), also at 1.6 ms. Pressure values reported here are in the hydrostatic sense, i.e., are defined by the equation $p = \sigma_{kk}/3$ where σ_{ir} is the stress tensor. Figures 18.6 and 18.7 give pressure/time plots of

Table 18.3 Assigned material properties of the constituent tissue of the human head and neck

EASE2 material number	Material Description	Density (g/ml)	Young's modulus, E (kPa)	Young's modulus, E (psi)	Poisson's ratio
1	Dura (membrane)	1.133	31 500	4 570	0.45
2	Dura (membrane)	1.133	31 500	4 570	0.45
3	Brain	1.04	66.7	9.68	0.499
4	Skull	1.41	$4.46(10^6)$	647 000	0.21
5	Disc	1.065	7 585	1 100	0.499
6	CSF	1.04	66.7	9.68	0.499
7	Vertebra	1.41	$4.46(10^6)$	647 000	0.21
8	Cord	1.04	66.7	9.68	0.499
9	Jaw	1.41	$4.46(10^6)$	647 000	0.21
10	Skull (thin shell)	1.41	$4.46(10^6)$	647 000	0.21

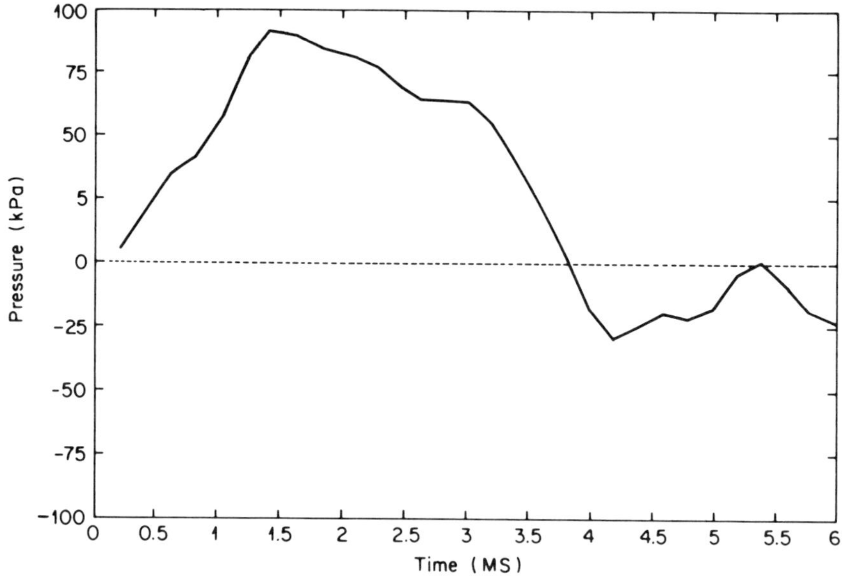

Figure 18.6 Pressure/time plot of occipital brain element 139 (coup region)

occipital brain element 139 (coup region) and frontal brain element 97 (contre-coup region). These elements in their initial configuration, i.e., the configuration prior to the application of the contact force, are illustrated in the sagittal view, Figure 18.1, and the transverse sectional view, Figure 18.3. Figure 18.8, which is defined by the same sectional cut as Figure 18.1, illustrates a sagittal pressure

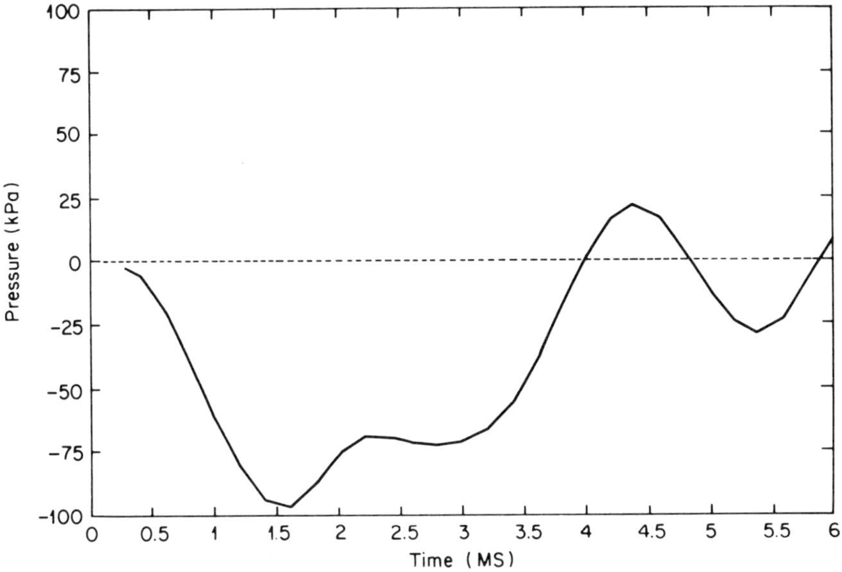

Figure 18.7 Pressure/time plot of frontal brain element 97 (contre-coup region)

distribution at 1.6 ms. The elements lying inferior to the tentorium are contiguous to the mid-sagittal plane; the elements superior to the tentorium, i.e., the brain elements, are very near the mid-sagittal plane.

Observe, from Figure 18.8, that the peak positive pressure in the region of the coup decreases linearly and eventually becomes negative as the region opposite the location of the contact force is approached. Also, the point of zero pressure is roughly located over the anterior portion of the foramen magnum. These observations, including the recognition that the maximum absolute value of the negative pressure in the contre-coup region is higher than the maximum positive pressure in the region of coup, were predicted in the paper by Engin (1969).

Figure 18.9 illustrates a transverse pressure distribution at 1.6 ms. Figure 18.3 can be used to identify the element numbers. The centrally located elements in this transverse section of the brain lie immediately superior to the corpus callosum, and the occipitally located elements are in the region of the coup. The peak pressures in the frontal and occipital regions of the distribution appear to be decreasing linearly in the lateral direction (away from the mid-sagittal plane), but a program rerun with pressure output of more elements would have to be made to verify this statement. If so, however, then the conclusion of Engin (1969) using the spherical shell model has again been verified.

The pressure/time plots of brain element B245, which is immediately superior to, but not adjacent to, the tentorium, and cerebellum element B344, which is immediately inferior to, but not adjacent to, the tentorium are given in Hosey

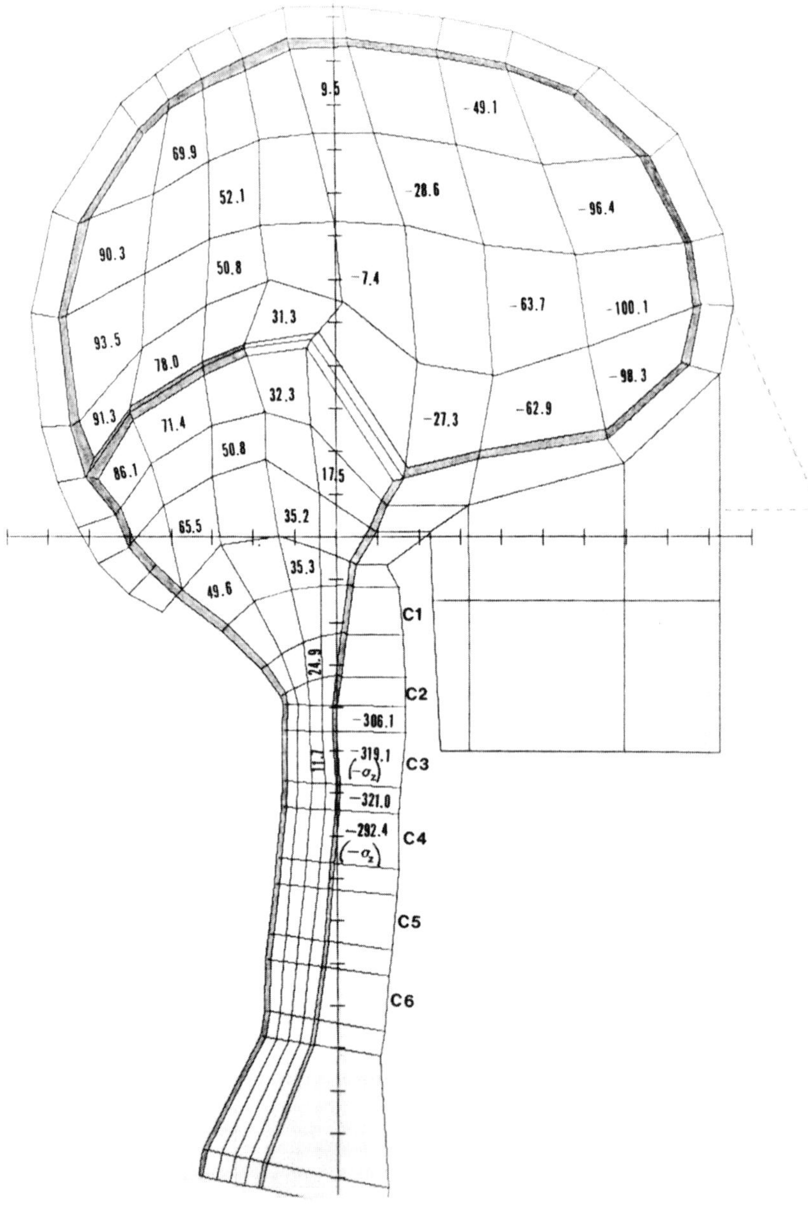

Figure 18.8 Sagittal section J–J, pressures at 1.6 ms, 6000 N peak input force

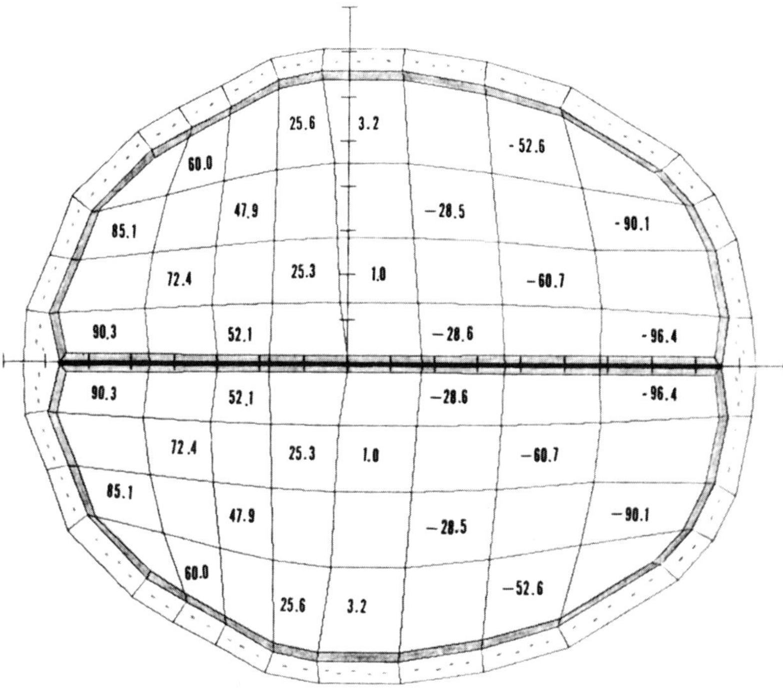

Figure 18.9 Transverse section A–A, pressures at 1.6 ms, 6000 N peak input force

(1980). Figure 18.1 gives the visual locations of these elements and Figure 18.4 their locations relative to the tentorium. The peak pressures of 78.0 kPa (11.3 psi) for the brain element superior to the tentorium and 71.4 kPa (10.4 psi) for the cerebellum element occur at 1.6 ms, and they are shown in the sagittal pressure distribution of Figure 18.8. The maximum pressure difference across the tentorium is 8.2 kPa (1.2 psi) and it occurred at 1.4 ms. The output of the EASE2 program gave a maximum in-plane compressive stress (maximum absolute value of the principal stresses) of 92.9 kPa (13.5 psi) for the tentorium element over the 6 ms of program execution. Since this value is at least two orders of magnitude below the ultimate strength of dura, destruction of the tentorium during the contact phase of the impact is highly unlikely.

The pressure/time plots of medulla oblongata element 402 (immediately superior to the foramen magnum) and spinal cord element 455 (immediately inferior to the foramen magnum), given in Hosey (1980), show a positive pressure throughout most of the contact phase of the input force with a maximum value of approximately 35.2 kPa (5.1 psi) occuring at the usual 1.6 ms as shown in Figure 18.8. Near the conclusion of the contact force, the pressure becomes negative with peak negative pressures of −40.7 kPa (−5.9 psi) for the medulla oblongata element and -56.8 kPa (-8.2 psi) for the spinal cord element, each occuring at 4.6 ms. The peak positive pressure of spinal cord element 479, which is posterior to the cervical disc between the C1 and C2 vertebrae, is 24.9 kPa (3.6 psi) and is again shown in Figure 18.8. The pressure curve of spinal cord element 479 is similar to that of the superiorly located spinal cord element 455 except that the peak positive pressure at 1.6 ms has been moderately damped, but such damping is expected in the caudal direction of the spinal cord. However, the peak negative pressure of spinal cord element 479 is -78.2 kPa (-11.3 psi) at 4.6 ms, which is approximately 38 % more severe than the superiorly located spinal cord element 455. If the peak applied force had been a little larger, say 7900 N (1776 lb), which was the value of the peak load used in the cadaveric experiments of Nahum *et al.*, (1977), then the estimated peak negative pressure would have been as low as -103 kPa (-14.9 psi), which is well into the range of cavitational pressure, possibly resulting in tissue destruction. Finally, the program output data shows damping in both positive and negative peak pressures of spinal cord elements which are located caudally to spinal cord element 479. As an example, Figure 18.8 shows that the peak positive pressure of spinal cord element 524 (located posteriorly to the C3 vertebra) became 11.7 kPa (1.7 psi) at 1.6 ms. At 4.6 ms, the pressure became -34.1 kPa (-4.9 psi).

The C3–C4 region of the cervical spine is a frequent site of spinal column injury in head impact, which appears quite plausible considering its central location relative to the massive head and the nearly fixed torso. The axial stress/time plot of cervical disc element 531, located between vertebrae C3 and C4, as shown in Figure 18.10, yields a peak compressive axial stress of -635.4 kPa (-92.2 psi) which occurred at 4.0 ms and a peak tensile axial stress of 1175.7 kPa (170.6 psi) which occurred at 5.8 ms. These peak pressures are well within the ultimate strength of the cervical discs since their values are at least an order of magnitude below the Young's modulus.

Figure 18.11 is the stress/time plot of the C4 vertebral element. A sagittal view of the C3–C4 region of the cervical spine in the initial configuration is illustrated in Figure 18.1 as brick elements B519, B531, and B543. From the stress history curve of Figure 18.11, one sees that the axial stress, σ_Z, is the most severe of the three normal stresses over the 6 ms of program execution. A peak compressive axial stress occurs at 4.0 ms in the C4 vertebra and has the value of -586.1 kPa (-85.0 psi). The peak tensile axial stress value for C4 is 1095.4 kPa (158.9 psi) occurring at 5.8 ms. The corresponding stress/time history of the C3 vertebra (brick element 519) is very similar to Figure 18.11 and has been omitted for the sake of brevity. Suffice it to record that its peak compressive and tensile stresses occur at 4.0 ms and 5.8 ms and has values of -654.9 kPa (-95.0 psi) and 1151.7 kPa (167.1 psi), respectively. Observe that the shape of the axial stress curve in Figure 18.11 is identical to the shape of the axial stress curve, Figure 18.10, of the inter-vertebral disk element 531. Note that the peak compressive and tensile stress values of -635.4 kPa (-92.2 psi) and 1175.7 kPa (170.6 psi) for this

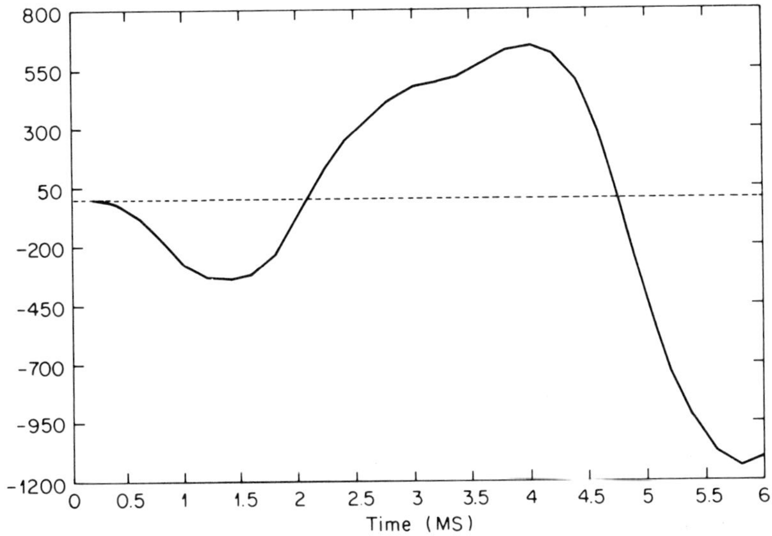

Figure 18.10 Pressure/time plot of element 531 (disc C3–C4). Pressure denotes negative axial stress, $-\sigma_z$

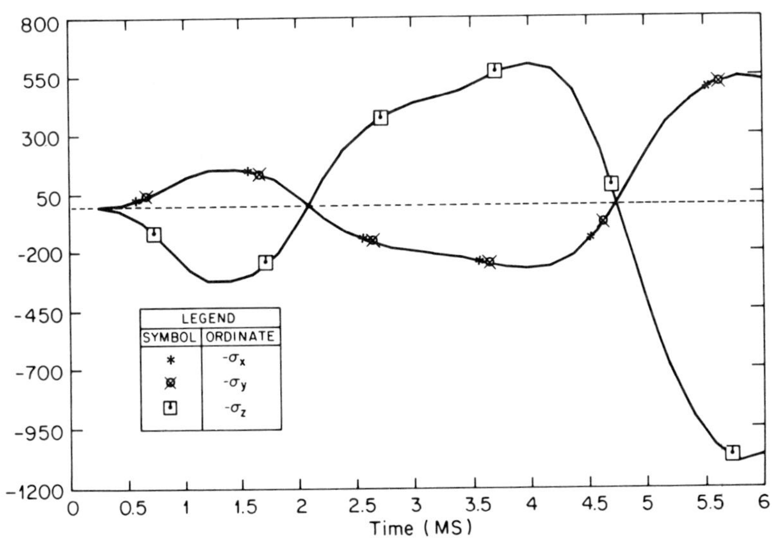

Figure 18.11 Pressure/time plot of element 543 (C4 vertebra). Pressure denotes stress values $-\sigma_x$, $-\sigma_y$, and $-\sigma_z$, respectively

cervical disc element lie between the peak compressive and tensile axial stress values, respectively, of the two vertebral elements.

Displacement output of skull nodal points by the EASE2 program shows that the rigid-body motion of the head is in the expected direction of the input force throughout the 6 ms duration of program execution with very little displacement in the inferior-superior (vertical) direction. Maximum horizontal displacement of frontal skull node 164 was 1.35 cm at 6.0 ms, but the magnitude of vertical displacement was not more than 0.2 mm for the skull nodes. The amount of head rotation was extremely small. The magnitude of the linear acceleration of frontal skull node 164 had a peak value of 139 250 cm/s^2 (142 g) at 2.2 ms. This value can be used as an approximation to the peak rigid-body linear acceleration of the head since node 164 is located on the skull opposite the site of the coup.

18.4 DISCUSSION OF RESULTS

The input force reported in Nahum *et al.*, (1977) was approximately 6 to 7 ms in duration, and the shape was typical of those reported in the literature, i.e., the shape was not unlike a half-sine wave. Also, the shape of this typical input force curve had most of its 'area under the curve' in approximately the first 4 ms, and therefore most of the impulse is delivered to the system during these 4 ms. For these reasons, the input force chosen for the head injury model is a half-sine wave of 4 ms duration where the peak applied force occurs at 2 ms. However, the peak pressures in the brain occur between 1.4 and 1.6 ms, as shown in Figures 18.6 and 18.7, and one would expect the peak applied force and the resulting peak pressures to be in phase for an elastic system. A two-fold explanation is required. First, the change in the force magnitude from 1.6 ms to 2.0 ms is very slight for the half-sine wave curve, i.e., for a sinusoidal curve of period 8 ms. Secondly, the displacement output of the program shows that the frontal and occipital portion of the brain has already begun a series of relative minimum and maximun displacements with respect to the skull by the time 1.6 ms is reached. As an example, the frontal brain nodes near the skull begin their first 'spring-back' around 1.2 ms, followed by an almost immediate rebound at 1.6 ms, even though an extension between nodes (and hence negative pressure in the brain elements defined by these nodes) is maintained throughout the load period. If the partition of thin brick elements, located in the anatomical region of the CSF space between the brain and the skull, had been assigned a value of Young's modulus somewhat less than that of brain material, then the additional cushioning effect would have delayed the initial 'spring-back'; and the phase difference between the pressure distribution and the input force would have been smaller. This last statement was verified in an earlier run using the linear finite element program, SAP4.

The shape and magnitude of the pressure/time curves of Figures 18.6 and 18.7 compare favourably to the results of Nahum *et al.*, (1977), who impacted seated, stationary, cadaveric subjects by a rigid impactor travelling at a constant velocity. The blow was delivered to the frontal bone in an anterior-posterior direction. In one of the experiments, an impact with a peak load of 7900 N (1780 lb) and a duration of approximately 6 to 7 ms produced a frontal peak pressure of 141 kPa (20.5 psi) beneath the impact site. The magnitude of the pressure at the posterior fossa, the point opposite the impact site due to the inclination of the skull when the blow was delivered, was − 60.3 kPa (− 8.75 psi). Lincar interpolation predicts a peak pressure of 107.1 kPa (15.5 psi) in the coup region of the cadaver for a frontal blow of magnitude 6000 N (1349 lb). When these values are compared with the peak pressure, 93.5 kPa (13.6 psi), in the coup region of the model, the correlation is good if one considers the increased geometrical damping experienced by an occipitally generated pressure wave from the close proximity of the tentorium, where the cranium has relatively less curvature.

A peak head acceleration of 204 *g* was obtained in the experiment by Nahum *et al.*, (1977). Again, linear interpolation predicts a head acceleration of approximately 155 *g* from a frontal blow of magnitude 6000 N (1349 lb). The peak linear acceleration of the model was 142 *g* at 2.2 ms. The amount of head rotation of the model was negligible, which is in agreement with the experimental results of Nahum *et al.* These investigators have shown that dynamic pressure changes within the skull practically cease before significant rotation of the skull occurs. From the graphs presented in Nahum *et al.*, a head rotation greater than 4° occurs only after 6 ms, i.e., subsequent to the time duration of this run.

Nakamura *et al.*, (1968, 1974) have reported on an autopsy study of 63 unselected clinical uses of head injury and found:

1. In 29 occipital impacts, contre-coup injuries were observed in 83 % of the population whereas coup injuries occurred in 10 % of the population. The remaining 7 % had both coup and contre-coup injuries.
2. In 14 cases of frontal impact, coup injuries occurred in 79 % and contre-coup injuries occurred in 7 %. Coup and contre-coup injuries were found in the remaining 14 % of the population.
3. In 20 cases of temporal impact, 50 % had contre-coup injuries and 40 % had coup injuries. The remaining 10 % had both.

Thus, clinical data suggest that contre-coup injuries from frontal impact should be less frequent as compared with occipital impact. However, frequency of coup and contre-coup injuries as a result of left or right temporal impact should approximately be the same. Additionally, the frontal region of the brain, i.e., the region furthest from the foramen magnum, should be the most frequently injured site. Lofgren and Zwetnow (1973) showed, in dogs, that the spinal cord cavity, which is joined to the rigid skull cavity via the foramen magnum, serves as an expansion vessel to moderate the intracranial pressure as the CSF volume is increased. The peak pressure of the model described in this dissertation, i.e., − 100.1 kPa (− 14.5 psi), also occurred in the frontal region of the brain. This peak value of the contre-coup pressure is below the cavitational pressure of the cerebrospinal fluid (CSF) of the brain, − 95 kPa (− 13.8 psi), and, although this computed value is theoretically impossible to observe experimentally, it does define a region of potential brain damage.

The two additional predictions of the present model i.e., (1) the approximately linear pressure gradient in the brain, and (2) the location of the point of zero pressure above the foramen magnum, agree with the experimental results of Roberts *et al.*, (1966) and the theoretical results of Shugar (1975, 1977). Since the location of the foramen magnum is slightly closer to the occiput than the forehead, the two above observations agree with the implication from the experimental data of Nakumura *et al.*, (1968, 1974) that brain injuries occur more frequently in the frontal brain. Therefore, it appears, at least theoretically, that a frontal impact has to be more severe than an occipital one in order to induce a high enough rarefaction to cause cavitation in the contre-coup region. The above conjecture needs to be substantiated by exercising the model with a series of frontal impacts.

REFERENCES

Becker, E. B. (1972) Measurement of mass distribution of anatomical segments, *Proc. Sixteenth Stapp Car Crash Conf.*, Soc. Auto Eng., 160–185.

Caveness, W. F., and Walker, A. E. (eds.) (1966) *Proc. Head Injury Conf.*, Univ. of Chicago, Lippincott.

Engin, A. E. (1969) Axisymmetric response of a fluid-filled spherical shell to a local radial impulse-a model for head injury, *J. Biomech*, **2**(3), 325–341.

Engin, A. E., and Liu, Y. K. (1970) The axisymmetric response of a closed fluid-filled spherical shell in free vibrations, *J. Biomech.*, **3**(1), 11–22.

Gurdjian, E. S., Lissner, H. R., Evans, F. G., Patrick, L. M., and Hardy, W. G. (1961) Intracranial pressure and acceleration accompanying head impact in human cadavers, *Surg., Gynec. Obstet.*, **113,** 185–190.

Hardy, C. H., and Marcal, P. V. (1971) Elastic analysis of a skull, *ONR Report No. N00014–0007/8*, Nov.

Hosey, R. R. (1980) A homeomorphic finite element model of impact head injury, Ph.D Dissertation, Tulane University, New Orleans, La.

Jacob, S. W. and Francone, C. A. (1965) *Structure and Function in Man.*, W. B. Saunders Co.

Liu, Y. K. (1978) Biomechanics of closed head impact, *J. Eng. Mech. Div., Ame. Soc. Civil Eng.* **104**(EM1), 131–152.

Löfgren, J. (1973) Pressure–volume relationships of the cerebrospinal fluid system–an experimental analysis in dogs, Goteborg, Sweden, Dept. of Neurosurg., of Goteborg.

Löfgren, J., von Essen, C., and Zwetnow, N. N. (1973) The pressure–volume curve of the cerebrospinal fluid space in dogs, *Acta Neurol. Scand.*, **49,** 557–574.

Löfgren, J., and Zwetnow, N. N. (1973) Cranial and spinal components of the cerebrospinal fluid pressure–volume curve, *Acta Neurol. Scand.*, **49,** 575–585.

McElhaney, J. H., Melvin, J. W., Roberts, V. L., and Portnoy, H. D. (1973) Dynamic characteristics of the tissues of the head, *Perspectives in Biomedical Engineering* (ed. Kenedi, R. M.), Macmillan Press Ltd., London, pp. 215–222.

Nahum, A., Smith, R., and Ward, C. C. (1977) Intracranial pressure dynamics during head impact. *Proc. Twenty First Stapp Car Crash Conf.*, Soc. Auto Eng., 337–366.

Nakamura, N. (1974) The dynamic study of head injury, *Neurol. Surg.* (Tokyo), **2**(2), 97–105.

Nakamura, N., Hirakawa, K., Masuzawa, H., Hashizume, K., and Sano, K. (1968) Dynamic mechanisms in closed head injuries, *Brain Nerve* (Tokyo), **20**(12), 1204–1212.

Ommaya, A. K., Hirsch, A. E., Flamm, E. S., and Mahone, R. H. (1966) Cerebral concussion in the monkey: an experimental model, *Science*, **153,** 211–212.

Pontius, U. R. (1975) The effect of the cervical neuromusculature on the dynamics of whiplash, Ph.D. Dissertation, Tulane University, New Orleans, La.

Roberts, V. L., Hodgson, V., and Thomas, L. M. (1966) Fluid pressure gradient caused by impact to the human skull, *Biomechanics Monograph*, Ame. Soc. Mech. Eng., 223–235.

Shugar, T. A. (1975) Transient structural response of the linear skull-brain system, *Proc. Nineteenth Stapp Car Crash Conf.*, Soc. Auto. Eng., 581–614.

Shugar, T. A. (1977) A finite element head injury model, I and II, *TR-R-854-1* and *R-854-II*, Civil Eng. Lab., Port Hueneme, Ca. USN.

Walker, L. B., Harris, E. H., and Pontius, U. R. (1973) Mass, volume, centre of mass, and mass moment of inertia of head and head and neck of the human body, *Proc. Seventeenth Annual Stapp Car Crash Conf.*, Soc. Auto. Eng., 525–537.

Ward, C. C. (1974) A dynamic finite element model of the human brain, Ph.D. dissertation, UCLA, Los Angeles, Ca.

Ward. C. C., Nikravesh, P. E., and Thompson, R. B. (1978) Biodynamic finite element models used in brain injury research. *Avia. Space Environ. Med.* **49**(1), 136–142.

Ward, C. C., and Thompson, R. B. (1975) The development of a detailed finite element brain model, *Proc. Nineteenth Stapp Car Crash Conf.*, Soc. Auto. Eng., 641–674.

Index